*The Earth on Show*

# The Earth on Show

Fossils and the Poetics of Popular Science, 1802–1856

Ralph O'Connor

*The University of Chicago Press* | *Chicago and London*

RALPH O'CONNOR is lecturer in Irish-Scottish Studies in the Department of History, University of Aberdeen. He is the author of *Icelandic Histories and Romances* (2002) and has published articles on nineteenth-century popular science, Romantic poetry, and the mediaeval literatures of Iceland and Ireland.

The University of Chicago Press, Chicago 60637
The University of Chicago Press, Ltd., London
© 2007 by The University of Chicago
All rights reserved. Published 2007
Printed in the United States of America

16  15  14  13  12  11  10  09  08  07      1  2  3  4  5

ISBN-13: 978-0-226-61668-1     (cloth)
ISBN-10: 0-226-61668-1         (cloth)

Library of Congress Cataloging-in-Publication Data
O'Connor, Ralph.
    The earth on show : fossils and the poetics of popular science, 1802–1856 / Ralph O'Connor
        p.   cm.
    Includes bibliographical references and index.
    ISBN-13: 978-0-226-61668-1 (cloth : alk. paper)
    ISBN-10: 0-226-61668-1 (cloth : alk. paper)
    1. Geology—Great Britain—History—19th century.   2. Geology—Social aspects—Great Britain—History—19th century.   3. Geology in literature.
    4. Literature and science—Great Britain—History—19th century.   5. Literature and history—Great Britain.   6. Science—Philosophy—History—19th century.
    I. Title.
    QE13.G7035     2007
    550.941'09034—dc22
                                                                    2007028061

*For Clémence*

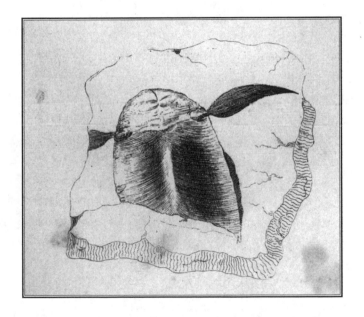

# Contents

# Acknowledgements

Since I began working on this project eight years ago, a great many people have given up their time and resources to help it on its way. Although it is customary to leave the spouse till last, I should like to begin by thanking my wife, Clémence. Despite the considerable demands of her own book projects, she read draft after draft of my work, giving it the full benefit of her literary-critical acumen. That this book ever reached completion is due not only to her apple cakes, but above all to her matchless support, encouragement, and remarkable patience. This book is for her.

Most of the research for this book was undertaken while I was a member of St John's College, Cambridge, first as a doctoral student, then as a Junior Research Fellow. It was funded initially by the Arts and Humanities Research Board, and then by the Master and Fellows of St John's College. I am extremely grateful to the College for supporting me so generously, for giving me the time to pursue my research in a congenial and stimulating setting, and for contributing towards the costs of the book's illustrations. I was subsequently appointed to a lectureship at the University of Aberdeen, where I was able to complete the book thanks partly to my employers' policy of offering first-time lecturers a generous reduction in initial teaching workload; I am grateful to my colleagues in the Departments of History and Celtic for shouldering the shortfall. The College of Arts and Social Sciences and the School of Divinity, History and Philosophy also made substantial contributions towards the costs of illustrations.

I am deeply grateful to the many friends and colleagues who have helped me to lick this book into shape. I owe particularly weighty debts to Anne Barton, Jim Secord, and Mike Taylor, all of whom have read more drafts of my work than I can remember. Anne Barton first encouraged me to undertake research in this area; I was very lucky to have her as my Ph.D. supervisor, and to have had the benefit of her critical insight, encyclopaedic learning, and moral support. Jim Secord provided much help from an early stage with practical advice, many perceptive suggestions for large-scale improvements, and an infectious enthusiasm for Victorian saurians. Mike Taylor not only responded to countless requests for obscure information, but also went through the whole manuscript twice with a fine-toothed comb, each time scrutinizing the beast with a palaeontologist's eye for detail and offering many valuable suggestions. I am grateful to Mike for sharing some of his research materials on Thomas Hawkins and Hugh Miller (his new biography of Miller is eagerly awaited). Exceptionally helpful and constructive comments on the entire manuscript were provided by Martin Rudwick, Nigel Leask, Boyd Hilton, Anne O'Connor, and above all the three anonymous referees for the University of Chicago Press. Santanu Das and Rob Macfarlane gave several chapters the benefit of their literary judgement and learning. I am particularly indebted to Harriet Truscott: her long-standing support and detailed comments on numerous draft chapters were crucial in getting this project off the ground in the first place.

For passing on valuable nuggets of information, offering practical assistance, and helping me shape my project, I am grateful to Alison Alexander, Iain Beavan, Paul Bibire, Patricia Boulhosa, Patrick Boylan, Richard Brasher, Tony Brook, Christopher Burlinson, Vicky Carroll, Paul Clasby, Peter Cochran, Phil Connell, Simon Conway Morris, Peter Davidson, William Delafield, David Duff, John Fuller, Tony Howe, Humphrey Hudson, Ralph Hyde, Elizabeth Irvine (of Maggs Bros., London), Simon Jarvis, Melanie Keene, John Kerrigan, Simon Knell, Julian Luxford, Donald and Ruth Lynden-Bell, Sherrie Lyons, Peter McCaffery, Ben Marsden, Subha Mukherji, Elizabeth Neswald, Mark Nicholls, Máire Ní Mhaonaigh, David Norman, Jane Partner, Erich Poppe, Katy Price, Michael Roberts, Ray Ryan, Tom Sharpe, Marianne Sommer, Bill and Sandra Truscott, Andrew Wawn, Paul White, R. Derek Wood, and especially Hugh Torrens and Anne Secord, whose generosity in this respect has been unstinting. Christopher Moule gave me essential instruction in the mysteries of Scottish geology, and the late Peggy Truscott offered some much-needed perspective on the whole project by sitting on it from time to time. I wish to thank my friends and family for bearing with me through the many 'hermit' phases I and my book have gone through. Above all, I should like to thank my parents for encouraging my interest in books (and extinct animals) at an early age, and for continuing to do so ever since. Their support has been incalculable.

ACKNOWLEDGEMENTS

I have benefited greatly from discussions arising from my presentations on this subject at Cambridge's Department of History and Philosophy of Science (the Cabinet of Natural History and the Science and Literature Reading Group); the History of Geology Group of the Geological Society of London (HOGG); the Twenty-sixth International Byron Conference, Nottingham; the Annual Conference of the British Society for the History of Science, Leeds; the Cultural History Seminar at the University of Aberdeen; and symposia commemorating William Buckland and Hugh Miller at the Oxford University Museum of Natural History and Cromarty respectively. I am especially grateful to the organizers of HOGG for their unstoppable energy in promoting research into the history of geology, and for maintaining such a stimulating and convivial atmosphere for discussion. My thanks to the participants of these various events for their helpful comments.

I have stretched the tolerance of many librarians over the past eight years. In particular I should like to thank Clare Welford, Stella Clarke, and Helen Hills in Rare Books, Cambridge University Library, for their unfailing patience and alacrity in the face of my increasingly unreasonable demands for large piles of books and photocopies at short notice. Michelle Gait and June Pirie in Special Collections, Aberdeen University Library, were extremely helpful during the final stages of reference-checking. I also wish to express my gratitude to the staff of the St John's College Library (Cambridge); Aberdeen University Library; the Bodleian Library, Oxford (in particular Julie Anne Lambert, curator of the John Johnson Collection); the British Library; Cambridge University Library; Cambridge's Whipple Library and English Faculty Library; the Devon Record Office, Exeter; the Guildhall Library, London; the Laing Art Gallery, Newcastle; the Lyme Regis Museum; Robin Darwall-Smith, the Archivist at Magdalen College, Oxford; the Oxford Museum of Natural History (in particular the Archivist, Stella Brecknell, and the Director, Jim Kennedy); Sir John Soane's Museum, London; and the University of St Andrews Library.

Many of these institutions have allowed me to use material in their collections, and they are credited formally on the credits page at the end. However, I wish to express particular gratitude to the following individuals for their patience in dealing with my more complicated requests: Marian Minson and Tim Lovell-Smith at the Alexander Turnbull Library, Wellington, New Zealand; Rory Lalwan at the City of Westminster Archives Centre, London; Alex Edouard at the Bridgeman Art Library; Julia Nurse and Angela Roche at the British Museum; Gerry Bye and the photographic staff at Cambridge University Library; Elizabeth and Roderick Gordon, owners of the Buckland Papers, formerly held in Exeter; and Tom Sharpe at the National Museum of Wales. In an age when exorbitant sums are often charged for granting copyright permission to the authors of scholarly publications, I wish to express my heartfelt gratitude to those people and institutions who waived their fees or reduced them to within the means of a humble university lecturer: the Alexander Turnbull Library, Wellington; the Bill Douglas Centre, University of Exeter; the Bodleian Library, Oxford; Cambridge University Library; the Department of Geology, National Museum of Wales; the Guildhall Library, London; Roderick Gordon; Lyme Regis Museum; Magdalen College, Oxford; National Galleries of Scotland; the National Portrait Gallery, London; John O'Connor; the Oxford University Museum of Natural History; and Martin Rudwick. I am also very grateful for the gifts of nineteenth-century books and photographs from various friends and family members, which have helped my work in a number of ways.

It has been a pleasure and a privilege to work with the University of Chicago Press. Particular thanks are due to Catherine Rice for taking on this project in the first place, and to my subsequent editors Alan Thomas and Karen Darling for seeing it through. Their commitment and enthusiasm are greatly appreciated. Many thanks also to my copyeditor, Barbara Norton, for her rigorous attention to detail and for bearing with the idiosyncrasies of British English; to Pete Beatty and Michael Koplow for their unfailing helpfulness throughout the editing process; and to the many other individuals working behind the scenes on this project.

Finally, I should like to thank you, the reader, for picking up this book. I hope you will not put it down too quickly.

Featherfield
Aberdeenshire
22 October 2006

# Abbreviations

ANH — *Archives of Natural History*
AS — *Annals of Science*
BJHS — *British Journal for the History of Science*
DNBS — *Dictionary of Nineteenth-Century British Scientists*, ed. Bernard Lightman (Bristol: Thoemmes, 2004)
ESH — *Earth Sciences History*
HS — *History of Science*
JVC — *Journal of Victorian Culture*
NRRSL — *Notes and Records of the Royal Society of London*
ODNB — *The Oxford Dictionary of National Biography*, ed. H. C. G. Matthew and Brian Harrison (Oxford: Oxford University Press, 2004); an online, updated version is available to subscribers at www.oxforddnb.com
OED — *The Oxford English Dictionary*, 2nd ed., ed. J. A. Simpson and E. S. C. Weiner (Oxford: Oxford University Press, 1989)
PGA — *Proceedings of the Geologists' Association*
SIR — *Studies in Romanticism*
WI — *Word & Image*

References to articles in *DNBS* and *ODNB*, two invaluable biographical dictionaries, are included in the footnotes for purposes of background information. I have not named the authors of these articles except when drawing attention to their views.

# Introduction:

## *Science as Literature*

It is familiar knowledge that the earth we inhabit is millions of years old; that we have not inhabited it for very long; and that it has been the theatre of successive geological changes, a shifting backdrop for strange and ancient creatures which have flourished and vanished in their turn. Museums, television programmes, and books have told this story thousands of times, presenting the "panorama of life on earth" as a "compelling drama of survival and extinction" or, with pardonable exaggeration, as "the most magnificent play ever enacted in the universe".[1]

But this story was not always familiar; nor was it always so theatrical. In fact, these two qualities go together. In the early nineteenth century—especially in Britain, on which this book focuses—the new science of geology was publicized in spectacular and theatrical forms which enabled it to gain the cultural authority it enjoys today. This rise to authority has been much studied, particularly in relation to social and political history, religion, and controversies within and between scientific and

---

1. Špinar 1972, 6; Gould 1993a, inside front dustjacket; Augusta 1961, 7.

educational establishments;[2] but the crucial enabling roles of spectacle and (in particular) of literature are only beginning to come into focus.[3] Performance was central to the public face of earth science. Its adherents pulled off an imaginative coup—a *coup de théâtre*—by giving their public tantalizing glimpses of an earth history far longer and stranger than the story of a literal six-day Creation which had held sway over much of that public at the turn of the century.[4]

This was no easy task. The new narrative had to compete not only with the Book of Genesis, but also with centuries of sacred-historical tradition, of which John Milton's epic poem of Creation and Fall, *Paradise Lost* (1667), was only the most prestigious expression.[5] Rather than assaulting this potent body of narrative head-on, proponents of the new science turned it to their own ends, "justify[ing] the ways of God to men" by forging a new Creation-myth for an imperial age.[6] Popularization took many forms, including lectures, exhibitions, and even custom-built geological museums. Public platforms such as the British Association for the Advancement of Science were initially dominated by genteel geologists: science's high status in the Victorian era owed much to the amateur theatricals of William Buckland and the stirring oratory of Adam Sedgwick.

But it was in their literary productions—in books, journals, magazines, and newspapers—that these geologists and their followers reached most of their increasingly variegated public. In an age marked by debates over the dangers of imagination and the deceptive allure of cheap romances and sensation novels, geology was marketed as the key to true facts which were nonetheless more marvellous and sensational than fiction. In his lavish volume *Illustrations of the Geology of Sussex* (1826), the

2. Some examples: Porter 1977; Morrell and Thackray 1981; Rupke 1983b; Rudwick 1985; J. Secord 1986; Desmond 1989; Topham 1992; Knell 2000; Rudwick 2005b; 2008 in press. On Victorian science and cultural authority more generally see Cannon 1978; Turner 1978; Yeo 1984; 1993; 2001.

3. Orange (1975) and Morrell and Thackray (1981, 157–63) have discussed the "carnival of science" in the 1830s and 1840s, focusing on the British Association for the Advancement of Science. Merrill (1989) has examined the rhetoric of wonder in (mostly later) natural-history writing. Rudwick (1992) has surveyed pictorial restorations of the ancient earth. Forgan (1994 and 1999) and Yanni (1999) have examined the architecture of geological display. J. Secord (2000a, 437–70) has explored cosmological works of the 1840s and 1850s in relation to popular spectacle, and more recently (2004) the Crystal Palace Gardens monster display. See also Sommer 2003 and Heringman 2004 for explorations of geology's relations with "Romantic" poetry. Of the popular accounts, the best are Cadbury 2000 and McGowan 2001, though these are focused more on fossil discoveries than on popularization (on which see Freeman 2004). For some insightful reflections on science and theatricality more generally see Lindqvist 1992, 84–93.

4. On biblical literalism see chapter 3.

5. On Milton's importance to the nineteenth-century scientific imagination see Beer 1983, 29–36; 1996, 210–15.

6. *Paradise Lost* I.26 (Milton 1971, 44).

Sussex surgeon Gideon Mantell promised his knot of wealthy readers that "the realities of Geology far exceed the fictions of romance", while in *The Old Red Sandstone* (1841) the Scottish newspaper editor Hugh Miller assured his own rather larger readership that "no man who enters the geological field in quest of the wonderful, need pass [. . .] from the true to the fictitious."[7]

These texts, too, were performances. Their authors mastered what James Secord has called a "rhetoric of spectacular display", inscribing themselves into their writing as omniscient authorities of an almost supernatural calibre—yet also friendly and approachable, politely showing the astonished reader around the wonders of the antediluvian world. They can seem rather like Raphael in Milton's *Paradise Lost,* the "affable archangel" who obliges Adam with scientific information from on high.[8] Yet not everything was revealed at once: tantalizing glimpses kept the reader enthralled. Suddenly, unexpectedly, a mass of hard facts about a rock stratum resolves into a stunning moment of epiphany in which the distant past is dimly glimpsed—and then, just as suddenly, the window mists over, and the reader is left gazing again at the blank rock-face. These bursts of vision drew on a range of literary and iconographic traditions. The new mythology of the ancient earth was stitched together with strands from all ages: the dreams and visions of Dante and Virgil rubbed shoulders with the refined topographical verse of James Thomson, and monsters straight out of Hieronymus Bosch mingled with the angelic beings of a more modern mysticism. Geology's popularizers were not necessarily hampered by contradictions between these worlds. Rather, they manipulated, even cultivated them as creative tensions within their writing, to titillate readers as much as to instruct them. In this book I will show how the power of the word was wielded to stage the story of life before man.

Nowadays, the staging or restoration of extinct monsters tends to lean heavily on pictures or models, often moving ones.[9] Steven Spielberg's film *Jurassic Park* (1993), Tim Haines's BBC documentary *Walking with Dinosaurs* (1999), and the animatronic creatures found in numerous museums all make some use of words and texts, but in a subordinate position to the visual material. Dinosaur books for the general reader are lavishly illustrated, and one can often learn much from these books without reading the main text. The situation was different in the early nineteenth century. The imagination was popularly seen as an organ which "pictured"

7. Mantell 1827, 78; Miller 1841, 45.

8. J. Secord 2000a, 439, 98.

9. In this book I use the term "restoration" to denote a representation of a creature or landscape in its living state, and "reconstruction" to denote the (re)assembling of a skeleton, with or without mock-up bones to supply gaps. However, these two words were (and are) often used interchangeably.

things, a kind of internal vision; but, despite some dissenting voices, the written word was widely felt to be the most reliable vehicle for calling up those pictures in the mind's eye. A good description, making use of visual tropes, was thought to be more effective at conveying vivid images than the act of looking at the object described.[10] This hierarchy was even reflected in urban visual culture which, for all its pictorial novelty, was rooted in reading: viewers "read" pictures with reference to well-known poems or novels, and upmarket exhibits were accompanied by detailed guidebooks.[11] Pictures of past worlds were introduced only gradually and cautiously into geological publications and displays. Even when "scenes from deep time" had become an established pictorial genre, the main burden of communication still fell on words.[12] So, when picturing the remote past, readers' imaginations were expected to work harder than they are now. This is something we tend to forget when we picture the Victorians in our own image, as a nation obsessed by new and spectacular visual displays from the diorama theatre to the glass shop front. This portrait has much truth in it, yet, even at the panorama, the Victorian mind's eye was given more exercise than is its present-day equivalent.

It is not surprising that some today find it difficult to see what was so appealing about nineteenth-century geological writings, for all that they were bestsellers in their day. Their poetic passages may seem like islands of evocative prose separated by oceans of fossil descriptions, lengthy discussions of strata, lists of minerals, and osteological analyses. To nineteenth-century readers, however, the moments of epiphany were not just romantic interludes; they were windows into a brave new world richer than romance. Through them we can look behind the stony details and glimpse the indescribable sensations underlying the whole, as the geologist—even the most articulate of geologists—gropes for the appropriate language.

Hugh Miller once described such a vision as having come upon him one summer's day in 1844, after a picturesque sea journey weaving among the Inner Hebrides. Alighting at Oban, he and his friend saunter off to inspect a nearby cliff, where by chance a section has recently been exposed. There they find a classic geological unconformity: the base of the cliff is made of slate, its strata upturned by "some long anterior convulsion", while on top of the slate lies the Old Red Sandstone conglomerate, a rock made up of pebbles loosely jumbled together—including bits of the original slate, broken and water-worn. Gazing at these two rock

---

10. On this hierarchy see W. Mitchell 1986.

11. See chapter 7.

12. The phrase "scenes from deep time" is taken from Rudwick 1992. "Deep time" was coined by John McPhee (1980, 20).

formations, Miller projects himself back into the period when the conglomerate was formed. In that world, strange fish sported in the depths; trilobites crawled the ocean floors; the land was barren, and even the reptile was a thing of the far future. Yet in that "incalculably remote period", reflects Miller, the slate must still have presented the same water-worn, pebble-chafed appearance as it does today: "it was in every respect as ancient a looking stone then as in the present late age of the world." His imagination soars away in a widening spiral which draws in the isle-studded seascapes of his journey—until language finally fails, and the bare rock comes back into view:

> a bit of fractured slate, embedded among a mass of rounded pebbles, proves voluble with idea of a kind almost too large for the mind of man to grasp. The eternity that hath passed is an ocean without a further shore, and a finite conception may in vain attempt to span it over. But from the beach, strewed with wrecks, on which we stand to contemplate it, we see far out towards the cloudy horizon many a dim islet and many a pinnacled rock, the sepulchres of successive eras,—the monuments of consecutive creations: the entire prospect is studded over with these landmarks of a hoar antiquity, which, measuring out space from space, constitute the vast whole a province of time; nor can the eye reach to the open shoreless infinitude beyond, in which only God existed: and—as in a sea-scene in nature, in which headland stretches dim and blue beyond headland, and islet beyond islet, the distance seems not lessened, but increased, by the crowded objects—we borrow a larger, not a smaller idea of the distant eternity, from the vastness of the measured periods that occur between.
>
> Over the lower bed of conglomerate, which here, as on the east coast, is of great thickness, we find a bed of gray stratified clay, containing a few calcareo-argillaceous nodules.[13]

Narrative—as "voluble" as the pebbles in view—leaps out from a rock formation, then abruptly leaps back in. Miller presents this passage as a fragment of the "sermons [. . .] which, according to the poet, are to be found in stones",[14] implying that similar visions can be seen wherever the bare materials present themselves. The sudden shift from seascapes of the mind to "the lower bed of conglomerate" may seem bizarre to some modern readers; but, from another perspective, the conglomerate itself

13. Miller 1858, 6–7. Miller's language of deep temporal perspective (on which see Pointon 1979, 89) harks back to Playfair 1805, 72–3 (discussed below, p. 56). Miller's book *The Cruise of the Betsey* has recently been reissued in a superb annotated facsimile edition (Miller 2003).

14. Miller 1858, 7, alluding to William Shakespeare, *As You Like It*, II.i.17 (Shakespeare 1993, 696). Compare Plate 4.

now appears charged with numinous potential. Miller's prose reveals the glory locked within the earth's structure.

By examining how and why this kind of narrative developed, I hope to illuminate how science worked in nineteenth-century public culture. Literary criticism might seem remote from the interests of today's historian; even among literary scholars, the aesthetic qualities of a piece of writing are often treated as a trivial affair, unworthy of serious attention because divorced from the concrete "realities" of politics or economics. Some historians have sought to displace the traditional scholarly obsession with textual evidence, focusing instead (with noted success) on the roles played by museums, public lectures, natural-history collecting, and industrial technology in the developing cultures of science.[15] Of those historians who do focus on the literature, most have steered clear of the words themselves, treating books and articles either as sources for scientific ideas (the traditional approach) or, more recently, as physical bibliographic objects, springboards for the study of readers, publishers, and printers. This newer approach is encapsulated by Jonathan Topham's dictum: "books are far too important to be treated merely as texts."[16]

Yet an understanding of narrative strategy is fundamental when examining a science like geology, whose dimly lit visions of the past sparked off a cacophony of competing versions of earth history. Visual and aural media certainly played a vital part in generating these narratives, but texts remained central, whether interacting with other media or standing alone. If we are to appreciate how and why geology appealed to its nineteenth-century public, we need to engage imaginatively with these texts. Here literary criticism, tempered by a historically informed sensitivity to aesthetics, comes into its own as a historical tool. As one Edinburgh periodical, the *Presbyterian Review,* put it in 1841:

> No man will maintain that the existing zeal for geology results entirely from the pure love of science. A few practical men cultivate it for its economical advantages. A few love it for its own sake. A very few study it that they may consecrate it. Many study it because they wish to be collectors [. . . .] But probably the greater number of its votaries are converts to what may be called the *literature* of the science.[17]

My analysis of "the literature of the science" will not explain the whole "conversion" process, even within a single country. Secord's magisterial reception study, *Victorian Sensation,* reveals just how much information

---

15. On this move see Cooter and Pumfrey 1994, 255.

16. Topham 1998, 262.

17. Anon. 1841–2, 210–11. The italics are original. Unless otherwise stated, all italicized words within quotations in this book are original.

must be assimilated and interpreted before the impact of even a single text can be convincingly recovered.[18] When dealing with a whole realm of scientific thought, embodied in numerous texts and other media, we can hardly hope to recover a full account. But before a reception history can even be dreamed of, some basic groundwork needs to be done on precisely what these people were trying to communicate in the first place, and how they meant to do so. Martin Rudwick has provided a comprehensive account of how geology became a historical science among the elite community of savants in Europe and America;[19] my emphasis will be on how its British popularizers projected this geology into the public realm. These insights can, in turn, suggest what kinds of audiences the authors were reaching out for, and what responses they expected. In short, we need to start reading.

The historian may still feel that this emphasis on authorial intention is misplaced. In promoting a less elitist account of popularization, Roger Cooter and Stephen Pumfrey have called for more scholarly focus on readers than authors, pointing out that audiences for science did not passively imbibe the information the would-be experts gave them, but actively transformed its meanings.[20] The self-confident scientific treatises of the 1830s are certainly ripe for this revision of intentionalist approaches: Topham's reception study of William Buckland's *Geology and Mineralogy* reveals how meanings directly opposed to the book's intended lesson were created by different readers in different contexts.[21] Nevertheless, this fluidity had its limits: literary culture was not simply a playground for the celebration of interpretative freedom.[22] We may celebrate the emancipation of a mythical "common reader" from authorial tyranny, but authors were no more dead then than they are now. Mantell's books, for instance, were advertised as being "by the author of 'The Wonders of Geology'"; and Lord Byron played fast and loose with the illustrious name of Georges Cuvier in his cosmic drama *Cain*.[23] Bringing extinct monsters to life was difficult and dangerous work, liable to be criticized as fantasy. This situation led to a persistent conservatism in depictions of the ancient earth: the words and authority of geology's better-known literary figures were routinely borrowed by other writers to do this work

18. J. Secord 2000a. On the problems involved in constructing a reception history which is both readable and properly representative see *ibid.*, 518–22.

19. Rudwick 2005b traces this development between 1787 and 1822; its sequel (Rudwick 2008 in press) takes the story up to about 1845. See also the essays collected in Rudwick 2004 and 2005a. For an alternative account see Oldroyd 1979.

20. Cooter and Pumfrey 1994, 249.

21. Topham 1998.

22. J. Secord 2000a, 521.

23. Mantell 1849, title-page; Byron 1986, 882 (on which see R. O'Connor 1999). On scientific authorship and reputation see Fyfe 2004a, 199–201.

INTRODUCTION

for them. The writings of these better-known figures will therefore take up a significant part of my discussion.

Yet even the most authoritative of authors did not necessarily intend a single monolithic "meaning" to enter the public realm. Most writers were aware of the heterogeneous nature of their public, and played to this plurality. As Topham puts it, "we have not only to contend with *readers* who actively multiplied the meanings of these works, but also with *authors* who intended their meanings to be multiple"—authors like Buckland, who aimed his treatise *Geology and Mineralogy* at geologists, clerics, and laymen alike.[24] These differences come across vividly in the spectacular rhetoric employed: Buckland opened up his science to multiple meanings by presenting pterodactyles as, simultaneously, monstrous Miltonic "fiends" and "beautiful" exemplars of the Creator's wisdom and benevolence.[25] Jon Klancher has suggested that the "foundational texts of British Romanticism" characteristically acknowledge the possibility of "colliding readings" in the "intense friction of their language", and Gillian Beer has shown how Charles Darwin struggled with conflicting metaphors in his heroic attempt to forge a satisfactory language for talking (and thinking) about the transmutation of species.[26] As we shall see, the histories of Creation written by Darwin's older contemporaries are shot through with similar tensions.

This, then, is not a reception study as such, but a study of a science's literary projection. The sources, of course, overlap: things we tend to see as evidence for reception (diaries, reported conversations, letters, reviews, advertisements) were cultural performances in their own right, while the usual evidence for projection (papers, lectures, exhibitions, books) also shows how a particular idea has been received by the author or designer. In this kind of analysis, which charts the changing meanings of old and new concepts, it is necessary to have a clear sense of the conceptual tools employed. For practical purposes, this means defining terminology and the ways in which it is to be used.

## Thinking about the Past

As geological writers repeatedly found, it is impossible to write about the past without infusing it with the concerns of the author's own time. For the historian of science, however, it is important to attempt some level of empathy with past perceptions, however outlandish the latter might

24. Topham 1998, 239.
25. W. Buckland 1836, I, 222–5.
26. Klancher 1987, 177; Beer 1983.

seem today. To this end I have retained nineteenth-century terminology wherever practicable, "translating" words into present-day analogues (or coining new formulations) only where this serves the analysis. Although some anachronisms will be inevitable, historians of science are generally agreed that words like "scientist" and "dinosaur" should not be used when referring to periods before these words were in use. They are not neutral labels, but reflect categories of thought quite different from their previous analogues. To call Mantell "the discoverer of the dinosaurs", as if this conceptual category were sitting in the Tilgate quarry waiting to be "discovered", is to miss the literally unspeakable strangeness Mantell experienced as he grappled with the notion of a herbivorous lizard larger than the largest elephant.[27]

In the world of science, moreover, new terms rarely passed into common use as soon as coined. "Scientist" and "dinosaur" (coined in 1833 and 1842 respectively) were rarely used before the late nineteenth century, while the word "prehistoric", coined in mid-century, was not widely used in the sense of "prehuman" until the early twentieth.[28] At a cosmetic level, retaining the most current early-nineteenth-century spelling of each animal discussed ("hyaena", "pterodactyle") helps preserve a sense of their strangeness which we have perhaps lost. In today's science books the names of extinct genera are always italicized (*Iguanodon*, *Plesiosaurus*), marking them off textually as the intellectual property of science; but the Victorian saurians jostled in public culture alongside dragons and leviathans, and their names were only occasionally italicized.[29] To avoid introducing spurious present-day meanings, I have also left place-names in their early-nineteenth-century forms, used gender-specific phrases such as "the creation of man", and reproduced the original spellings of

27. On this philosophical problem see Torrens 1999, 188.

28. William Whewell, who coined the term "scientist," was nervous about using it even in 1852 (Whewell 1852, 33). The fullest analysis of categories of extinct animals is Taylor 1997, xxxvi–xxxix. See also Torrens 1999; J. Secord 2004, 164–5. The commonest adjectives connoting "prehuman" in the nineteenth century were "ancient" and "antediluvian" (literally "before the Flood"), reflecting the lack of a clear and agreed boundary between human and prehuman history. This confusion increased when geologists in the second quarter of the century came to view the so-called geological deluge as a distinct event from the biblical Deluge: the term was already too widely used to be dropped from popular scientific discourse, resulting in further blurrings. The categories "saurian" and "dragon" included marine and flying reptiles as well as crocodiles and lizards; "monster," while problematic, was commoner still and included extinct mammals, fish, and big birds.

29. Species names (*Iguanodon mantelli*) were, however, increasingly italicized from the 1840s on as the scientific community tightened up the taxonomic system. Furthermore, with ongoing research, species once ascribed to a single genus are often now assigned to several genera; for instance, *Ichthyosaurus* now comprises a much narrower range of species than in 1840 (Michael Taylor, pers. comm.).

quoted passages without the help of that pedantic little word *sic*.[30] The past is, after all, a foreign country.[31]

In this spirit I have also avoided mentioning current thinking on the specific geological questions discussed, except in specific cases where such a perspective seems useful or stimulating. Contrasting the state of a nascent science with "what we now know" offers a wonderful boost of cultural self-esteem; but it is no part of my purpose to point up Mantell's mistake in placing the Iguanodon's spiked thumb on its snout, or to observe that today's archaeologists interpret the "Temple of Jupiter Serapis" at Pozzuoli (made famous by Charles Lyell) as a marketplace. Readers interested in finding out about the "real" dinosaurs or Romans are encouraged to refer to the many excellent textbooks available in these areas. This book, by contrast, subjects the whole idea of a "real" Iguanodon to critical and historical analysis. It examines how imaginative restorations were constructed so as to convey what Roland Barthes called the "reality effect" to an audience hungry for simulations of distant objects.[32] This was the essence of successful popularization.

But what was popularization? In the late eighteenth century this term meant simply "making something widely known". With the rise of new print technologies and the broadening of the reading public in the 1820s and 1830s, the term began to be used in a more specialized sense, referring to the "diffusion" of knowledge from a new breed of self-styled "experts" down to a passive public. This model for public science was promoted by certain gentlemen from the 1820s onward and is exemplified by the writings of Lyell, although the work of communication was increasingly left to professional science writers (disparagingly termed "hacks" since the 1820s), many of whom were not scientific practitioners. Because this diffusionist model of knowledge production continues to enjoy wide currency in the public science of our own time, the word "popularization" is often casually used in the same restricted sense, with the same powerfully hierarchical assumptions—reflected, for example, in its frequent pairing with those weasel words "mere" and "only".[33]

However, many of these so-called hacks were not (and still are not) neutral mouthpieces for scientific practitioners, but pursued their own agendas and made new meanings out of the science they communicated, even as they wielded the words of established authorities. The docile audience, too, proved to be more of an ideal than a reality. Clearly,

30. Obvious errors of spelling in quoted passages are, however, corrected using square brackets. Italics are original unless otherwise stated.

31. Some of these problems are discussed in Lowenthal 1985.

32. Barthes 1986, 141–8.

33. For discussion see Whitley 1985; Cooter and Pumfrey 1994; Topham 2000; R. Williams 1983, 237.

the diffusionist model cannot account for the totality of nineteenth-century public science. As a result it is not suitable for use by historians as a neutral analytical tool; it is itself a cultural formation which needs to be put under the microscope. For our purposes a more inclusive concept is required to cover the forms by which science manifested itself. In this book, rather than inventing a new term, I reclaim the word "popularization" by using it in its older, more inclusive sense. This inclusiveness extends to all the forms and genres in which earth science was communicated: "geological popularization" was not restricted to books or museums, but included any medium by which geology of any kind was propagated beyond an immediate circle of initiates, whether in utilitarian or imaginative forms. This produces a daunting array of materials in literary, oral, visual, architectural, and other media. Even with a primary focus on texts, some degree of selection is called for. In this book I focus on what contemporaries would have seen as the "imaginative" end of the spectrum, on texts which restored visions of vanished worlds before the mind's eye—even if these restorations were only fleeting, or marginal to the main body of the book in question. What we are exploring is not a genre or generic system, but a cluster of modes and techniques by which writers interested in geology painted the deep past.

Then there are the problematic concepts of "popular culture" and "popular science".[34] Too often, the term "popular" carries an implication of a simple oppositional relation to a ruling establishment. Whether "popular culture" is construed negatively (ignorance or bad taste untouched by enlightenment) or positively (resistance against tyranny or marginalization), the term arguably begs too many questions to do any useful work purely as a term. This is not to say that the promoters of earth science avoided using socially divisive terms themselves: this could help present them as authorities and flatter their chosen readers. Thomas Ashe's guidebook to the fossils at William Bullock's museum addresses itself in its subtitle to "ladies and gentlemen" and later makes disparaging references to "the vulgar"—defined as those people who are not interested in fossils, and are therefore not reading this guidebook. At the same time Ashe used a highly sensational rhetoric, sounding variously like an auctioneer calling for bids, a circus ringmaster, or the narrator of a chapbook romance. Despite posing as a genteel naturalist, Ashe was no gentleman, but a blackmailer and con-man whose vicious and picaresque career left him little time to acquire much scientific knowledge.[35]

Such cases are complex, and this Irish adventurer was by no means

34. On the historical problems associated with these concepts see Shiach 1989; Anderson 1994, 7–12; Cooter and Pumfrey 1994.
35. Ashe 1806, title-page, separately paginated "Introduction", 12. For discussion see chapter 1.

alone in simultaneously cultivating and repudiating "vulgar" sensationalism: we find the same mixture in the work of an Oxford don, a Ross-
shire stonemason turned accountant, a Sussex surgeon, and the son of
a Somerset cattle-dealer.[36] Social boundaries blurred in other ways, too.
Elements we might associate with popular culture, such as the singing of
ballads, enlivened the festive side of genteel geology in the 1840s; at the
same time, the ostentatious use of Latin and Greek scientific terminology
was keenly taken up by artisan naturalists.[37] Then, as now, popular culture was a contested category rather than a historical fact: in the tapestry
of nineteenth-century public science, overt oppositions often concealed
underlying continuities and exchanges. Of course, several of the differences between competing interest-groups ran deep, and their clashes can
often be mapped onto larger socioeconomic divisions: the question of
whether (and how far) science should be "made available" to the lower
classes was bitterly fought over during the early nineteenth century, and
the politics of popularization forms a central thread in the story I am going to tell. But these issues tend to be obscured rather than illuminated
when the terminology of a uniform popular culture is wheeled in as an
explanatory device.[38]

To allow for a clearer perspective on ideologies of popularization,
then, I shall be using words like "popular" and "popularization" as neu-
trally as possible, without any implicit social assumptions. In this book,
"popular" refers simply to the presence (real or imagined) of a non-specialist public, whose identity and constitution varied.[39] Knowledge has
always been socially distributed, but the disparity in incomes across the
social spectrum was far greater in the early nineteenth century than it
is today. The top slice of church livings earned their incumbents over
£1,000 per annum (perhaps equivalent to £175,000 or $347,000 today),[40]
but many curates earned as little as £50 per annum (= *c.* £8,750 or $17,350
today), while tradesmen and servants were lucky to earn half that. The

36. Respectively, William Buckland, Hugh Miller, Gideon Mantell, and Thomas Hawkins.

37. These artisans did not necessarily know the English meaning of the Latin terms, still
less the Latin language more generally (A. Secord 1994b, 292–3).

38. For similar cautionary remarks on the subject of botany see A. Secord 1994b.

39. On this sense of the word "popular" see Shiach 1989, 27–8. On the concept of a "public"
for science see Shapin 1990; Cooter and Pumfrey 1994.

40. My ratio of 1:175 for pounds sterling was arrived at by the crude means of comparing
today's median average British full-time annual salary (*c.* £21,840) with Topham's (1992, 400 n.
19) lower estimate of typical "comfortably middle-class" British incomes in the 1830s, 48s per
week. Rudwick (1985, 461) has suggested 1:40 from a more genteel perspective, while Taylor
and Torrens (1986, 145–6) have suggested 1:200 from a farm labourer's perspective. As all these
scholars have noted, however, such ratio calculations are only of value as a very rough guide.
For further discussion see chapter 6. My equivalent sums in U.S. dollars have been calculated
on the basis of today's median average American full-time annual salary (*c.* $43,300, yielding a
ratio of £1:$347). On nineteenth-century currency see the appendix.

usual price for visiting a "typical" early-Victorian exhibition such as the panorama was 1s—a drop in the ocean for the high-earning vicar, but over half a day's wages for the tradesman. Phrases like "the era of cheap print" should not lead us to think that these writings were equally accessible to all classes: it was not until the 1860s that working-class readers could afford to buy a wide range of up-to-date scientific publications.

"Popular geological writing" in this book includes not only cheap penny periodicals and illegal unstamped newspapers, but also books, which were prohibitively expensive for all but a minority of the population. As Richard Whitley has noted, popularization includes "all communication to non-specialists which involves transformation" (this last qualification being something of a tautology).[41] By definition, to publish a text is to place it before a public, though the nature and breadth of that public varied during the half-century under consideration. Before the 1860s most writing on geology was to some extent "popular", and "popularizers" cannot be cordoned off from "scientists" as though they were two separate groups.[42] Periodical production kept pace with book production until the 1850s (when books were overtaken), and periodical reviews and miscellanies became increasingly significant in the developing culture of public science. The present analysis will focus primarily on books, whose greater scope, permanence, and authorial presence enabled large-scale literary strategies to be developed with more freedom.[43]

The word "literature" also needs divesting of some of its present-day associations. One of my central claims in this book is that science writing was an integral part of nineteenth-century literary culture—not that science writing and literature enjoyed a fruitful relationship, but that scientific writing *was* literature. Today, the terms "literature" and "literary" are most often used in a sense which excludes non-fiction: they denote a limited range of usually fictional works including novels, poems, and plays. In the eighteenth and nineteenth centuries, however, "literature" was a more inclusive concept, incorporating scientific and historical writing as well. As disciplinary specialization intensified, some branches of this unitary literary culture began to bud off as separate concerns, resulting in new boundaries and self-images which demanded new labels or the narrowing down of old labels. By 1800, for example, the word "science"

41. Whitley 1985, 25.

42. This dichotomy (modulated into "natural history" versus "science") informs the otherwise excellent analysis in Merrill 1989. It is perhaps less out of place when discussing the late nineteenth century (e.g. Lightman 1999, 2–4), when the professionalization of both writing and science had gained momentum; yet Thomas Huxley continued to be a fully paid-up member of both groups, and one could cite many more recent examples (see the essays in Shinn and Whitley 1985).

43. On the relative paces of book and periodical production see Dawson *et al.* 2004, 10.

was often used in a sense which excluded theology, once the "queen of the sciences", and this exclusion was reinforced in the 1830s by the new coinage "scientist", designed to include only the natural sciences.[44] Likewise, by the late nineteenth century a move was afoot among a subgroup of literary professionals to redefine "literature" in terms of a new concept of the "creative imagination", excluding non-fiction: they took their cue from their readings of certain early-nineteenth-century poets, for whom they coined a new collective label, "the Romantics".[45] Some scholars have attributed this redefinition to the Romantics themselves: Klancher, for instance, cites it as a "fact" that the word "literature" meant "one thing in 1780—the whole array of educated genres from natural philosophy and history to poetry and drama—and something very different after 1820 (the restricted category of imaginary genres we know today)".[46]

The question is, to whom did this word "literature" mean "something very different"? To a tiny handful of self-styled literary reformers, or to the middle- and upper-class reading public as a whole? This question is just as pressing when applied to the late-Victorian period, where the redefinition of "literature" is supported by more concrete evidence. Citations in the *Oxford English Dictionary* reflect the continuing inclusiveness of the term in the teeth of this redefinition, and many nineteenth-century journals and anthologies which claimed to contain only "literature" or "literary" material included scientific, historical, and travel writings under this heading.[47] On the other hand, one need only look at the above quotation from the *Presbyterian Review* of 1841, commenting on "what may be called the *literature* of the science", to see that some writers were slightly uncomfortable referring to scientific writing as "literature". Ever since the seventeenth century, natural philosophers had bolstered the truth-claims of their practices by defining them against "fictions" of the imagination. By 1800, vigilance against the deceptive potential of language and imagination constituted one of the foremost tropes of scientific truth-claims.[48] Conversely, several poets of the period were promoting their own claims to philosophical or spiritual truth against the myopia of mere fact-gatherers, a position which many of their Victorian (and later) admirers misread as hostility towards science in general. Later in the century, this specific opposition fed directly into the more generalized debates among educationalists over the relative importance of Classical literature and modern science, and the political ramifications of

44. Ross 1962; Yeo 1993, 32–3.
45. On the term "Romantic" see St Clair 2004, 211–13 (especially n. 10).
46. Klancher 1994, 524. See also Siskin 1998, 6; Heringman 2004, 7.
47. OED, s.v. *literature* 3a. For further discussion see R. O'Connor 2005a and chapter 6 below.
48. Christie and Shuttleworth 1989b, 1–2; Daston 2001.

this debate led ultimately to a perception of a still more generalized gulf between "two cultures", as C. P. Snow would put it in 1959.[49]

This rhetoric of a fundamental opposition between "science" and "literature" was (and is) a vital, constitutive element in public debates on science's place in culture. Yet, while we cannot ignore it, we must not take it at face value: to remain sensitive to these shifting fields of meaning, we need once again to attend to underlying continuities as well as public oppositions. Scientific popularization is a case in point: science, a group of practices defined against the imagination, was promoted in literary texts using imaginative techniques. The tension between practice and representation is particularly evident in writings on earth history, because—as one present-day palaeontologist has acknowledged—"the past is always a fiction".[50] History requires narrative: in exploring how the fossil past was represented, we shall see how dramatic techniques and poetic images flooded in to give life to the bare bones. My aim is not to expose scientific truth-claims as illusory, or to assert that science can be somehow "explained" in purely literary or narrative terms. Rather, I aim to show how the truth-claims of public science have been supported by (and expressed within) structures which we are used to thinking of as fundamentally opposed to scientific procedure.

By examining science *as* literature, rather than science *and* literature, I hope to complicate some of the oppositions to which the latter duality has given rise. To put the case crudely, nineteenth-century savants' anxieties about objectivity and poets' concern for subjectivity cannot simply be mapped onto a disciplinary divergence of "science and literature". Much poetry, and still more prose fiction, continued to be valued for its factual content; equally, scientific writing continued to have an aesthetic dimension, both self-consciously and by default. Some recent scholarship has worked to restore these consonances,[51] but there is a risk of perpetuating the old polarities if we continue to treat scientific writing purely as a vehicle for communicating scientific facts and ideas, while treating "literature" purely as a space for their subjective expression. The polarization becomes still more acute when "science" is represented only by a canon of well-known scientific thinkers, and "literature" only by well-known poets, novelists, and dramatists. In these pages, I shall be using the term "literature" in its most inclusive sense, denoting any written text. Writers are qualified for discussion if they simply sought to fire up British readers' imaginations on the subject of earth history—not only in scientific treatises, but also in pamphlets, magazine articles, epic

49. On the late-Victorian debates see Beer 1990, 786–7; P. White 2005. On the "two cultures" see Snow 1993 and Collini 1993.

50. Fortey 1997, 253; compare C. Cohen 2002, 220.

51. See in particular J. Smith 1994 and Heringman 2004.

poems, autobiographies, children's stories, travelogues, and sermons. By approaching science writing from this angle, it will become possible to redraw the literary landscape.

These writers included not only pioneering geologists like Buckland and Lyell, but the whole spectrum of scientific expertise: local fossil collectors, specialists in other sciences, proponents of a biblically literalist young-earth geology, and writers who never touched a hammer in their lives. Whether they derived their information from the field, the library, or their own imagination, my focus will be on how they communicated it. Terms like "amateur" and "professional" cannot be introduced into this discussion without confusion, because their present-day connotations worked the other way round in the early nineteenth century.[52] Those whose profession was geology—civil engineers such as William Smith who practised it for a living—were looked down on by their social superiors: the gentlemanly geologists, exemplified by the leading lights of the Geological Society of London, had the leisure to explore and synthesize the whole field and thus came to win middle- and upper-class support as the final arbiters of scientific truth. They styled themselves as an elite, a term often used by historians today.[53]

But while these gentlemen may be seen as having won the public ear, their message reached that ear in forms which sometimes differed considerably from what was being talked about within the rooms of the Geological Society. Furthermore, this elite was neither fixed nor universally recognized: its unspoken membership constantly shifted, and for many members of the British public its authority was as nothing compared to longer-established elites such as Classical scholars and learned clergymen. Joining the clergy, indeed, was the high road to scientific authority in the early nineteenth century, and the many "Reverends" among the gentlemen of science are not to be dismissed as clerical dabblers. This negative image has become so prominent in popular accounts, and clerical labels such as "the Rev." or "Dean" are now so often used as a sly shorthand for "amateurish" or "prejudiced", that I have decided in this book to omit such designations altogether, discussing professional commitments only where relevant.[54]

What of the word "geology" itself? This word became current among men of science in the last quarter of the eighteenth century, but its se-

---

52. On this semantic minefield see Torrens 2006b.

53. Porter 1978a; Morrell and Thackray 1981; Rudwick 1985 (especially 418–28); J. Secord 1986. On the social composition of the Geological Society see Wennerbom 1999; on the practical men see Torrens 2002.

54. On the persistence of the epithet "Dean Buckland," see Boylan 1984, 512.

mantic range was more expansive than it is today.[55] Then, as now, it connoted the study of the earth's structure and history; but then it was primarily associated with Enlightenment cosmology and "theories of the earth", a learned literary genre in which the world's physical origins were authoritatively narrated.[56] In the 1780s and 1790s this speculative literary discipline became associated with other established practices: biblical exegesis, natural history, antiquarianism, mineralogy, surveying, and mining. These components were variously reconstituted and realigned in the struggle to define geology's conceptual boundaries.[57]

Like Frankenstein's monster in Mary Shelley's novel, this unstable composite science lurched into the world shortly before the French Revolution ended centuries of absolute monarchy in France and replaced it with the First Republic (proclaimed in 1792). Conflicts within and without brought extremists to power, and in 1793–4 the bloodbath known as "the Terror" alienated many of those in Britain and elsewhere who had initially supported the Revolution. These developments provoked a strong reaction against Enlightenment cosmological speculation, especially from abroad. French anticlerical *philosophes,* who had revived Aristotle's idea that the earth had always existed, were blamed for fostering a godless materialism and plotting the Revolution; so when the Edinburgh savant James Hutton claimed to find in the strata no evidence of historical direction ("no vestige of a beginning,—no prospect of an end"), his ahistorical *Theory of the Earth* was widely vilified by conservative critics as an "infidel" production.[58] All such speculation about earth history came under suspicion. In these turbulent times, people who doubted the Bible's literal inerrancy risked being popularly labelled "atheists", and while most geological writers did not subscribe to Hutton's ahistorical eternalism, most did not accept a six-day Creation either. After a coup in 1799, the First Republic became an Empire under Napoleon Bonaparte, and when Napoleon began waging his expansionist wars between 1803 and 1815, British Francophobia reached new heights. Geologists continued to be seen as a dangerous species, particularly because several of them were former radicals, and many were Dissenters, worshipping outside the Established Church.[59] The nation's spiritual health seemed in jeopardy.

55. On the history of this concept see Rudwick 2005b (especially 133–5, 345–8, and 644–8); see also Dean 1979 and Rudwick 1996.

56. Rudwick (2005b, 134) has called the "theory of the earth" a "scientific *genre,* just as landscapes, operas, sonnets, and novels were artistic genres," to which it might be added that the "theory of the earth" was *also* a literary (and hence artistic) genre.

57. Porter 1977.

58. Hutton 1788, 304; Gould 1987, 61–97; Hole 1989, 152–6; Dean 1992. See also Brooke and Cantor 1998, 195–200.

59. Gillispie 1951; Garfinkle 1955; Porter 1978b, 435–6.

To avoid controversy in this climate of paranoia, many geologists ostentatiously isolated their science not only from politics and theology, but from any form of speculation. Geology was to comprise strict empirical induction, focusing upon the relative positions of the strata; and fossils became increasingly important tools in this stratigraphic project. In England, several practitioners dissociated themselves from the delusions of "theory" in a conscious break with the science's past, sharpening the anti-literary rhetoric discussed above. When one group of genteel geologists formed the Geological Society of London in 1807, they deliberately narrowed the semantic range of the term "geology", defining it against the cosmological speculation with which it had been synonymous. They cultivated a historical myth according to which their predecessors had been romantic theory-mongers while they themselves were sober men of science. In this context, by the mid-1810s, terms like "romance" and "poetry" amounted to a standardized rhetoric of dismissal.[60]

Yet public interest was required, and geology had to avoid the opposite taint of being boring. This was never a problem with "theories of the earth", which continued to be read and admired for their imaginative power. Coleridge found Thomas Burnet's *Sacred Theory of the Earth* (1680–9) to be "poetry of the highest kind",[61] while the Comte de Buffon's *Époques de la nature* (1778) vied with Burnet's work in the poetic force of its visions of former and future worlds.[62] Although they dissociated themselves from the content of these works, the new geologists wanted to maintain their sense of excitement: the Geological Society retained the word "Geological" partly for its romantic associations with a manly outdoor lifestyle and heightened sensibility.[63] The literary techniques of the old genre would exert much influence on later geological writing, and the Victorian drama of earth history owed much of its force to the providential historicism of sacred chronology and universal history, another parent discipline from which many early geologists were quick to distance themselves.[64] So, while the term "geology" was appropriated and publicly defined against its older, imaginative connotations, the lat-

60. For examples see Porter 1977, 204–8; Gould 1987, 23–4; Hamblyn 1996, 201–3; Daston 2001; Heringman 2004, 270–1. On the Geological Society generally see Boylan 1984, 305–39; Rudwick 1985, 18–27.

61. Burnet 1965; Coleridge 1985, 318. Compare [Rennie] 1828, 11–12. On Burnet and his reception see Nicolson 1963, 184–270; Dean 1968, 80–4.

62. Buffon 1988. For one early-nineteenth-century British response to Buffon see the letter by Percy Shelley in F. Jones 1964, I, 499 (and Leask 1998b, 194–6).

63. On the "romantic" dimension of early geology see Rudwick 1963, 328; Dean 1968; D. Allen 1976, 52–72; Sommer 2003. On the quite different "romanticism" of German geology see Rupke 1990; Ziolkowski 1990, 18–63.

64. On the relations between chronology, exegesis, and earth science see Rudwick 1986; 2001; 2005b.

ter were tacitly turned to its advantage. Imagination was at once the site of fatal delusion and vital energy.

At the same time, the evocative word "geology" was also claimed by many writers who held that the earth had been made in six solar days a few thousand years previously. Most early writers on earth science, evangelical or otherwise, did not adopt this particular form of biblical literalism, but it was widespread in Genesis commentaries well into the nineteenth century and enjoyed considerable public currency.[65] Commentary on the nineteenth-century debate between proponents of young-earth and old-earth geologies has traditionally been founded on an inherent opposition between "geologists" and "biblical literalists", and in this book I use the term "literalist" as a shorthand to refer to the proponents of a young earth; but it should be remembered that some of these literalists saw themselves as geologists, and some were acknowledged as such by their opponents, while many proponents of an old earth felt that this did not violate their own "literal" interpretations of Genesis 1. In its broadest outlines, the early history of geological spectacle is the story of how a new intellectual community—the old-earth geologists and their supporters—laid exclusive claim to the term "geology" against the counterclaims of other communities promoting young-earth cosmologies. In a sense, the latter maintained the old genre of "theories of the earth" in its more conservative form.[66]

For large sectors of the reading public then and now, the new science succeeded, and the word "geology" became roughly synonymous with an earth history in which man was a relatively recent arrival. But "geology" was still a contested term. If we use it in a sense that automatically excludes the literalist side of the debate, we become complicit in a partisan view which—however justifiable from the standpoint of a present-day geologist—is inappropriate for the historian. In this book the words "geology", "geological", and "geologist" will refer to both schools of thought, unless qualified by the context.

## The Literary Background

Narratives of earth history had been current long before the nineteenth century. This literary tradition can be traced back to Classical times, and it had flowered in the great synthetic narratives of the eighteenth-

65. For some examples of commentaries taking this view see Mortenson 2004, 40–5.
66. Strictly, the terms "young-earth" and "old-earth" are anachronistic: they were not coined until the late twentieth century. Literalist geologies are discussed in more detail in chapters 3 and 5.

century "theories of the earth". However, a distinctively modern form of this tradition was born in early-nineteenth-century Europe and North America, when the old techniques united with new ways of thinking about fossils and the strata, and with a new model of scientific authority to address the expanding reading public. In this new narrative, fossils proved overwhelmingly popular for the purpose of promoting geology's appeal as a spectacle of deep history: they revealed the presence of living creatures, actors within the vanished scene. The term "palaeontology" was coined in 1822 to denote this branch of geology, but the word "geology" continued to be used in a very broad sense well into the Victorian period. For many people around 1850, "geology" simply meant "a scientific way of talking about fossils". My emphasis naturally falls on this aspect of the science.[67]

Stories of life before man took a bewildering variety of forms, differing according to the author's (and audience's) nationality, social class, regional identity, and intellectual or religious background. In order to give some sense of this variety while also allowing enough space for proper discussion, this book's scope is limited to a single country. Paris may have been renowned as the scientific hub of the world, but Britain's accelerating urbanization and cutting-edge print technology helped to place her in the vanguard of popular scientific literature, particularly in the second quarter of the century. Nevertheless, spectacle as well as science was a fully international practice: the same forms of display in which geological popularization was rooted, such as panoramas and biblical paintings, were current in most of the major urban centres in Europe and North America, including many of the same shows and canvases.

Our story begins in 1802, when the British public first set eyes upon the complete skeleton of a large extinct animal. This skeleton, on loan from Philadelphia, was bound ultimately for Paris. But books were easier to transport than monsters, especially before the consolidation of international copyright laws. The nineteenth century saw a constant traffic of scientific writings across the English Channel, the North Sea, and the Atlantic Ocean. Foreign works in British translations or imprints must therefore be considered part of the British literary scene, from Cuvier's *Theory of the Earth* (so called in Britain) to *The Religion of Geology* by the New England savant Edward Hitchcock.[68] By 1857 a Glasgow imprint of Hitchcock's book was being sold in Britain's railway stations for only eighteen pence, making it much more widely available to British readers than many home-grown products.[69]

---

67. For an outline history of vertebrate palaeontology see Buffetaut 1987.
68. Cuvier 1813; Hitchcock 1851 (on which see Guralnick 1972, 537).
69. Anon. 1857b, 317.

*Figure 0.1.* The anatomy of the landscape: Gideon Mantell's "Plan of the Stratification of the South Eastern part of Sussex", drawn by his wife, Mary Ann, and coloured by hand. Mantell 1822, plate 3, detail.

Although I have been speaking of a literary tradition, these narratives of earth history did not occupy a single genre, however broadly we might define that term. They often occupied only small parts of works belonging to other, much better established genres. This reflected geology's disciplinary origins as a composite science which drew on several older practices. Depending on the author, geology could be seen as a new branch of any of these practices. It was housed in biblical exegesis on the one hand, and in the overlapping literatures of landscape aesthetics, local antiquarianism, and natural history on the other. These links are manifest in the literary and iconographic conventions which came to dominate later geology. Despite the new science's ostensible self-definition against picturesque tourism, for example, drawings of stratigraphic sections often paraded their kinship with picturesque conventions of landscape depiction (Figs. 0.1 and 1.4): [70] Such drawings aimed to please as well as to inform.

In the eighteenth century there had been nothing controversial in the idea that scientific ideas could appeal to the imagination. Joseph Addison's "On the Pleasures of the Imagination" is a case in point. This essay, which remained a firm favourite in reprints throughout the nineteenth century, was first published in separate numbers of his middle-class literary periodical *The Spectator* in 1712. After praising the Roman historian Livy's consummate ability to bring the past to life, Addison granted a still higher place in the literary pantheon to "the authors of the new philosophy, whether we consider their theories of the earth or heavens", for

70. On the development of this visual genre see Rudwick 1976 (especially 175–6 and fig. 7); compare Rudwick 2005b, fig. 8.6. On geologists' anti-touristic rhetoric see Hamblyn 1996, 201–3.

their ability to "gratify and enlarge the imagination". But such writings are seen to elicit an even greater pleasure in showing the inadequacy of the imagination:

> when we survey the whole earth at once, and the several planets that lie within its neighbourhood, we are filled with a pleasing astonishment, to see so many worlds hanging one above another [. . . .] If, after this, we contemplate those wild fields of Æther, that reach in height as far as from Saturn to the fixed stars, [. . .] our imagination [. . .] puts itself upon the stretch to comprehend it. But if we rise yet higher, and consider the fixed stars as so many vast oceans of flame, that are each of them attended with a different set of planets, and still discover new firmaments and new lights that are sunk farther in those unfathomable depths of Æther, [. . .] we are lost in such a labyrinth of suns and worlds, and confounded with the immensity and magnificence of nature.[71]

The words "astonishment", "confounded", and "lost" imply that reason has temporarily fled, generating a "pleasing" sensation. By the mid-eighteenth century, this delight in awestruck confusion had come to be categorized under the umbrella term "the sublime".[72] Such writing often claimed to excite wonder at God's works, but more sensory pleasures were rarely absent: devotional reflection and sensational gut reaction were two sides of the same coin.[73] This dichotomy would soon be put to creative use by writers on geology, who exhibited freakish yet perfectly designed extinct animals and expressed the thrill of deep time in markedly similar ways to Addison's astronomical raptures.

Nor should it be forgotten that verse, as well as prose, had long been an important vehicle for scientific popularization. James Thomson's descriptive poem *The Seasons* (1726–30) was still widely read and reprinted in the mid-nineteenth century, and like the innumerable Newtonian odes of its time it was intended as a serious contribution to philosophy as well as to letters.[74] Didactic verse enjoyed a new lease of life in the hands of the celebrated doctor Erasmus Darwin, whose encyclopaedic botanical poems such as *The Loves of the Plants* (1789) enjoyed an international vogue, with their epic sweep and their steamy depictions of plants' love-lives. Darwin was widely considered a great poet until the conservative backlash of the 1790s, when his evolutionary speculations fell into dis-

71. [Addison and Steele] 1805, II, 208–9 [No. 420].
72. Ashfield and de Bolla 1996. For alternative approaches see Nicolson 1963; Heringman 2004.
73. Wilton 1980, 29–30.
74. On eighteenth-century scientific verse see Nicolson 1946; 1963; W. Jones 1966.

favour (although his techniques lived on).[75] In these scientific poems, it is extremely difficult to distinguish clearly between poetry which uses scientific ideas and verse popularizations of science: the two categories merged, and a canny publisher could easily transform the one into the other. By the 1800s, however, prose was becoming increasingly favoured as a vehicle for scientific popularization.[76] The elegant but flexible style of Addison was an inspiration for such writers as Buckland, Lyell, and Miller.

Geological writing, then, emerged at the beginning of the nineteenth century from a confluence of disparate genres. This allowed popularizers to channel pre-existing enthusiasms into a new and unfamiliar science. But eliciting enthusiasm and wonder was not the only aim of this eclectic literature, and in some treatises it played a distinctly minor part in terms of textual bulk. Consequently, although the integrity of each work must be kept in view, we shall necessarily end up examining parts more often than wholes.

This kind of dissection is invited by the literature's compilatory tendencies. As may be seen in review articles, anthologies, and periodical miscellanies, quotation was not just an occasional gesture; it lay at the heart of nineteenth-century literary culture.[77] Geology, springing from such a variety of disciplines, was particularly eclectic. Its authors made full use of neighbouring texts and genres by quoting them, often at length. At one level, this technique is unsurprising: you do not necessarily have to produce original prose in order to communicate information or articulate an argument. But writing in someone else's words also had its defensive uses for a suspect science. Buckland's inaugural lecture was stuffed with vindicatory quotations from Classical and theological authorities, while Mantell's first treatise was theologically insulated, sandwiched between a prologue provided by a clergyman and a final peroration culminating in devotional poetry.[78] Most of all, scientific writers quoted from themselves and from each other. Important facts and effective rhetoric were recycled, leaping out of their original context and doing service in another.

This compilatory attitude contributed to the stylistic instability of scientific writings. This instability is particularly clear in works aimed at

---

75. On the backlash see Garfinkle 1955; on his international importance see Boime 1990, 432–4, 477–9. On Darwin's poetry see McNeil 1986; Heringman 2004, 191–227. The continuing popularity of didactic verse in the early nineteenth century is discussed in Duff 2001.

76. Compare W. Jones 1966, 200–9.

77. On these three compilatory forms see Butler 1993, Price 2000, and Topham 2004a, respectively. On compilatory strategies more generally see Dean 1968, 16–38; Crawford 2000.

78. W. Buckland 1820; Mantell 1822, 1–13, 304–5.

a wide readership, which often display considerable stylistic variety in order to capture readers' imaginations. Quoting from a poem caused a sudden lurch into a new form of discourse, the break clearly marked by indentation. This lurch could be just as sudden when the writer launched into lyrical descriptive prose, as in the example from Miller's *Cruise of the Betsey* quoted above. Such demarcation was encouraged by the wider literary culture of the time, dominated as it was by review articles and anthologies which excerpted the most attractive or striking passages ("beauties") for readers with no time to read entire books. As Leah Price has shown, early-nineteenth-century readers developed an eagle eye for choice passages and an ability to skim the surrounding matrix. Novelists, for all their protestations to the contrary, often encouraged this kind of reading by stylistically marking passages of special significance, using rhetorical equivalents of a dotted line and scissors-symbol.[79] Writers were constantly aware of how their work might be excerpted, and science writing was not exempt from such practices.

For popularizers of science, the need to appeal to both specialists and non-specialists represented a more pressing reason for these rhetorical swerves. Hence Miller's apology in the preface to *The Old Red Sandstone:*

> My facts would, in most instances, have lain closer had I written for geologists exclusively, and there would have been less reference to familiar phenomena. And had I written only for general readers, my descriptions of hitherto undescribed organisms [. . .] would have occupied fewer pages, and would have been thrown off with perhaps less regard to minute detail than to pictorial effect. May I crave, while addressing myself, now to the one class and now to the other, the alternate forbearance of each?[80]

But the two modes cannot be neatly mapped onto distinct readerships. Miller's plea for "forbearance" is addressed not to the totality of his readership, but only to its learned and unlearned extremes—geologists on the one hand, and readers intimidated by the "hard words" of science on the other. Between these two extremes lay the large body of non-specialist readers of all classes who relished the profusion of "minute detail" and scientific precision offered by natural-history writing, as well as the more "pictorial" or "familiar" passages. Miller had no need to apologize to this middle group: he knew they would enjoy both modes. Even for many working-class practitioners, Greek and Latin species names and anatomical terms made up a seductive language of fine distinctions, difficult and satisfying to master, granting its users a sense of authority

79. Price 2000.
80. Miller 1841, vii–viii.

over the objects described. Specialist terminology was appropriated by non-specialists precisely because it made them feel like experts.[81] Miller's readers valued a detailed scientific description—working in tandem with minutely accurate plates and diagrams—as a kind of virtual specimen, its vital statistics gathered together with consummate precision, allowing readers to apprehend it as a particular piece of the natural world. At the same time, they valued particularly vivid descriptions (here presented as something separate from "minute detail") and "reference to familiar phenomena" for teaching them how to look through the fossil as a window on a former world.[82]

The "swerve effect", then, was due only in part to the perceived demands of two kinds of reader. It also reflected changing ideas about what kinds of language were appropriate to communicating knowledge about nature. The period we are examining saw renewed suspicions concerning fiction and sensational rhetoric among many savants, and a growing mistrust of utilitarian agendas among some poets. This fuelled ongoing attempts to distinguish stylistically between "Poetry and Matter of Fact, or Science", as William Wordsworth formulated it in 1802.[83] The distinction between two "languages"—fact versus feeling, objective versus subjective[84]—cannot easily be sustained on closer examination: the dry enumeration of a monster's vital statistics could achieve as sensational an effect as a comparison with one of Milton's demons. But the *idea* of such a natural distinction was, and still is, tenacious.[85] During the nineteenth century this perception was particularly widespread, as in Miller's unquestioning contrast between "minute detail" and "pictorial effect". The literature of Victorian science is studded with these overdrawn dichotomies. Passages intended to be especially "pictorial" were introduced with a new note of authorial self-consciousness. They were often separated out from what was seen as the scientific exposition proper or, in children's books, presented as preliminaries to the real thing. Such passages are the main focus of the present study. Because they were both intended and received as "imaginative", "poetic", or "pictorial" in a quite different sense to their textual surroundings, this artificial dichotomy plays a large part in my reconstructions of authorial strategies; but this does not mean that I personally subscribe to such a neat division between fact and feeling.

These "pictorial" passages tended to cluster around beginnings and endings (whether of a chapter or of the whole text), areas under special

81. Merrill 1989, 46–7.

82. See chapters 9 and 10.

83. W. Wordsworth 1944–72, II, 392 n. 2. On the prehistory of "facts" see Daston 1991–2.

84. These two words were coined in the early nineteenth century (Daston 2001, 85; see also Daston and Galison 1992).

85. For insightful discussion see J. Smith 1994; Daston 2001.

rhetorical pressure. A common result was what has been called a "sand-wich structure", a format shared by many kinds of non-fiction writing from the seventeenth century to the present day[86]—though if we imag-ine the reader progressing through the text, a more nourishing analogy would be that of a three-course meal. Non-specialists' intellectual ap-petites are whetted by the prefatory matter and introduction, in which they are given a tantalizing foretaste of what is to come and why it is important. They are led gradually from the familiar to the unfamiliar, bringing them to a state where they (like those already initiated) will be able to digest difficult information. Then comes the main course, the bulky heart of the text. Here specialists and non-specialists share the feast, and an author sensitive to possible indigestion on the part of his uninitiated readers will often scatter occasional garnishes—homely, re-ligious, or spectacular—to retain in the minds of these readers a lively sense of the subject's nature, relevance, and impressiveness. Wonder and mental enlargement often dominate the conclusion, which acts in part as a digestive—but unlike real-life dessert, it does so not by providing a refreshing contrast to earlier fare, but by intensifying the flavour of what has already been eaten and explaining why it is good for you. Visionary restorations of ancient landscapes are most often found in these conclud-ing portions, as narrative "retrospects".

Writers associated with the new school of geology at first made a point of not indulging in such spectacle, tainted as it was with eighteenth-cen-tury cosmologies. Rebranding their science as disinterested and empiri-cal meant focusing on the structure, rather than the story, of the strata.[87] Their visions of former worlds were made available only to strictly lim-ited audiences, whether in the poems they circulated among themselves (maintaining Erasmus Darwin's techniques on the quiet), in lectures to the young gentlemen of Oxford and Cambridge, or in occasional brief passages in their expensive publications. This cautious attitude persisted until the 1830s, when the literary market was flooded by a new wave of popular-science writing which embraced the techniques of theatrical display beloved of the urban middle classes. In this way, the grand narra-tives of eighteenth-century cosmology returned in a new guise to bring the deep past before the eyes and imaginations of a wider readership.

Accordingly, this book is divided into two parts. Part I tells the story of how, why, and for whom geological writers developed a rhetoric of spectacular display before the 1830s. In this project, high-profile gentle-

---

86. For recent examples see Eger 1997. In the field of earth history, one deliberate exploita-tion of the "sandwich structure" is the "Cambridge Sandwich" which structures Simon Conway Morris's recent treatise *Life's Solution* (Conway Morris 2003, xi–xvi).

87. This remained the dominant strand in nineteenth-century geological practice, al-though historical visions dominated its popularization (J. Secord 1986, 29–35).

men like James Parkinson, John Playfair, Buckland, and Lyell made common cause with humbler practitioners like Mantell and Robert Bakewell: they borrowed techniques from the less exalted world of commercial exhibition, developing narrative possibilities and self-images for the science either in private or for very select audiences. The so-called scriptural geologists—biblical-literalist writers on earth history—used similar devices to promote a cosmology whose opponents soon found they had to combat publicly. Lyell duly stepped into the breach in 1830: the first volume of the *Principles of Geology* was, among other things, a powerful assertion of genteel geology's unique authority to tell the story of Creation.

The *Principles* ushered in a golden age for popular geological writing, to which we turn in Part II. Here old masters like Buckland and Mantell were joined by younger and still more poetically inclined authors such as Miller, Thomas Hawkins, and George Richardson. The new wave was characterized by a confident, almost cinematic deployment of poetic and spectacular techniques which drew directly on cinema's own ancestors, contemporary forms of theatrical display such as panoramas and dioramas. Part II begins by considering the new drive to promote geology among a wider middle-class readership, and examines the publication formats, literary genres, and narrative frameworks which housed these productions. In the light of the panoramic displays, with their encyclopaedic fusions of text and image producing heady visions of empires past and present, the final chapters explore the literary forms and cultural meanings of geological spectacle between 1830 and 1856, by which year all but one of the authors named above were either dead or no longer interested in geology.[88]

These panoramic visions of the ancient earth blended human, sacred, and natural history to point towards a single cosmic pageant, dimly discerned behind the visible universe. They represent a major literary achievement, adapting ancient narrative structures to enact the supreme authority of modern science. To this day, our mental images of the prehistoric world ring the changes on themes set out by the poets of old. Examining these transformations may not help palaeontologists in their ongoing quest to learn what the remote past was really like. Yet, by illuminating the science's rise to celebrity, it may help us to understand more fully our need for such a quest.

88. Lyell was the exception, dying in 1875.

## Part I

Building the Story

# Enter the Mammoth

1

In the autumn of 1802, the first reasonably complete mammoth skeleton to be shown in Europe was exhibited in London's Pall Mall by a young American artist with the improbable name of Rembrandt Peale.[1] The monster stood eleven feet high and seventeen and a half feet long (Fig. 1.1), and could be viewed for the princely sum of two shillings. Along with two other specimens, its bones had been dug up from a marl pit in New York State by Rembrandt's father, Charles Willson Peale.[2]

Peale senior had long had an eye for spectacle. Since the late 1780s, the paintings on display at his Philadelphia home (later renamed the American Museum) had been joined by natural curiosities: he pioneered "naturalistic" ways of displaying stuffed animals, fossils, rocks, and minerals. In the United States, natural-history collections were taking on special

1. On the Peale mammoth see Semonin 2000, 315–40; Rudwick 2005b, 400–2. I use the term "mammoth" here in its colloquial, generic, and contemporary usage; more precisely, this was a mastodon, and was named as such by Cuvier in 1806.

2. The standard biography of Charles Willson Peale is Sellers 1947. On his collection see also Altick 1978, 126 n.; Sellers 1980; Brigham 1995, 13–67; Yanni 1999, 28–30; Bedell 2001, 8–11.

significance in polite society as symbols of cultural autonomy for a land which had, until recently, been ruled by the British crown. As the naturalist and statesman Thomas Jefferson had discovered in the 1780s, the gigantic bones of the so-called "American *incognitum*" (Latin for "unknown creature") were particularly useful for refuting French claims that the New World's natural productions were inferior to those of the Old. This beast was bigger than its European counterparts, and was fast becoming an emblem of America's natural grandeur.[3]

So when Charles acquired three entire specimens in 1801, he quickly reassembled the best one and mounted it in his museum. As the first such fossil reconstruction ever undertaken in North America, it crowned his collection, conferred a grand historic perspective on his portraits of

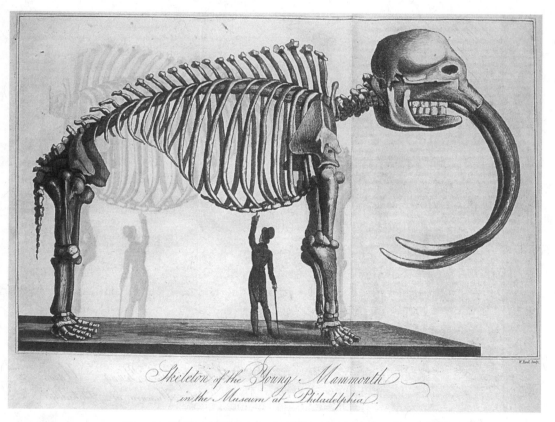

*Skeleton of the Young Mammouth in the Museum at Philadelphia.*

Figure 1.1. Skeleton of the mastodon (the "Young Mammouth") unearthed by Charles Willson Peale, as displayed in his Philadelphia museum in 1821. On the significance of the downturned tusks, see below, p. 34. Montulé 1821, plate opposite p. 9.

3. The mammoth's political and cultural meanings are explored in Semonin 2000 (focusing on eighteenth- and early-nineteenth-century America) and C. Cohen 2002 (surveying concepts of the mammoth from prehistoric cave art to the present day).

American patriots, and gave him the chance to style himself as a national hero.[4] Visitors to his museum normally had to pay twenty-five cents, but they were charged double to see the "Skeleton of the Mammoth" in its separate room. After all, as the advertisement put it, this "ANTIQUE WONDER of North America" was "the LARGEST of *Terrestrial Beings!*"[5] It soon earned him both fame and money.

Fired by this success, Charles sent his sons Rembrandt and Rubens with the second mammoth specimen to England, where Rembrandt was to exhibit it to help meet his tuition expenses at the Royal Academy.[6] To accompany the exhibition, Rembrandt also published a ninety-eight-page *Historical Disquisition on the Mammoth* early in 1803, and both book and beast had a considerable impact in polite society.[7] Word spread abroad rapidly: extracts from the book and pictures of the fossil appeared in learned journals across Europe. In 1806 Rembrandt's figures of this specimen would enable the French comparative anatomist Georges Cuvier to draw his authoritative distinction between the mammoth, the mastodon, and the modern elephant. The brothers wanted to take their "pet" to Paris and sell it to the Muséum National d'Histoire Naturelle, where Cuvier was based, but sadly the Napoleonic Wars got in their way. Later in 1803 they exhibited it in Reading and Bristol; but the costs of their travels (and of living in London) were now too great to be met by such precarious means, and in November the three of them returned to the United States.

Showmanship did not always pay. In this chapter, however, we shall see how it underpinned geology's public appeal in the first two decades of the nineteenth century, whipped up in guidebooks such as Peale's *Historical Disquisition* that presented fossils as sublime relics of a legendary past. Practising geologists may have repudiated such rhetoric as a rule, but even they had their uses for it when promoting their science as worth the attention of polite society. This will be illustrated in the final section of this chapter, which turns to the English translation of a work by Cuvier in which the grand historical vistas of Enlightenment "theories of the earth" were recast in a new and approved guise.

4. This imagery is embodied in the oil painting Peale produced to celebrate this reconstruction, *The Exhumation of the Mastodon*, reproduced and discussed in Semonin 2000, 325, 357–61; Bedell 2001, 10–11.

5. Broadside, *c.* 1802, reproduced in Semonin 2000, 329. The word "mammoth" took on its now-fossilized adjectival sense of "huge" in response to Peale's discoveries (Sellers 1947, II, 142–3; Semonin 2000, 328–30).

6. The third son was called Raphaelle.

7. Peale 1803; on Rembrandt's conception of the mammoth see Semonin 2000, 334–9. Rembrandt also wrote two briefer, more sober accounts for the *Philosophical Magazine* in 1802.

## Mammoths on Show

The Peale mammoth exhibition was unusual in Britain for two reasons. Mammoth bones themselves were hardly unique: among other London cabinet collections, the Hunterian Museum at Lincoln's Inn (see Fig. 5.1) had long boasted an impressive array of fossils, as had Sir Ashton Lever's museum on Leicester Square.[8] For several decades mammoths had been known about and their table manners speculated on. But their bones were typically presented as part of a collection, and with little or no direct textual accompaniment. The Peale mammoth was an independent exhibit, accompanied by a book which presented it within the context of a controversial idea. In 1796, in a widely-reported public lecture in Paris, Georges Cuvier had made the strongest case yet for the reality of species extinction: in his hands the Siberian mammoth proved the existence of a prehuman world long since destroyed.[9] Rembrandt's *Historical Disquisition* opened with detailed references to Cuvier's work and made the idea of catastrophic extinction vividly clear. Although Cuvier's achievements were attracting some notice in Britain, many people would have learned about extinction for the first time from Rembrandt.[10] The subject was calculated to excite "curiosity"—a word which, in the early nineteenth century, implied a much greater emotional and intellectual investment than it does today.[11]

Furthermore, while individual mammoth bones were not new, a complete skeletal reconstruction was unheard of in Britain.[12] Collectors usually arranged fossil bones in their cabinets as separate, singular objects, and in any case complete specimens of the larger vertebrates rarely came their way. Not only was the Peale mammoth almost complete, but the missing bones were supplied by wooden mock-ups. For the first time, those who could afford to pay had the opportunity to *see* the mammoth, its tusks pointed downwards to emphasize its supposedly carnivorous character.[13] Such "visualizing moments" helped to focus and define the distant past: the mammoth had suddenly been brought to life. This iconic status made it a perfect vehicle for the humorist and the metaphor-

8. Altick 1978, 29–32.

9. Rudwick 1997, 13–24. On late-eighteenth-century concepts of species extinction see Rudwick 2005b, 239–376.

10. Cuvier had, however, also presented a paper on the extinct fossil mammal Megatherium in 1796 (Rudwick 1997, 25–32), of which a shorter English version appeared in London's *Monthly Magazine* (Rudwick 2005b, 360).

11. On the changing meanings of "curiosity" see Daston 1995; Leask 2002.

12. The only European precedent was the complete Megatherium skeleton from Spanish South America, sent to Madrid in 1789 and displayed in the Real Gabinete (Royal Cabinet). See Rudwick 1992, 27–32; 2005b, 356–60.

13. On the tusks see Semonin 2000, 334–7; Rudwick 2005b, 401 fig. 7.16.

monger. There was already ample precedent for turning the mammoth into a political animal: jokes about Jefferson and mammoths were rife in the United States, and lately Napoleon Bonaparte had fallen victim to the inevitable puns on his surname.[14] When war with France broke out in May 1803, Charles Dibdin the younger seized the opportunity for musical satire on the London stage:

> Of all the sights in London town, that take the folk's attention,
> The Mammoth is a wond'rous form, of monstrous huge dimensions;
> 'Twas brought from North America, and on it wags are smart, Sir,
> For as it is a skeleton, they call it Bonypart, Sir.
>     Bow, wow, &c.[15]

The third line suggests that the pun was already somewhat worn, but Dibdin develops the analogy with enthusiasm. The next four verses describe how the real "Bonypart" circumvents his countrymen's proposition to buy the mammoth, declaring his own intention to carry it off as a "lawful prize" when he invades England (echoing an episode in 1794 when the French Revolutionary armies had carried off a massive fossil "crocodile" skull as booty from the Dutch town of Maastricht).[16] The personification of England, "John Bull", is unimpressed by Boney's tactics, but—

> Then little Boney, to look big as Mammoth, stretch'd his wizen, Sir,
> And ordered all the English folks at Paris into prison, Sir;
> But we'll return the compliment, if Bonyparte we meet, Sir,
> And with a Habeas Corpus serve his worship from the Fleet, Sir.
>     Bow, &c.[17]

The Dibdins were experienced actor-dramatists, and they knew exactly what their audience wanted. These allusions to the Peale mammoth indicate how much it had caught the public imagination.[18]

The Peale mammoth fostered further interest in fossil bones in cabinet collections around the country. One such collection—a typical mixture of natural and artificial curiosities—had been opened in Liverpool in 1795 by the jeweller William Bullock.[19] Around 1806 Bullock acquired

14. The Peales' departure for England in 1802 had been marked by a festive dinner inside the mammoth skeleton, concluding with a toast wishing "Success to these *boney parts* in Europe" (Sellers 1947, II, 145).

15. C. Dibdin 1807, 168 ("Mammoth and Buonaparte"). The rest of the refrain is not recorded.

16. For a judiciously demythologized account of this famous story, and of the "Maastricht animal" itself (later christened the Mosasaurus), see Bardet and Jagt 1996.

17. C. Dibdin 1807, 170.

18. See also T. Campbell 1904, 108.

19. Altick 1978, 235.

some large North American fossil bones, deriving from mastodons and a mysterious clawed beast, from the Irish adventurer Thomas Ashe. In his melodramatic and deeply untrustworthy autobiography, Ashe describes how he amassed this collection in North America after taking a sudden and passionate interest in fossils: he spent all his money (£1,100) in the process and meant to make his fortune by establishing a museum "of the first distinction in the world" in London, in which these fossils "were to take the principal lead". But, on arrival at Liverpool, "all was dreadful in the extreme". He was unable to pay the hefty customs duties, and his property was seized by the officers. Here Bullock enters the story: having learned of this opportunity, he agreed to extricate Ashe from his "pitiable and condemned state" by paying the necessary duties, on condition that Ashe sell him the fossils for "the miserable and contemptible sum" of £200.[20]

This disaster left Ashe in "a state of mental darkness" as he went on down to London. Deprived of the opportunity to set up his magnificent museum, he earned a living by blackmailing a succession of society ladies and authoring a scurrilous and anonymous biography of Princess Caroline. His autobiography, *Memoirs and Confessions*, was written in 1815, ostensibly to warn readers against falling into the state of depravity into which its unfortunate author had lapsed, which he was now able to abjure thanks in part to the characteristic generosity of Lord Byron, to whom Ashe (a complete stranger) had appealed for money in 1813.[21] However unreliable Ashe's autobiography might be, it is revealing that he chose to engage the reader's sympathy by presenting himself as a heroic fossil-collector who dreamed of displaying his hard-won antediluvian treasures and basking in their reflected glory. The assumption underlying Ashe's narrative is that fossils were captivating enough to earn their discoverers a lasting reputation.[22]

In 1806 Ashe published a cabinet-by-cabinet guide to Bullock's fossils, combined with reflective "memoirs" on the two species. His seventy-two-page book, addressed ostentatiously to the leisured classes and published in Liverpool, rejoiced in the title of *Memoirs of Mammoth, and Various Other Extraordinary and Stupendous Bones, of Incognita, or Non-Descript Animals*: Ashe claimed to have composed it while crossing the Atlantic, intending it as a guidebook for his own London museum.[23] In the event it was Bullock who took the fossils to London when he moved there with

20. Ashe 1815, II, 192–215. On Ashe and Bullock see *ODNB*.

21. Marchand 1973–94, III, 198; IV, 15. Byron was one of the dedicatees of Ashe's autobiography (Ashe 1815, I, v–viii).

22. Compare the ambitious collectors discussed in Knell 2000.

23. Ashe 1815, II, 209. "Incognitum" and "non-descript" were technical terms denoting species not yet formally described or named.

*Figure 1.2.* William Bullock's museum, 22 Piccadilly, known as the "Egyptian Hall" for its striking façade. Aquatint by Rudolf Ackermann, *Ackermann's Repository of Arts,* 1815.

his collection in 1809; and in 1812 the collection came to occupy a bizarre and fashionable building known as the "Egyptian Temple" or, later, "Egyptian Hall" (Fig. 1.2).[24]

Like any cabinet museum, this architectural monstrosity displayed organic and artistic relics together. Bullock was keen to acquire "any uncommon production of Art or Nature" to display for the entry price of a shilling per exhibition.[25] In 1814 he exhibited what appeared to be a fossil dragon, unearthed by the Dorset fossilists Joseph and Mary Anning;[26] after 1815, Napoleon's carriage was displayed there, prompting Byron to have a replica built for himself. The animal department or "Pantherion" was noted for the educational value of its arrangements, which in some respects mirrored those of Charles Peale's Philadelphia museum. Bullock, too, pioneered the use of realistic habitat displays to make his mounted animals look more like the real thing (Fig. 1.3). He transported visitors in imagination to Staffa, designing a passage between two rooms as a gigantic simulacrum of Fingal's Cave. The Egyptian Hall soon became a popular venue for artists like John Martin, whose apocalyptic painting *The Destruction of Pompeii and Herculaneum* was exhibited there in 1822.[27] Here, as in many other venues, edification was inseparable from theat-

24. On this museum see Altick 1978, 235–52.
25. Bullock 1812, ii, v.
26. Torrens 1995, 259–60. This creature was named Ichthyosaurus ("fish-lizard") in 1817.
27. [J. Martin] 1822; Feaver 1975, 55–9.

BULLOCK'S MUSEUM,
22, Piccadilly.

*Figure 1.3.* Lifelike animal displays in Bullock's museum, perused by well-dressed visitors, including some children examining a suit of armour to the left. Aquatint by Rudolf Ackermann, *Ackermann's Repository of Arts,* 1810.

ricality. One might say the same of Ashe's book, which found a second readership when the mammoth bones moved to London.

The guidebooks (to use the term very loosely) which Ashe and Rembrandt Peale wrote to accompany their displays reveal their desire to maintain a sense of novelty and to generate dramatic or narrative interest in the bones. Both men staked out their personal connections with the fossils, incorporating autobiographical anecdotes on the bones' discovery and making bold excursions into the world of comparative anatomy, presenting deductions of their own which differed from those of better-known authorities. The story of how the Peale mammoth was exhumed had been told many times in Philadelphia, and Peale here repeated it for his British audience: his own role alongside his father in unearthing these particular bones was attested by other sources. For Ashe, craftier techniques were called for:

> I abandoned Canada, [. . .] patiently and laboriously visiting and exploring the regions where the object of disquisition and dispute was said to abound. I trod those plains he once devastated.—I drank of the water of those lakes in which he had once quenched his thirst.—I crossed the Apellactian and the Alleghany.—I descended the Ohio, the Illinois, the Wabash, the Missouri, and the Mississippi. I traversed the depths of the deepest vallies, and

the summits of the highest mountains [. . . .] and I at length obtained the full completion of my wish, the ardent object of my prayer; a grand collection of stupendous bones [. . . .][28]

The details are suspiciously vague: nowhere does Ashe describe actually digging out the bones. This is no surprise when we find from other sources that he purloined them. Their original discoverer was a Cincinnati physician named William Goforth, who wanted to sell his collection in New Orleans. According to a recent account, Ashe persuaded Goforth to employ him as agent, then promptly absconded with the entire collection across the Atlantic.[29] But this only makes Ashe's narrative stance all the more striking. Discovering fossils was the stuff of heroism: Ashe's rhetorical authority is based on legwork and fieldwork, like that of later geologists such as Hugh Miller ("I have been an explorer of caves and ravines, a loiterer along sea-shores, a climber among rocks, a labourer in quarries").[30]

In their guidebooks, Peale and Ashe attempted to present the current state of research, with lengthy excerpts from other authorities; they also provided narrative backgrounds for the specimens on display. The anatomical content is balanced by a melodramatic "Indian Tradition" which both writers linked with the mammoths' extinction, directing their readers' romantic interest in doomed races towards the huge fossils on display. This "Indian Tradition" was based on a Shawnee monster-legend published by Jefferson in his *Notes on the State of Virginia* (1785); an embellished version had been disseminated more widely in the United States in the December 1790 issue of the *American Museum,* a monthly miscellany.[31] This second version was printed on the advertisement to the Philadelphia mammoth exhibition, and now appeared in both Peale's and Ashe's guidebooks.

According to the story, a race of vast animals—"huge as the frowning precipice, cruel as the bloody panther, swift as the descending eagle, and terrible as the angel of night"—had once threatened to annihilate the Indians. They were so large that "the pines crashed beneath their feet [. . .] and whole villages, inhabited by men, were destroyed in a moment." The "good Spirit" intervened by destroying the creatures with thunderbolts, "and the mountains echoed with the bellowings of death." But the biggest bull of the herd proved indestructible. Like King Kong surrounded

28. Ashe 1815, II, 203–4; compare Ashe 1806, 39–40.
29. Semonin 2000, 345.
30. Miller 1841, 13.
31. Anon. 1790. On these stories see Semonin 2000, 182–3, 285–7.

by aeroplanes on top of the Empire State Building, this lone survivor of a giant primeval race defied the "artillery of the skies":

> He ascended the bluest summit [. . .] and, roaring aloud, bid defiance to every vengeance. The red lightning scorched the lofty firs, and rived the knotty oaks, but only glanced upon the enraged monster. At length, maddened with fury, he leaped over the waves of the west at a bound, and this moment reigns the uncontrouled monarch of the wilderness, in despite of even Omnipotence itself.[32]

Peale comments that the "Indian mode of description [. . .] is always highly poetical, and much in the stile of Ossian".[33] Ossian was the Homer of the North, a mythical Scottish bard whose sorrowful epics had been "reconstructed" by the upwardly mobile Highlander James Macpherson in the 1760s and caused a great sensation across the Western literary world.[34] Ossian was Jefferson's favourite poet; when Peale was writing, however, the authenticity of the Ossianic poems was widely disputed. Peale voiced doubts about the "Indian Tradition" as well, finding its language "perhaps a little too highly dressed"; but he credited the underlying ideas as being "truly Indian".[35] The anonymous author who embellished Jefferson's account clearly saw the implied Indian reciter as a potential Ossian, recounting ancient victories over forces of unimaginable destruction. Like Ossian, the reciter is aware that his own race has not long to survive: his tale is set in a glorious past, "long before the pale men, with thunder and fire at their command, rushed on the wings of the wind to ruin this garden of nature".[36] The firearms of the invading white settlers prefigure the supernatural "artillery of the skies": as the latter had once destroyed the monsters, so the former are now destroying the Indians. The thrill of species extinction, of long-lost monsters ruling an untamed wilderness, was here built on a substructure of human conquest and racial extirpation, filtered and romanticized through the lens of Ossianic sensibility.[37]

But Peale and Ashe were more interested in "stupendous *incognita*" than in Indians: their treatment of the "Tradition" makes the "last monster" into the hero. His destructive excesses are narrated within a context of appalled admiration. In both accounts, the curtain drops on his

---

32. This is Peale's version (1803, 88); Ashe's (1806, 7–8; 1815, II, 196–8) are largely identical, though he replaces the beast's Ossianic "fury" with a more Napoleonic "disdain".

33. Peale 1803, 86.

34. On Macpherson and Ossian see F. Stafford 1988.

35. Peale 1803, 88. The historicity of the tradition was much debated by American naturalists (Semonin 2000, 308–9).

36. Peale 1803, 86–7; Ashe 1806, 7.

37. Semonin 2000. On the "last of the race" myth more generally see F. Stafford 1994.

Promethean posture of defiance in a kind of pantomime tableau. Here, again, is that tension between sensationalism and higher reflection so characteristic of the sublime. On the one hand, this creature is part of the "chain of nature", created by the "Author of Existence [who is] wise and just in all his works"; [38] on the other hand, he is "monstrous" and "tyrannic", the object of divine wrath in the form of a "deluge" or some such "sudden and powerful cause". [39] The dissonance comes across most clearly in Ashe's account, to which we now turn.

Ashe clearly relished playing the authoritative guide, showing the reader around each box:

*No. 4,*

Contains an object of inexpressible grandeur and sublimity. It is the foot of a clawed animal [. . . .] The animal to whom it appertained [. . .] must have been the terror of the forest and of man. This monument stands alone. It has no competitor. It is the first and only one of such exorbitant magnitude ever discovered, or probably that ever will be. [40]

It is almost as if Ashe were trying to sell it. Elsewhere a higher tone creeps in:

The huge leg and thigh bones, how monstrous, how massive! [. . .] And the fragments of ribs! how admirable their construction! [. . .]

But above all, I beg your attention to the claw [. . . .] 4 ft. long by 3 ft. wide [. . . .] a beautiful mechanism [. . . . allowing it] to seize its prey, rend, and annihilate it.

From this rapid view of these majestic remains *it must appear, that the creature to whom they belonged was nearly 60 feet long, and 25 feet high!* [41]

Here, besides the rhetoric of terror, is a response reminiscent of William Paley's recently published *Natural Theology* (1802), in which each creature's perfect fit to its environment was presented as evidence for a benevolent Creator. For Ashe, the monster's claw has been designed so perfectly for its horrific purpose that it is "beautiful". This seeming dissonance is conventional enough for its time, being one of several possible developments of the argument from design in an age when the existence of animals long before man was becoming increasingly familiar among the learned. Briefly, the new argument ran thus: death is a necessary evil

38. Peale 1803, 75; Ashe 1806, 41.

39. Ashe 1806, 43, 50; Peale 1803, 74, 90.

40. Ashe 1806, "Introduction", 7–8. The phrase "the terror of the forest and of man" was taken from G. Turner 1799, 518, as was much of Ashe's information.

41. Ashe 1806, 43–4.

in a world of plenitude, and a quick death is a small price for an animal to pay for a happy life—therefore, efficient claws are a mark of divine benevolence in the grand scheme of things. Natural theology was built on the Baconian conceit that nature was the "Book of God's Works" and could be read like the King James Bible. This textual analogy was reinforced by the antiquarian habit of viewing rocks and fossils in a continuum with human remains, as the "monuments", "archives", or "medals" of nature.[42] Natural theology represented a dominant structuring principle in science popularization throughout the nineteenth century, and it had the advantage of appealing across the theological spectrum. Deists, Catholics, Protestants, and Jews could all respond to such arguments, and the implied response was often framed in these writings as a thoughtful appreciation of "beauty".[43]

The theological aesthetic of Ashe's display is reinforced by his choice of two illustrative poetry quotations, both highly visual: the spectacular description of the giant beast Behemoth from the Book of Job, and a long passage from Mark Akenside's physico-theological poem *Pleasures of Imagination* (1744). This second passage enjoins "man, poor short-sighted man" to "Once more search, undismay'd, the dark profound / Where Nature works in secret"; having scanned the chaotic immensity of creation, the reader is invited (like Job) to trust in its ultimate order, its mysterious "symmetry of parts".[44] The providential design of each bone makes these fossils into holy relics, which "must be touched with a trembling and a pious hand, by him who can admire the wonderful greatness and wisdom displayed in the operations of nature".[45] At the same time, Ashe's intense focus upon each singular object, particularly the nastier-looking ones, makes it hard for the reader to draw back and take Akenside's larger view. Ashe's self-conscious exclusion of "the vulgar" from these exalted reflections cannot disguise the fact that it was the monstrosity of these creatures rather than their evidence of divine wisdom which fascinated him. As monsters, they would captivate all but the most idealistic of spectators.

Accordingly, Ashe gave the lurid "Indian Tradition" a greater interpretative weight in his argument than the principles of scientific reasoning could justify. Objects with stories attached were more "curious" than objects alone, so it was in Ashe's interests to insist that this Ossianic ro-

---

42. On the early history of this analogy see Rudwick 1972; Rappaport 1982; 1997. On its constitutive importance in geology from the late eighteenth century on see Rudwick 2005b.

43. On the language of natural theology see Brooke 1979; 1991; Corsi 1988; Robson 1990; Topham 1993; Brooke and Cantor 1998, 141–243. On Quaker and (the surprisingly few) Jewish uses of natural theology see Cantor 2005, 233–42, 308–14.

44. Ashe 1806, title-page, 19–23, 50. The Book of Job, with its references to giant animals and other wonders of creation, remained a common reference-point for later popularizers.

45. Ashe 1806, "Introduction", 12.

mance contained a kernel of truth. According to Ashe, the fossil remains suggest that "However clouded the sublime tradition [. . .] may be with fiction", it has "a considerable degree of truth [. . .] in it" [46]—a claim which in turn allows the "Indian Tradition" to tell the story of the fossils. Peale had a similar narrative need to take the "Tradition" seriously, which in his case was satisfied by quietly disagreeing with Cuvier and stating that "the Mammoth was exclusively *carnivorous*".[47] This conclusion led to a colourful speculation on its lumbering method of hunting:

> the Mammoth probably fed [. . .] on such animals as could not well escape him, and which would not require much artifice or speed to be caught, nothing more being necessary than his long tusks and some powerful pro-truberant cartilagenous instrument, for the purpose of taking up his prey, whether, like the Elephant, it was a nose elongated; or like the Walrus, it was a large and powerful lip; or like the Ant-eater, it was a long and powerful tongue.[48]

Ashe had no need to posit a carnivorous mammoth, since he had another beast to hand: the *"Second Incognitum"*, bristling with claws. Here Ashe freely plundered the publications of American naturalists. Jefferson had named this beast the "Great-Claw or Megalonyx", deducing that it was a kind of lion three times the size of the present-day African species but denying that it was extinct.[49] This interpretation had since been scotched by Cuvier and retracted by Jefferson. However, Ashe now appropriated it (pretending it was his own) as the master key to his realization of the "Indian Tradition":

> I early discovered, that the description pointed at some stupendous voracious animal; cruel, fleet, and capable of bounding suddenly on his prey. Furnished with carnivorous teeth to consume, and with claws to rend and destroy: in short, a monster of the tiger line, endowed with every bloody and malignant property, and differing in every character but bulk from the mammoth [. . . .][50]

Such a monster Ashe now claimed to have found. He speculated that if the Megalonyx had existed at the time of man, either God would have had to destroy it for mankind's sake (as in the "Tradition"), or else men must then have been giants—and "if so, [. . .] 'how have the mighty

46. Ashe 1806, 37.
47. Peale 1803, 48.
48. Peale 1803, 80.
49. Greene 1959, 102; Semonin 2000, 303–310; Rudwick 2005b, 373–5.
50. Ashe 1806, 37.

fallen!'"[51] This recalls William Hunter's comment on the Ohio mammoth in 1768: "if this animal was indeed carnivorous [. . .] though we may as philosophers regret it, as men we cannot but thank Heaven that its whole generation is probably extinct".[52] In Victorian times the same sentiments, alive with the pleasing sensation of danger averted, would be expressed repeatedly of the great fossil saurians.

One final method by which Ashe demanded open-mouthed attention for his voracious mammals was to insist that they were constantly at war with each other. Since the bones of mammoths and Megalonyx had been found mingled together in a single site, "There is no question, but that the mammoth was his perpetual rival, and avowed adversary. Wherever they met, they fought; and wherever they fought, one or both fell."[53] This ritualistic, seemingly pointless duelling is reminiscent of the outdated knightly heroes of late-mediaeval romance, but Ashe may have been applying Ossianic analogies again. This image endured: each era in the history of palaeontology seems to have had its typical pairs of bitter enemies, culminating in standoffs between *Tyrannosaurus* and *Triceratops*. In Ashe's case we might recall Dibdin's comic comparison of the mammoth to Napoleon "Boney Part". In the early years of the war, British attitudes towards Napoleon were characterized by just this kind of ambiguous indulgence: even while fearing and hating him, many found it hard not to admire his audacity and military success.[54]

The Megalonyx was Ashe's Napoleon of the ancient world. Ashe concluded his guidebook with a retrospective attempt to "form to ourselves some idea of his character" in an anthropomorphic vision of his clawed hero. To launch this passage, Ashe used that perennially evocative form of scientific rhetoric, the enumeration of vital statistics:

> His length 60 ft. his height 25; his figure magnificent; his looks determined; his gait stately; his voice tremendous! In a word, his body must have been the best model of deadly strength, joined to the greatest agility. [. . . .] Accustomed to measure his strength with that of all other animals he used to encounter, the habit of conquering must have rendered him haughty and intrepid!
>
> Having, perhaps, never experienced the strength of man, as the power of his arms, instead of discovering any signs of fear, he would disdain and set an army at defiance! Wounds might irritate, but they could not terrify

---

51. Ashe 1806, 50. Both these ideas were combined in a highly original manner, with the help of Cuvier, by Ashe's benefactor Byron in *Cain* (1821).

52. Hunter 1768, 45 (echoed in Playfair 1802, 464–5).

53. Ashe 1806, 48.

54. On attitudes towards Napoleon see Lean 1970; Boime 1990; Bainbridge 1995.

him; and after a violent and obstinate engagement, should he find himself weakened, he would retreat fighting, always keeping his face to the enemy, looking proud, great, and ferocious.

THE END.[55]

Ashe did not mention that two years before, Cuvier had shown the Megalonyx to have been an inoffensive ground sloth which wandered the forest floor, quietly chewing on leaves. But Ashe had not yet exhausted his hero's Napoleonic potential. In 1815, in the dedication to his autobiography, he complained that "the world" had treated him as "a sort of wild beast which could neither be dragooned nor caressed into tameness; [. . .] as an old lion in a cage, ever apt to leap into his natural wildness".[56] He took up this comparison when telling the story of his defeat at the Liverpool docks, dramatizing himself as the greatest lion that ever lived:

> Haughty, intrepid, and mad, I attended the Custom-house daily. I disdained and set all its officers at defiance; nay, after violent and obstinate engagements, which induced them to force me out of doors, I retreated fighting and blaspheming, always keeping my face to the enemy, and looking proud, great, and ferocious as the Mammoth, when known to be the terror of the forest and of man.[57]

Though Ashe mentions "the Mammoth", his stirring description largely reproduces his earlier eulogy on the Megalonyx, while the last eight words reproduce his account of the Megalonyx's bones in Box 4.[58] This self-identification with a fossil beast may seem bizarre, even infantile;[59] but in itself it is no more unusual than striking the pose of a Napoleon. The history of extinct animal impersonation has yet to be written, but it is clear that such posing was not confined to children, as (we like to think) it is today.[60]

The "Indian Tradition" is prominent in both guidebooks: Ashe put it near the beginning, Peale at the end. Peale concluded by suggesting that this story reproduced folk memories of a past race of mammoths, which "must have filled the human mind with surprise and wonder."[61] Both writ-

55. Ashe 1806, 59–60.

56. Ashe 1815, I, vii–viii.

57. Ashe 1815, II, 214.

58. See above, p. 41.

59. On the popularity of extinct saurians as self-images for twentieth-century children see, for instance, Haste 1993; W. Mitchell 1998.

60. See chapter 2 on Buckland's example.

61. Peale 1803, 91.

ers invited their prospective audiences to imagine themselves in the deep past, gazing at these primitive, dauntless tyrants with the same admiration that they directed at the imposing antiquities of Egypt, more recent relics of a tyrannical but hugely impressive empire whose decayed remains Napoleon's army of savants was now busy plundering.[62] Peale and Ashe presented the world of savage primeval man and his gigantic mammalian adversaries as one more scene from ancient history to awe the public.

These exhibitions played a vital role in shaping new knowledge about fossils for the polite classes in the first two decades of the nineteenth century. At this time most geological practitioners were deliberately avoiding the "storytelling" possibilities of the strata, focusing instead on their structure and composition. But some, even at this time, engaged in popularization of a similar kind. Their occasional forays into poetry and sensationalism must be seen in the context of the shows we have just been looking at: in order to forge a scientifically aware audience from the wealthy throng that had gaped at the mammoth in Pall Mall, they had to tread on methodologically forbidden ground.

## Four Experts in Search of an Audience

The most successful public spokesmen for the new geology in the 1800s and 1810s operated, in different ways, outside the rigorously circumscribed empiricism favoured by the knot of British gentlemen who came together to form the Geological Society of London. This section treats four of these spokesmen: Robert Bakewell, James Parkinson, Humphry Davy, and John Playfair.[63] Bakewell was a freelance surveyor who never joined the society, though its members greatly admired his *Introduction to Geology* (1813).[64] Bakewell's peripheral position allowed him to recommend the practice of speculation, which, he insisted, "awaken[ed] curiosity and stimulate[d] inquiry".[65] Parkinson, a genteel physician, was one of the society's founders, but his luxurious three-volume treatise *Organic Remains of a Former World* (1804–11; Fig. 1.5) was peripheral to the society's practice in two ways. Written in the form of fictionalized "letters" like Gibert White's classic *Natural History and Antiquities of Selborne* (1789), Parkinson's book was squarely aimed at the general reader and collector. It also focused exclusively on fossils, which had not yet acquired constitutive importance in geology: Parkinson aimed to provide an overview,

62. Laissus 1998; Leask 2002, 102–56.
63. On these men see *ODNB* and (apart from Parkinson) *DNBS*. On Davy see also Knight 1992.
64. According to Hugh Torrens (*ODNB*), Bakewell was refused membership.
65. Bakewell 1813, 330; see Page 1963, 34.

since all the available British writings on fossils focused on individual areas or specimens.[66]

Davy was another founding member of the Geological Society: his work as a chemist had propelled him from humble origins to honorary gentility. He was already well known as a flamboyant lecturer, which gave him the freedom to indulge in occasional flights of fancy in the geology lectures he gave between 1805 and 1811 for fashionable Londoners at the Royal Institution.[67] Lectures were a tried and tested site for the union of earth science with spectacle, and this union could take unexpected forms. Robert Jameson enthused many budding geologists by instituting regular field lectures on mineralogy at Edinburgh University, where professorial wranglings gave earth science an added appeal: in the early nineteenth century a battle over the origin of basalt was raging between those who argued that it had been deposited under water (Neptunists) and those who argued for its formation by the earth's heat as molten rock (Vulcanists).[68] Jameson, Professor of Natural History and a disciple of the German geological giant Abraham Gottlob Werner, headed Edinburgh's large Neptunist contingent; the Vulcanists were led by Playfair, Professor of Mathematics and Natural Philosophy. Playfair sparked off a new phase in the debate when he promoted his own substantially altered version of James Hutton's theory in his *Illustrations of the Huttonian Theory of the Earth* (1802);[69] Jameson's own *Treatise on Geognosy* (1808), steeped in Wernerian terminology, was less accessible, but it fanned the flames of controversy. Nor was this battle confined to the lecture theatre. In 1812 Edinburgh's Theatre Royal staged a verse melodrama entitled *Helga and Her Lovers,* based on the mediaeval Icelandic *Gunnlaugs saga ormstungu* (*The Saga of Gunnlaug Wormtongue*). The author was the mineralogist George Mackenzie, recently returned from Iceland; *Helga* was the first adaptation of any Icelandic saga on the British stage. But Mackenzie was an outspoken Vulcanist, and on the opening night the theatre was packed with Neptunists, so the unfortunate *Helga* was booed into oblivion.[70] In this context, at least, science and literature did not mix well.

Geology's public appeal was thus reinforced by the whiff of controversy. Nor should we ignore economic factors during the difficult years of the Napoleonic Wars. The stratigraphic expertise of engineers like William Smith was crucial for landowners hoping to make the best

66. Parkinson 1804–11, I, v; J. Thackray 1976.
67. On Davy's lectures see Foote 1952; Yearley 1985; Golinski 1992.
68. D. Allen 1976, 61–3; Hallam 1989, 18–23.
69. Dean 1992, 102–25.
70. For details on *Helga* see Wawn 1982.

use of their lands, and Bakewell and Davy also took care to meet their needs.[71] Science was marketable, geology especially so. In London's Fleet Street, John Tatum lectured on chemistry, galvanism, mineralogy, and "natural philosophy", charging one shilling per lecture. According to the printed advertisement, Tatum's final lecture was "GEOLOGY.—A great variety of curious Minerals will be exhibited, and their constituent parts explained, with the theory of their formation." The small print added that "Only subscribers to the Course" would be "admitted to the last lecture":[72] this meant paying not one shilling but twenty. Geology was used as bait—but not for the lower or most of the middle classes. For them, twenty shillings (i.e. £1) was prohibitively expensive: it meant more than half a week's wages for the best-paid of the skilled workers, a week's wages for the average curate, and at least two weeks' wages for tradesmen or servants. Instruction in geology—whether from lectures or from books, which were similarly priced—was a luxury commodity in the 1800s and 1810s, aimed at the wealthiest 5 percent or so of the population. Its potential economic benefits were, as Tatum knew, well worth the outlay for his target audience.

Yet economic benefits were not everything. As Tatum made clear by signposting "the theory of their formation", many of these lecture-goers were attracted by the speculative elements of grand theory with which the idea of geology was still imbued. The need to admit these elements into the science's public manifestations is encapsulated by Davy's comments on Georges Buffon, an eminent aristocratic naturalist of pre-Revolutionary France.[73] Davy toed the party line by stressing that excessive "speculation" was a hindrance to knowledge, dismissing Burnet's *Sacred Theory of the Earth* as a "poetical romance" whose author's "philosophical opinion" was "wholly unworthy of attention";[74] but he could subvert the traditional vocabulary of disparagement to striking effect. His censure of Buffon's equally romantic geological epic *Les Époques de la nature* (1778) is followed by this encomium, which concludes the third lecture:

> it would be unfair to deny him the highest praise for order of arrangement, sp[l]endour of description, and brilliancy of style. He has adorned a subject of the most abstracted kind in the refined charms and elegancies of literature, and he has bestowed on it an interest which it had never possessed before. His merit in this respect is transcendent. There is nothing highly difficult in the task of delineating human character [. . . .] But to clothe the

71. On geology and landowners see Boime 1990, 178–82, and the essays collected in Torrens 2002. See also Siegfried and Dott 1980, 10–14; Bakewell 1813, title-epigraph.

72. Cutting in the Bodleian Library, Oxford, John Johnson Collection, Entertainments Box 7.

73. On Buffon see Roger 1997.

74. Siegfried and Dott 1980, 43.

dull and dead objects of a speculative branch of science in the forms of beauty and to infuse into them animation and life, this is indeed a work which the strongest genius only can accomplish. And it is a work in which Buffon has perfectly succeeded.[75]

It was now customary among practising geologists to dismiss Buffon as a "mere stylist",[76] but Davy here turns this rhetoric on its head, elevating Buffon's imaginative powers and ascribing to them a constitutive role in the progress of knowledge. Buffon's literary skills alone guarantee him a place in Davy's pantheon of "geniuses".[77] The pejorative or laudatory overtones of such language depended wholly on the context: Davy's unusual reputation gave him the freedom to step occasionally outside the new empiricism, and his success with the public both derived from and contributed to this flexibility.

To make the science attractive to the leisured classes, Davy, Parkinson, Bakewell, and Playfair drew on the range of geology's component discourses outlined in the previous chapter. Besides economic factors, they recommended geology as an "invigorat[ing]" outdoor pursuit which led its "votaries" into "alpine districts" and the "seeming magic scenery" of romantic caverns.[78] According to Bakewell, geology's combination of fresh air and intellectual stimulation offered geologists all the benefits of the Grand Tour, "conspir[ing] to invigorate the health, and give to the mind a certain degree of elasticity and freshness, which will enable them on their return to discharge with greater alacrity the duties of active and social life."[79] The frontispiece to the first edition of Bakewell's *Introduction* emphasizes this point by depicting a walker (perhaps Bakewell himself) sitting happily on a Welsh mountainside.

Pleasure in picturesque landscapes could be channelled into higher delights, as in Davy's introductory lecture:

In travelling he [the geologist] is attached to a pursuit which must constantly preserve his mind awake to the scenes presented to it. [. . . .]

The imagery of a mountain country, which is the very theatre of the science, is in almost all cases highly impressive and delightful, but a new and a higher species of enjoyment arises in the mind when the arrangements in it [. . .] are considered.[80]

75. Siegfried and Dott 1980, 45.

76. Rappaport 1997, 257–8.

77. On eighteenth-century debates over the concept of "genius" and the role of the subjective imagination see Schaffer 1990.

78. Bakewell 1813, 22; Parkinson 1804–11, I, 381. Compare Siegfried and Dott 1980, 13.

79. Bakewell 1813, 22.

80. Siegfried and Dott 1980, 13.

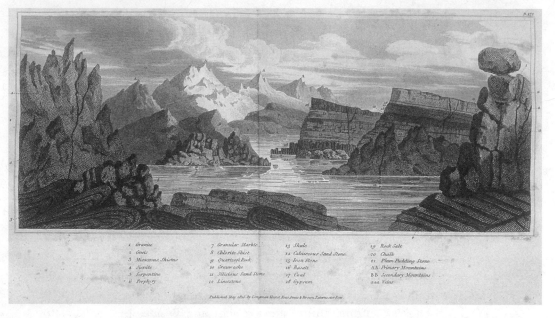

*Figure 1.4.* The structure of the sublime. Keyed diagram of rock types, painted by Thomas Webster and engraved by F. C. Bruce for Davy's *Agricultural Chemistry*: a striking blend of picturesque and scientific visual conventions. Davy 1813, Fig. 16.

Davy is hinting at a different kind of vision, requiring an alert mind rather than the passive consumption of a view. The modulation of one species of vision into the other is beautifully encapsulated in Thomas Webster's picture of the "General Appearance and Arrangement of Rocks and Veins" (Fig. 1.4) which illustrates Davy's firmly utilitarian *Agricultural Chemistry* (1813). This diagram shows all the different rock types, keyed and labelled, but arranged in a sublime fjord-like landscape: the geologist looks through as well as at the scenery.[81] Playfair makes use of a similar movement from passive delight to active study in his *Illustrations,* imagining the geologist placed for the first time on Alpine heights: "as soon as he has recovered from the impression made by the novelty and magnificence of the spectacle before him, he begins to discover the footsteps of time [. . . .] He sees himself in the midst of a vast ruin".[82] And the first chapter of Bakewell's *Introduction* eases this movement by means of verse quotation:

> No inanimate objects on the terrestrial globe excite in the mind such sublime emotions as the immediate presence of lofty and precipitous mountains that rear their heads in awful silence amidst the clouds, and secure

81. Davy 1813, 172. On this form of vision in relation to the developing mountaineering vogue see Macfarlane 2003, 35–6. Compare Fig. 0.1.
82. Playfair 1802, 110–11.

in their immensity remain unchanged, while successive generations and empires pass away from the earth.

> "Surely there is a hidden power that reigns
> 'Mid the lone majesty of untamed nature
> Controlling sober reason ———"

To the geologist such scenes are peculiarly interesting; for it is here he must receive his most instructive lessons, and learn the antient and present condition of the globe.[83]

Bakewell here quotes from William Mason's poem *Caractacus,* in which a Roman general surveys the awesome cliffs surrounding a Druid haven on Anglesey and intuits the presence of higher powers than reason can contemplate.[84] These lines are appropriated by Bakewell to imply, without forcing the case, that the contemplation of mountains is "sublime" because it hints at God's "hidden power".

The language of antiquarian spectacle appears most often in Parkinson's *Organic Remains.* Here the words "curiosity" and "curious" occur frequently, usually in connection with some historical speculation.[85] The objects described may have been sitting isolated in cabinets, but, like human monuments, they were objects with stories attached.[86] The collector's eclectic attitude surfaces occasionally, as when a discussion of pentacrinites is interrupted by the "curious inquiry, perhaps worthy of the heraldic antiquary [    ] how far the origin of the heraldic symbol, the mullet, has depended on the frequent discovery of these pentagonal vertebrae".[87] Parkinson declares that "a view of the remains of a former world" necessarily excites an "eager curiosity".[88] His task was to direct that curiosity towards a more philosophical interest in the antediluvian world. The literary content and epistolary form of *Organic Remains* might seem "diffuse" and "old-fashioned" to a modern eye;[89] but when one con-

83. Bakewell 1813, 21–2.

84. Caractacus was the king of the Silures, an ancient British tribe later to be immortalized in the name of a geological "system" which Roderick Murchison would make his own. Murchison's delight at being crowned a "second Caractacus, a 'King of Siluria'" (J. Secord 1982, 440) recalls Mason's drama, which remained popular well into the Victorian period.

85. For example: Parkinson 1804–11, I, v, 8–10; II, 6; III, 133.

86. Compare Carroll 2004, 49–53.

87. Parkinson 1804–11, II, 254.

88. Parkinson 1804–11, I, v.

89. Rudwick 1992, 18 (reading Parkinson within the context of elite scientific discourse). The book was popular enough for its plates to be reprinted in 1850 with geological notes by Mantell, who also excerpted most of Parkinson's first chapter and praised its "pleasing style" (Mantell 1850, 14–16). As for the outdatedness of the epistolary form, the similarly constructed *Natural History and Antiquities of Selborne* by Gilbert White (1789) enjoyed enormous popularity in the early nineteenth century (D. Allen 1976, 50–1), and the form was still being used in the 1830s (e.g. Nichol 1837).

siders the novelty and experimental boldness of the whole enterprise, one is inclined to admire the rhetorical techniques with which Parkinson achieved his aims. The epistolary form sets up the genteel context necessary for Parkinson to win a hearing among his intended audience: readers familiar with Gilbert White's *Selborne* and other popular books on natural history would feel immediately at home.

In Parkinson's scenario, he presents himself as an inquisitive gentleman as yet unversed in oryctology (i.e. fossils), writing a letter to an exceptionally learned friend. His letter describes how, with two other characters, Emma and Wilton, he spent a night in a house near Oxford. That evening the landlady showed them some "curiosities":

> "Here," said she, shewing us a stone of a conical form, "is one of the *fairies night-caps,* now also become stone[.] Do, madam," said she, addressing Emma, "pray observe; is it possible that lace-work, so beautiful as this, should ever be worked by human hands? This," said she, "and this, are pieces of the *bones of giants;* who came to live here, when the race of fairies was destroyed." These bones, she informed us, were frequently dug up in several parts of the country; as well as innumerable *thunderbolts:* some of which she also shewed us; stating, that these were the very thunderbolts, with which these people were, in their turn, also destroyed.[90]

"Parkinson" is then told that the university museum at Oxford contains whole cabinets of such curiosities, and he describes his frustration to his philosophical friend: "How mortifying will it be to have objects presented daily to my view, whose form alone renders them highly interesting; and whose history is probably fraught with entertainment; and to find myself totally ignorant of their origin". He asks his friend to send him "a regular, and systematic history" of these curiosities;[91] his friend duly obliges, and the rest of the work is taken up by the philosopher's letters back to "Parkinson".

This scenario serves several ends. First, by presenting all the information originating from a philosopher-persona in a particular narrative setting, Parkinson avoids the textbook's illusion that the knowledge imparted is permanent. In this way the discrepancy between volumes 1 and 3 concerning the interpretation of Noah's Flood (after Parkinson had become acquainted with Cuvier's latest research) seems less glaring. Second, Parkinson's inclusion of himself within the scenario as a neophyte

---

90. Parkinson 1804–11, I, 4.

91. Parkinson 1804–11, I, 5. On the "shame in ignorance" topos as an opening gambit in popular-science dialogues see Myers 1989.

rather than an expert is calculated to engage the uninitiated reader. More significantly, the informative landlady is presented with considerable sympathy: "Parkinson" and Emma smile at her "romantic" tales when she leaves the room, but Wilton (characterized, interestingly, as a sceptic) defends her story's claim to be taken seriously.

There is, of course, a strong congruence between the landlady's tales of racial extirpation and the philosopher's own accounts of giant mammalian extinctions. The folktale version is not presented as a useless product of superstitious ignorance, but (as in Ashe's and Peale's guidebooks) as a popular understanding of a "truth" to which the philosopher has closer access. In Britain as much as in the United States, fossil science rested on a thick and fertile substratum of folklore, and its nineteenth-century popularizers consistently upheld the imaginative continuity between the two. This continuity took several forms. According to many writers, folklore preserved memories of creatures now known only in a fossil state, representing them as legendary monsters coexisting with humanity (e.g. the Indian "Great Buffalo" or the biblical Behemoth and Leviathan).[92] Other writers alluded to this view in a more critical vein, implying that monster-legends derived ultimately from the discoveries of fossils by pre-scientific peoples:[93] Parkinson's book exemplifies this technique, and his employment of a Yorkshire legend adapted by Walter Scott in his verse-romance *Marmion* (ammonites being seen as serpents petrified by St Hilda's prayers) was recycled by many later writers.[94] In the Victorian period, however, the continuity between legends and fossils would be upheld most often on a metaphorical level for poetic effect, aligning extinct animals with elves,[95] goblins,[96] dragons,[97] demons,[98] or the biblical Leviathan.[99]

In all these ways, curiosity could be channelled into scientific interest. According to Parkinson, science shows these objects to have histories

92. Anon. 1790; [Rennie] 1828, 365; Thompson 1835; Best 1837, 13–15. A similar device is used in young-earth creationist popularization today (e.g. Ham 2001, 35–8).

93. This possibility has been argued recently, with respect to Classical monster-legends, in Mayor 2000.

94. The same lines from *Marmion* illustrate the ammonite in Parkinson 1804–11, III, 133–4; Mantell 1836a, 12; Milner 1846, 617; R. Hunt 1854, 355; G. Richardson 1855, 23–4; Miller 1859, 144–5.

95. Miller 1841, 59–60.

96. Miller 1858, 103; 1859, 136–7.

97. W. Buckland 1836, I, 74; Mantell 1838, 369; Hawkins 1840a; Anon. 1841b, 513; Broderip 1848, 326–80; Miller 1859, 149. See Figs. 2.10 and 6.5 below.

98. Hawkins 1834a, 51; W. Buckland 1836, I, 224–5.

99. W. Buckland 1836, I, 164; Milner 1846, 722; Goodrich 1849, 275; Mill [c. 1855], 79–80. These alignments will be explored more fully in chapter 10. See also Haste 1993, 353–4; C. Cohen 2002; Sommer 2003, 182–4.

"fraught with entertainment", no less entertaining for being true (another durable theme in subsequent literature).[100] His folklore allusions reinforce the practice of looking at curious objects with a historical eye to "see" the attached story; but Parkinson then gestures towards the stranger and truer, but dimly understood, histories that geology is now attaching to these objects. The rhetoric of spectacular display situates his reader within a metaphorical chamber of curiosities, and his engravings only hint at what the reader "will behold". What is to be viewed by the mind's eye is no mere cabinet, but the museum of nature herself: "You will behold her [Nature], incessantly labouring in the deep recesses of the earth [. . .] reducing to form and beauty, the mutilated wrecks of former ages." These wrecks are a frightening sight:

> innumerable beings have lived, of which not one of the same kind does any longer exist [. . .] enormous chains of mountains, which seem to load the surface of the earth, are vast monuments, in which these remains of former ages are entombed [. . .] laying thus crushed together, in a rude and confused mass [. . . .][101]

Order is imposed on this chaos by the geologist-guide, who leads the awed spectator among the exhibits:

> you will behold the bones of an animal, of which the magnitude is so great; as to warrant the conviction, that the bulk of this dreadful, unknown animal, exceeded three times that of the lion; and to authorise the belief, that animals have existed, which have possessed, with all the dreadful propensities of that animal, its power of destroying, in a three-fold degree. [. . .] In a word, you will be repeatedly astonished [. . . .][102]

Like Ashe, Parkinson places a purely sensational emphasis upon the "dreadful" character of these beasts—even though most of the fossils described are those of plants and other peaceful beings. He too resorts to the melodrama of vital statistics to convey the enormous size of the mammoth in the frontispiece to the third volume, an enormous drawing labelled "The back grinding tooth of the MAMMOTH OR MASTODON of Ohio.—weight 4 lb. 7 oz."[103]

The narrative background to the cataclysm which overtook all these extinct animals is vividly depicted in Richard Corbould's frontispiece

100. Parkinson 1804–11, I, 5.
101. Parkinson 1804–11, I, 10, 8.
102. Parkinson 1804–11, I, 11.
103. Parkinson 1804–11, III, frontispiece.

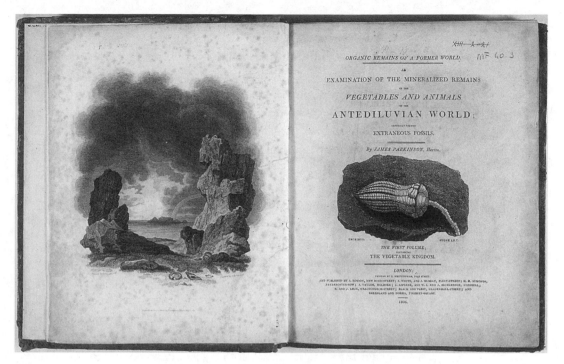

*Figure 1.5.* Frontispiece and title-page of the first volume of Parkinson's *Organic Remains.* The frontispiece, by Richard Corbould, shows the retiring waters of the Deluge: a tiny Noah's Ark can be seen in the distance, beached on some rocks. Parkinson 1804–11, I.

to the first volume (Fig. 1.5), showing the retiring waters of the biblical Deluge, a common theme of historical landscape art at this time. Noah's Ark is seen in the distance, with a group of fossil shells in the foreground.[104] The composition of this two-page spread compels curiosity: the most prominent words on the title-page are "ANTEDILUVIAN WORLD"; the frontispiece gives the broad narrative backdrop; and the title-page vignette of the "Stone Lily" (a crinoid) offers a captivating close-up. Iconography and rhetoric combine to draw the reader into a lost world of wonders.

Parkinson stopped short of providing full-scale imaginary voyages into former worlds.[105] These could only begin to appear when the specific antiquarian objects—the fossils—were united with a philosophical view of the landscape's structure, translating strata into periods of time; but stratigraphy lay outside the specific aims of Parkinson's book. The idea of travelling back along the strata was, however, vividly embodied by

104. Parkinson had abandoned the Deluge as a geological event by 1811, when the third volume was published. On Corbould's picture see Rudwick 1992, 18–20.

105. Percy Shelley, however, performed a visionary extrapolation of Parkinson's account in *Prometheus Unbound* IV.270–318 (Shelley 1989–2000, II, 629–33; for discussion see Heringman 2004, 178–86).

Playfair in 1803, in a brief reminiscence of a geological excursion which he, Hutton, and James Hall had made back in 1788. This passage was tucked into his posthumous biographical tribute to Hutton, delivered before the Royal Society of Edinburgh and printed in its *Transactions* two years later. As Playfair recalls, he and his friends had been boating at Siccar Point on the Berwickshire coast when they observed an outcrop where horizontal beds of sandstone overlay vertical beds of "schistus". Hutton interpreted this angular junction on the spot: as he saw it, the beds of "schistus" had clearly been uplifted by subterranean forces and subsequently eroded over the millennia to form a flat plane for the sand to settle on. Playfair later recalled the effect of Hutton's exposition:

> We felt ourselves necessarily carried back to the time when the schistus on which we stood was yet at the bottom of the sea, and when the sandstone before us was only beginning to be deposited [. . . .] An epoch still more remote presented itself, when even the most ancient of these rocks, instead of standing upright in vertical beds, lay in horizontal planes at the bottom of the sea, and was not yet disturbed by that immeasurable force which has burst asunder the solid pavement of the globe. Revolutions still more remote appeared in the distance of this extraordinary perspective. The mind seemed to grow giddy by looking so far into the abyss of time [. . . .][106]

Playfair's skill at communicating the sense of wonder opened up by the new geology is borne out by the fact that this passage is regularly quoted today to support the dubious claim that Hutton was the "discoverer" of deep time.[107] Hordes of geologists today make pilgrimages to Siccar Point to pay homage to Hutton, but in choosing that spot they pay unconscious tribute to the rhetorical gifts of his younger friend. In this passage Playfair extended the Enlightenment image of time as an abyss: listening to Hutton, he not only looks deep into this abyss, translating temporal distance into visual perspective, but feels himself transported back to those remote "epochs". This sense of transport is not found in Hutton's own reflections on the abyss of time, which seem almost ahistorical: "we find no vestige of a beginning,—no prospect of an end."[108] Hutton was concerned to show the earth as a stable, divinely-created system, not to

---

106. Playfair 1805, 72–3 (for a facsimile reprint see Eyles 1970). This passage almost certainly harks back to a piece of visionary narrative written in 1786 by Horace-Bénedict de Saussure (Saussure 1779–96, II, 339–40), recalling a view from an Alpine summit which stimulated him to imagine witnessing the successive "revolutions" undergone by the landscape. See Rudwick 2005b, 232–4.

107. The unfathomability of time was a commonplace among Enlightenment savants (Rudwick 2005b, 169).

108. Hutton 1788, 304 (for a facsimile reprint see Eyles 1970). On Hutton's views on time and extinction see Gould 1987; Rudwick 2005b, 169–72.

tell stories about past ages; but in Playfair's hands he became a sublime (and secular) historian.

Playfair's incorporation of extinct life-forms into his *Illustrations of the Huttonian Theory* would have made Hutton turn in his grave. Hutton had believed that present life-forms (including, it would seem, humans) had always existed, and extinction would have disfigured his perfect world-machine. Playfair had no such scruples, and in a short section on fossils at the end of his book he briefly used the language of visionary transport to point not only to the depth of geological time, but to the inhabitants of specific epochs: "Thus we are carried back to a time when many larger species of animals, now entirely extinct, inhabited the earth".[109] Yet this passage is very cautious by comparison with his Siccar Point epiphany, written three years later and referring entirely to inorganic processes. Playfair was not ready to make fossils into a sublime spectacle, or to fuse the latest advances in comparative anatomy with the kind of panoramic narrative Buffon had pioneered in *Les Époques de la nature*.[110] A brief attempt to do just this was, however, made by Bakewell at the end of his *Introduction,* snatching us up momentarily into a visionary realm where even exclamation marks are permitted:

> What various reflections crowd upon the mind, if we carry back our thoughts to the time when the whole surface of the globe was agitated by tumultuous and conflicting elements; or to the succeeding intervals of repose, when all was one vast solitude; and again to a subsequent period, when the deep silence of nature was broken by the bellowings of the great mastodon and the mammoth, who stalked the lords of the creation, and perished in the last grand revolution of the globe before the formation of man![111]

This vivid sequence of scenes illustrates the narrative power that fossils gained when combined with attention to the order of the strata. The last scene echoes the anthropomorphism of Ashe's tyrant beasts, ruling the world before the revolution came.

Such magnified views often prompted devotional reflections. As Davy put it,

> To the geological enquirer every mountain chain offers striking monuments of the great alterations that the globe has undergone. The most sublime speculations are awakened, the present is disregarded, past ages

109. Playfair 1802, 476.
110. Buffon 1988.
111. Bakewell 1813, 326–7.

crowd upon the fancy, and the mind is lost in admiration of the designs of that great power who has established order in which at first view appears as confusion.[112]

The crucial word is *lost:* although geology is presented throughout Davy's lectures as a triumphantly rational enterprise, its moment of highest validation comes in this momentary suspension of the rational faculties, whereby the "enquirer" can no longer enquire, but must wonder and adore. This was a different and more emotive kind of natural theology to that promoted by William Paley.[113]

What made natural theology so useful to geological popularizers was its flexibility: its rhetoric of wonder could fit into virtually any religious framework and (ostensibly) exalt the sensational into the sublime. The corresponding passage in Parkinson's *Organic Remains* is relatively restrained:

> By widening the views of the natural philosopher [. . .] and, by showing him a glimpse of other creations; more just, and more grand sentiments, also, must be excited, of the immensity of animated nature, and of the power of the great Creator of all things.[114]

Parkinson's philosopher does not "lose" his reason like Davy's rapt "enquirer"; but this momentary vision of the higher truths revealed by geology serves to validate the science in the same way. Parkinson's readers have, however, already been invited to suspend their rational faculties in quite a different manner. Gut sensation and pious adoration meet at this pleasurable point where reason temporarily flees. Their proximity shows what a difficult job educationalists would have in trying to encourage the booming new readership of the 1820s and 1830s to "relocate" their "sites of pleasure", as Anne Secord has put it.[115] It was always easier to appeal to spectacle in itself than as a high road to heaven, and many writers would opt for the former rather than the latter.

The movement from delight to piety, from physical phenomena to religious epiphany, can hardly be described as empirical induction; but such external validation was necessary for this difficult and time-consuming science to win public interest and support. Geology had, somehow, to appeal to the imagination. Both Bakewell and Parkinson aligned

112. Siegfried and Dott 1980, 13.
113. For another example see Herschel 1830, 4–5. On this "doxological" natural theology see Astore 2001.
114. Parkinson 1804–11, I, 12.
115. A. Secord 2002, 32.

their science with the popular physico-theological poetry which had nurtured polite science in the eighteenth century. Parkinson conveyed the wonders of mineralogy with twenty fulsome lines from Thomson's *Seasons*,[116] while Bakewell (like Ashe) employed Akenside's *Pleasures of Imagination* to argue that God's benevolence could be seen in his provision of "picturesque" landscapes.[117] *The Seasons* presents an inclusive vision of an endlessly curious, seemingly chaotic, indeed often disturbing universe; but, by a willed leap of faith, Thomson imposes scientific and religious unity on his vision by proclaiming the universe a "harmonious whole" in which human beings could be sure of their divine Governor.[118] Science writers were competing with cheap reprints of these very poems for the public's attention and cash. In quoting from such works, they claimed this imaginative territory for their own; but by this act they inherited all the metaphysical tensions of the genre.

However, only when fossils played a more constitutive role in geology could the historical vistas suggested by these writers flourish on a larger scale and, as Davy had said of Buffon, infuse the science's "dead objects" with "animation and life" and the "elegancies of literature".[119] It took some years for this to happen in earnest. Caution was still of paramount importance.

## Science with a Sting in Its Tail?

After the Battle of Trafalgar in 1805, the idea of a French invasion of England had become increasingly unlikely, and the excitement of Napoleon's further adventures hardly mitigated the troubles at home. Radical agitation, machine-breaking, rioting, and the still-fresh memory of the Terror in France resulted in a resurgence of post-Revolutionary anxiety. In 1815 the Battle of Waterloo brought the long war to an end and left thousands unemployed: bread was unaffordable for many of them because of the protectionist Corn Laws then in place, which provoked still more working-class discontent. There was worse to come: in the summer of 1815 Mount Tambora erupted in Southeast Asia. This was one of the most violent eruptions in recorded history, spewing clouds of ash into the atmosphere, but nobody in Europe knew this. All they knew was that the next winter was the coldest they had ever experienced, and that there was no summer to speak of in 1816. Crops failed all over Europe, rain

---

116. Parkinson 1804–11, I, 363–4.
117. Bakewell 1813, 95–6.
118. R. Cohen 1970.
119. Siegfried and Dott 1980, 45.

and frost predominated, the sky was often unnaturally dark, and people starved.[120]

A fresh wave of reform agitation and radical pamphleteering now swept Britain. The reading public itself was undergoing convulsive changes. In November 1816, to avoid the recently-imposed laws against seditious meetings (which included reading radical newspapers aloud in pubs), the former Tory William Cobbett issued his radical weekly, the *Political Register,* as a cheap twopenny pamphlet, so that working men could read it at home.[121] Its circulation soared to over forty thousand, more than any other periodical of the time. The next three years saw a dramatic rise in the activities of the radical press and the political awareness of working men. The near-tyrannical response to these developments by the ruling classes seems surprising today, but they were in a state of escalating desperation bordering on panic. In the space of a few years they had witnessed the appearance of a massive new urban public—angry, hungry, apparently unimpeded by any fear of God and, worst of all, articulate. Every new reader was a potential demagogue, and the conspiracy theories of the 1790s revived to stoke fears of revolution.

Consequently, Anglican sermons tended increasingly towards themes of social control, and the government clamped down on free thought wherever possible.[122] This often served to further the cause of the radicals they targeted: the bookseller Richard Carlile was arrested in 1818 for reprinting Tom Paine's *Age of Reason,* and the resulting publicity caused this inflammatory book to sell out.[123] Riots were now suppressed with increasingly draconian measures: in 1819 the so-called Peterloo Massacre, in which a peaceful pro-reform demonstration was violently put down by the cavalry, provoked furious popular outcries—to which the panicking government's response was, once again, to tighten its grip. That same year, new legislation revitalized the concept of "blasphemous libel" whereby any offence against the Church could be construed as sedition.[124] Deists like Carlile, who believed in a Supreme Being but not in the divinity of Christ, were routinely labelled "atheists". Scholars and savants might disagree with a literal interpretation of the Bible, but in this polarized political climate it was not advisable to publicize such views beyond their own learned circles.

120. Clubbe 1991.

121. Altick 1957, 324–6; on radical print in the 1830s see Hollis 1970. On the implications of the Seditious Meetings Act for scientific lecturing see Inkster 1979; Weindling 1980.

122. Hole 1989, 178–9.

123. On Carlile see Wiener 1983; *ODNB; DNBS.*

124. Schock 1995, 188–90; Marsh 1998, 18–77.

If the geologists had had to tread carefully before, they now had to tiptoe. It had been possible, in the early years of the Geological Society, to avoid controversy by insisting that their science now had nothing to do with cosmogony. An informal truce had existed by which exegetes and scholars of sacred chronology were expected to keep to their discipline, correlating scriptural with secular history, while geologists would keep to theirs, correlating the strata of different regions without claiming to tell tall stories about the deep past.[125] Now, however, many geological writers were finding it necessary to take a more defensive tone, inserting prefaces to argue that the new science was not atheistic.[126]

Yet geology won its wider audience precisely by presenting its objects as historical antiquities, romantically evocative of bygone eras. So it is no surprise that the best-known treatise of all before the 1820s was an English translation of a long essay (*Discours préliminaire*) by Georges Cuvier, in which the antiquarian approach to geology was developed with flair, erudition, and scientific acumen. It was edited by the great Edinburgh Neptunist, Robert Jameson.[127] At first glance, this book seemed to be directly opposed to everything the Geological Society stood for. It was entitled *Theory of the Earth,* and its author was one of the leading savants of a country with which Britain was still at war. It explicitly reunited geology with the study of ancient history, and furthermore its detailed use of textual evidence did not privilege the Bible over other documents. Pronounced with all the weight of his European authority, Cuvier's *Theory* risked breaking the fragile truce between geologists and more conservative interpreters of Genesis in Britain, and not everyone welcomed his views. Nevertheless, thanks partly to its editor's skilful marketing, this book sold extremely well: first published in 1813, it went into further editions in 1815, 1817, 1822, and 1827. Cuvier's work, mediated in this way, was definitive for the science's developing image and set the tone for subsequent popularization.[128] It is therefore worth first sketching Cuvier's own enterprise, before turning to the remarkable story of his *Theory*'s British reception.

Cuvier's cultural milieu differed sharply from Regency England. Paris was the scientific hub of Europe, and even during the reactionary 1800s and 1810s science was still perceived as one of the glories of the French state. In 1812, when his magisterial treatise on comparative anatomy was

---

125. Porter 1977, 202.

126. Conybeare and Phillips 1822, xlix–lxi; Mantell 1822, 1–13.

127. *ODNB; DNBS.*

128. On Cuvier's British reception see chapters 2 and 3 below; Page 1963; 1969; Rudwick 1972, 132–9; 1997; 2005b, 596–638. Some of the narrative strategies of the *Discours* itself, and its French reception, are examined in C. Cohen 2002, 105–24.

published (*Ossemens fossiles* or "Fossil Bones"), Cuvier was an internationally respected savant whose position in the French establishment was assured. His diplomatic skills had enabled him to weather the turbulent 1790s in positions of increasing authority: religion continued to swing in and out of political correctness, so Cuvier kept quiet about his own beliefs.[129] The *Ossemens* opened with a *Discours préliminaire* ("Prefatory Essay") whose aim was to set earth science on a new empirical footing. Like the spokesmen of the Geological Society, Cuvier set about debunking speculative "theories of the earth"; but, unlike them, he advocated a rigorously historical approach to fossils and chided mere "mineralogists" for ignoring them.[130] The earth science he envisaged, building on work done by other Continental savants during the 1780s and 1790s, would be a disciplinary blend of geology and comparative anatomy. His *Discours* put this case forward.

The essay begins by setting up Cuvier's relationship with the reader in conventional Enlightenment terms as a guide leading us on a metaphorical journey into the unknown. He introduces himself as a "new kind of antiquary", capable of "deciphering and restoring" the "monuments" of the earth's history before our wondering eyes, "gathering, in the darkness of the earth's infancy, the traces of revolutions before any nation existed".[131] The *Discours* functioned in part as a plea for future collaborative research which would, if carried out, unveil an extraordinary tale of periodic catastrophes and extinctions. For man to recover "the story of this world" would be no less than "to burst the limits of time" as Newton had burst those of space—the choice of analogy claims the prestige of celestial mechanics for the new subterranean science.[132] Cuvier's paean to the penetrative intellect is resumed at the very end:

> man, to whom a mere instant has been granted on earth, would have the glory of reconstructing the history of thousands of centuries which predated his existence, and of thousands of beings which have never been his contemporaries![133]

129. Cuvier's religious views and stances are discussed in Outram 1984, 141–60; Rudwick 1997, 83–8, 257–62; 2005b, 445–56.

130. Cuvier 1812, I, 25–35.

131. Cuvier 1812, I, 1–2 ("Antiquaire d'une espèce nouvelle"; "à déchiffrer et à restaurer"; "monumens"; "recueillir dans les ténèbres de l'enfance de la terre les traces de révolutions antérieures à l'existence de toutes les nations"). All translations in this book are my own. For an alternative translation of the whole essay see Rudwick 1997.

132. Cuvier 1812, I, 3 ("l'histoire de ce monde"; "franchir les limites du temps").

133. Cuvier 1812, I, 116 ("l'homme, à qui il n'a été accordé qu'un instant sur la terre, auroit la gloire de refaire l'histoire des milliers de siècles qui ont précédé son existence, et des milliers d'êtres qui n'ont pas été ses contemporains!").

The exclamation mark is quite deliberate, and rare in the *Discours* as a whole. The reader is asked to imagine a not-so-distant future in which the history of the earth (truer than the "romances" of writers like Buffon) will be available for all to read.

The implications of this new narrative disturbed notions of a Creator which many in Britain held dear. True, most savants and many evangelical divines already saw no difficulty in accepting the notion of species extinction: to those who knew their Origen and their Augustine, there were ways of reading the Genesis account of Creation alongside a belief in an "old earth" without doing violence to the principles of natural theology. Thanks largely to Cuvier's ongoing work, practising geologists were beginning to view the Creation as a providential sequence of events stretching over countless ages in which God had created increasingly advanced life-forms as the successive "tenants" of an increasingly habitable earth, culminating in man.[134] But such views were far from universal. Many clergymen and laymen were either ignorant of, or hostile towards, ideas of an earth ruled by beasts before man, and of death having entered the world before the Fall.[135] In this age of sensibility towards animal suffering, the idea of a God who extinguished entire species came under much criticism, not only from Anglicans but also from radical Deists like Elihu Palmer, who considered extinction to be a desecration of the order and harmony of the eternally existing universe.[136] Cuvier's fossil-oriented geology thus came with a sharp sting in its tail. He was by no means the first writer to raise the question of species extinction in Britain, but nobody before Cuvier had made the case so convincingly or made it such a central feature in their writings.

That such a potentially disruptive book won such a widespread following outside learned circles suggests that it must have been marketed effectively. Besides the force of Cuvier's rhetoric and his often delightful prose style, credit is due to the shrewd tactics of its now much-maligned editor Jameson. Of course, many critics on both sides of the Tweed felt that Cuvier's so-called *Theory of the Earth* was just another speculative attempt to undermine the authority of textual scholars in telling the story of Creation.[137] Even the Scottish evangelical Thomas Chalmers, who scorned the idea that geology "nurture[d] infidel propensities" and eagerly supported the new science, warned against its over-enthusiastic

---

134. On Cuvier's importance here see Rudwick 2005b, 507–8.

135. For an example of potential difficulties even within the Royal Society see Rudwick 2005b, 126–7.

136. Palmer 1819, 56–60.

137. "F. E---s" 1816; Boyd 1817; Gisborne 1818, 30–6.

encroachment upon the Bible's domain.[138] Jameson's editorial matter was designed to allay these fears.[139] His preface began:

> Although the Mosaic account of the creation of the world is an inspired writing, and consequently rests on evidence totally independent of human observation and experience, still it is interesting [. . .] to know that it coincides with the various phenomena observable in the mineral kingdom.[140]

This effectively reverses Cuvier's insistence on the pre-eminent authority of the natural "monuments" themselves. At this point the reader has not yet read Cuvier's own textual-critical observations, so Jameson is able to mould the way in which they will be perceived. By casting Cuvier in advance as a defender of Genesis, Jameson reassures his readers that the Bible's account of Creation—traditionally held to have been written by Moses—will not be challenged. The practice of geology (which just happens, like some pagan traditions, to coincide with Genesis) is rescued from superfluity by an appeal to antiquarian curiosity: geology is "interesting", another word which bore much more force in the nineteenth century than today. A handful of additional footnotes by the translator, Robert Kerr, refer some of Cuvier's observations to a framework of biblical chronology, discreetly maintaining an editorial presence in the main body of Cuvier's text.

Overall, Jameson's ploy worked. His success is best demonstrated by the response of his old enemy Playfair, whose review article in the *Edinburgh Review* might be expected to seize on any hint of a cover-up. But although Playfair sniped at the conservative tone of Jameson's preface, he happily accepted that "defending the Mosaic chronology" was Cuvier's own aim.[141] The book was attacked in the London-based *Philosophical Magazine* from a more conservative standpoint by H. S. Boyd, who likewise assumed that Cuvier, rather than just Jameson, was (falsely) attempting to reconcile Genesis and geology.[142] Both reviewers played into Jameson's hand: their Cuvier is definitely "Jameson's Cuvier". As a

138. Hanna 1849, I, 80–1; Chalmers 1836–42, XII, 361–2. For Chalmers's support of geology see Chalmers 1816, 175–85; 1836–42, XII, 369–70. His 1804 statement that *"The writings of Moses do not fix the antiquity of the globe"* (Hanna 1849, I, 80–1) was widely quoted by later popularizers.

139. Elsewhere Jameson had made very few references to the Deluge or the Mosaic account of Creation (see Page 1969, 260). It is possible that the reason he decided to import Cuvier's essay (and entitle it *Theory of the Earth*) was to provide a "directionalist" counterblast to the "Huttonian Theory of the Earth", against which his loyal Neptunians were still battling in 1813.

140. Cuvier 1813, v.

141. [Playfair] 1813–14, 468.

142. Boyd 1817, 375–8.

separate publication, the *Discours* was available only in Jameson's English edition until a German one came out in 1822.[143] In 1825 Cuvier authorized a separate printing of the *Discours,* but before then, if you wanted to read the original you had to obtain the enormous four-volume *Ossemens fossiles,* a tall order in wartime. "Jameson's Cuvier" set the tone for the adoption of Cuvier's ideas into British geology.

The *Theory of the Earth* brought geology to many for the first time. The Irish novelist Maria Edgeworth reported in 1816 that, although Jane Austen had just sent her a complimentary copy of *Emma,* she really wanted to read Cuvier:

> we are [. . .] reading a book which delights us all, though it is on a subject which you will think little likely to be interesting to us, and on which we had little or no previous knowledge [. . . .]—Cuvier's *Theory of the Earth*. It is admirably written, with such perfect clearness as to be intelligible to the meanest, and satisfactory to the highest capacity.[144]

Jameson continued to adapt later editions to suit changing demands.[145] As he was preparing his second edition, hostilities between France and Britain reached a final peak, so Jameson patriotically altered Cuvier's nationality, changing his preface's first reference to "Cuvier" to "the illustrious German naturalist Cuvier". A footnote explained that "Cuvier is a native of Mumpelgardt in Germany".[146] Not only could Jameson manipulate Cuvier's supposed religious attitude to suit his purposes; he could also capitalize on the liminal status of his home town.[147] After Waterloo, this rather heavy-handed gesture could be dropped, but the third edition of 1817 also reveals that Jameson no longer felt his detailed observations on Genesis and geology were necessary or helpful. Although the new preface omitted references to Moses himself,[148] Jameson retained what was now a somewhat ambiguous emphasis on the Deluge, stating that "the subject of the *deluge* forms a principal object of this elegant discourse".[149] He also kept Kerr's footnotes on the biblical chronology. But the equation of the six days of Creation with six undefined epochs was more contentious than it had been in 1813: Chalmers had popularized

---

143. A German paraphrase had, however, appeared in 1816; see Rudwick 1997, 254–6.

144. Edgeworth to Mrs. Ruxton, 10 January 1816 (Barry 1931, 231–2).

145. He also adapted them to fit his own changing ideas about Lamarckian transformism (J. Secord 1991, 9–11).

146. Cuvier 1815, vii.

147. Montbéliard/Mumpelgardt, a Francophone border-town, was in the German duchy of Württemburg, which had been annexed by France during the Revolution (Rudwick 1997, 1).

148. Page 1969, 260–1.

149. Cuvier 1815, vii; 1817, ix.

an old scheme now known as the "gap theory", by which all the former worlds required by geologists could be shoehorned into the gap between Genesis 1.1 (the earth created) and 1.2–31 (the six days' work).[150]

To continue tempting the curious, Jameson incorporated into the third edition's preface a five-page account of the bizarre German "monster", the pterodactyle. This creature had been discussed in the previous edition, but he now chose to draw special attention to it.[151] By 1822, when the fourth edition came out, British writers were promoting geology's devotional role with more energy; accordingly, Jameson enlarged the preface with a meditation on geology's ability to fill the mind of the "antiquary" with "sentiments of piety and feelings of devotion".[152] Four years on, there was room for complacency about Cuvier's fame and the public status of geology, "now deservedly one of the most popular and attractive of the physical sciences", yet the tone is still noticeably defensive about the threat of conservative opposition.[153] Jameson had now removed all traces of the Deluge, sidestepping the escalating controversy over diluvial phenomena. Erratic boulders, U-shaped valleys, and gravels, associated with glaciation since the mid-nineteenth century, represented in the 1820s one of the most intractable problems in earth science. The Deluge also formed the focus for parallel debates on the relation between historical texts (including the Bible) and natural phenomena, and on the role of miracles in scientific explanation. Some geologists, such as William Buckland in Oxford, at first understood diluvial phenomena to be evidence for Noah's Flood, while others saw them as resulting from prehuman causes (whether gradual or catastrophic).[154] In Jameson's fifth edition a new section, "On the Universal Deluge", now explained diluvial phenomena in terms of natural causes, with no reference to Moses.[155] This section was apparently included in response to numerous requests from his friends, and reveals that Jameson—in his capacity as purveyor of popular geology—was now as willing to wrench Cuvier's theories away from the Bible as he had been to weld them together at first.[156]

This shifting prefatory matter in "Jameson's Cuvier" is a revealing index to the changing public projections of geology, filtered through the shifting viewpoint of one Edinburgh man of letters and geologist. Jameson's rhetorical flexibility allowed him to dovetail Cuvier's Enlightenment research scheme with the changing needs of the British intel-

---

150. Chalmers 1816, 184. On this and other exegetical schemes see Yule 1976, 332.

151. Cuvier 1817, xi–xv.

152. Cuvier 1822, iv.

153. Cuvier 1827, v.

154. On diluvial debates see Page 1963; 1969; Boylan 1984, 446–70; Herbert 1992; Rudwick 2005b, 557–638.

155. Cuvier 1827, 417–37.

156. Compare J. Secord 1991, 10–11.

lectual establishment. To many of the more liberal evangelical clerics of the day, Cuvier's work offered a useful means of publicly reinforcing scriptural authority (mainly via the Deluge) while providing a thrill of curiosity associated with fossils.

John Bird Sumner exemplified the cautious approval with which many learned evangelicals greeted the new historical geology of Cuvier.[157] The first appendix to his influential treatise *The Records of the Creation* (1816) is entitled "That the Mosaic History Is Not Inconsistent with Geological Discoveries"—the double negative signalling the shift from Jameson's highly specific day-age equivalence to the more non-committal reassurance that geology was harmless to the faith.[158] But Sumner also valued the new geology as a sublime form of entertainment, and was particularly drawn to Cuvier's analogy with astronomy:

> the description of fossil remains of unknown genera, form[s] a curious and interesting subject of speculation, but can never interfere with the knowledge acquired from less disputable sources of information. The alternate revolutions which Cuvier supposes, are in the science of geology what systems beyond our own are in astronomy. They are matters of curious reflection and sublime interest; but they lead us beyond the regions of legitimate science or certain history, into those of vague speculation [. . . .]
>
> [. . . .] But we are not called upon to deny the possible existence of previous worlds, from the wreck of which our globe was organized, and the ruins of which are now furnishing matter to our curiosity. The belief of their existence is indeed consistent with rational probability, and somewhat confirmed by the discoveries of astronomy, as to the plurality of worlds.[159]

This last phrase refers to the theory that there were numerous inhabited worlds in space: this was widely credited between 1750 and 1850 and, generally speaking, excited controversy only in matters of theological interpretation.[160] Sumner here uses astronomical discoveries to familiarize Cuvier's proposal of a plurality of worlds in time. He warns that the geological narrative "supposed" by Cuvier is not "certain history" or even "legitimate science", but insists that such "speculation" is not dangerous if practised along historical lines (rather than in the eternalist sense employed by anticlerical *philosophes*).

157. Contrary to a common misconception, Sumner was not bishop of Chester when he wrote *The Records of the Creation*. He became bishop in 1828 and Archbishop of Canterbury in 1848 (Scotland 1995; *ODNB*).

158. Sumner 1816, I, 267–85.

159. Sumner 1816, I, 282–5. Forty-five years later, Sumner would use the same tone of circumspect approval in welcoming the latest "shocking revelation" of palaeontology, the antiquity of man (Brooke 1999, 27).

160. Crowe 1986.

Sumner's phrasing belies today's popular Whiggish myth according to which nineteenth-century clerics reluctantly "admitted" that geology was "true" after all and grudgingly bowed to the march of science, adjusting (and reducing) their religious beliefs accordingly. Sumner clearly felt, or wished to project, genuine enthusiasm for Cuvier's new science, in terms already expressed by the other writers examined in this chapter. Not only are the findings of Cuvier "interesting" and "sublime"; they are repeatedly described as "curious". The final paragraph suggests that Sumner actually wanted the notion of "previous worlds" to be proved correct. Elsewhere he enthuses about the "sublime discoveries as to the prospective wisdom of the Creator" opened up by astronomy, and adds that geology would do the same "if that science should ever find its Newton, and break through the various obstacles peculiar to that study".[161] This is surely a direct allusion to Cuvier's own exhortations.[162] That an evangelical bishop, concerned to safeguard the claims of Scripture, was willing to use such language indicates the irresistible attraction which geology seemed to hold; and so Sumner himself joined the ranks of its popularizers.

This was, of course, a very different kind of popularization to what we see taking place later in the century. In the 1800s and 1810s the reading public for the new science was restricted to the wealthiest sector of society. Working people could afford cheap pirated reprints of philosophical works sold by radical publishers like Carlile, but the cosmology of these works tended to be of the old ahistorical, eternalist kind beloved of eighteenth-century deism. New books and periodicals were prohibitively expensive, as was membership in the Geological Society of London. Many of the authors themselves, such as Bakewell and Smith, could not afford these luxuries. From the viewpoint of the genteel geologists, however, this restriction guarded their science against radical appropriation as they carved out a space for it in a culture of opposition. Before the idea of successive "former worlds" could command broader assent, it had to take root among the great and the good. So the geologists wooed polite society, both in person and in their elegant, comfortingly exclusive publications. They leavened the appeal to economic utility with glimpses of terrifying monsters drawn from the world of commercial fossil exhibitions; they scattered the occasional attractive and morally improving piece of verse around their calm and measured prose; and, in the example we have just been looking at, they reassured their readers that geology posed no threat to the Established Church.

From this position it would only take a slight shift in emphasis to pres-

161. Sumner 1816, I, 271.
162. Cuvier 1813, 3–4.

ent geology as a positive support for Scripture. This is precisely what happened at Oxford University, where, in 1818, the Prince Regent gave geology the royal seal of approval by endowing a new Readership in Geology for Cuvier's charismatic English follower William Buckland. Oxford was at the very heart of the Anglican establishment: if the new science could take root here, among the future leaders of the Church, its future would be assured. Buckland rose to the occasion. Within this socially circumscribed setting, he could afford to brush aside the polite conventions of Regency public science and get on with the show. His startling restorations of vanished worlds and their impolite inhabitants mark a turning-point in the history of the geological imagination. They call for a closer look.

# William Buckland:

*Antiquary and Wizard*

2

### Historical Geology at Oxford

Cuvier's science came at an opportune time for the beleaguered University of Oxford. Since the beginning of the century Oxford and Cambridge had been under increasing attack by periodicals such as the *Edinburgh Review* for their neglect of secular learning. Reform was required: at Oxford, progressive Fellows pressed for the introduction of new secular subjects to complement a Classical, clerical education. William Buckland, who had been giving optional lectures on geology since 1814, gained his Readership in this context: the curriculum was centred on ancient texts, so Buckland presented his science initially as a historical support for Holy Writ.[1] In his inaugural lecture of 1819 he announced an

> ingrafting (if I may so call it) of the new and curious sciences of Geology and Mineralogy, on that ancient and venerable stock of classical literature

---

1. On this process see Edmonds 1979–80; Rupke 1983b; 1997. Rupke's work is complemented by Boylan's (1984) superb biographical study of Buckland, which includes detailed analysis of his Oxford lectures. See also *ODNB; DNBS*.

from which the English system of education has imparted to its followers a refinement of taste peculiarly their own [. . . .][2]

Geology would be both comfortingly traditional and attractively novel. Buckland then enumerated its connections with established disciplines, which it would enhance by serving "a subordinate ministry in the temple of our Academical Institutions."[3] In this clerical milieu, Buckland "ingrafted" geology directly on to the more biblically aligned "theories of the earth", as well as to Cuvier's procedures, without drawing (as Cuvier had) a polemical distinction between the two. This attitude did not impress some of Buckland's colleagues in the Geological Society, who continued to dissociate themselves from such a speculative, text-centred perversion of their empirical principles. Yet the publicity which Buckland achieved for his science in the 1820s gave "Oxford geology" a prominent place among the competing versions of geology on offer to leisured non-specialists.

As a public vindication of geology, delivered first at Oxford and then (in print) before a wider audience, Buckland's inaugural lecture indicates the appeal geology was expected to exert in an academic, clerical context. In 1819 geology was still a science under surveillance, and Buckland must have felt the eagle eye of public opinion fixed on him as he spoke. Whether what he said was original or not—and much of its content was supplied by his friend William Conybeare—it was vital that he say it in such a way as would convince his patrons that geology was worthwhile.[4]

Most of the lecture's tropes recapitulate the sentiments of Parkinson, Bakewell, and Davy. Buckland first mentions geology's practical utility, but almost at once reaches further:

The human mind has an appetite for truth of every kind, Physical as well as Moral; and the real utility of Science is to afford gratification to [. . . .] this large and rational species of curiosity [. . . .]

He then claims an unmatched spectacular appeal for the "monuments of the mighty revolutions and convulsions" suffered by the globe,

convulsions of which the most terrible catastrophes presented by the actual state of things (Earthquakes, Tempests, and Volcanos) afford only a

2. W. Buckland 1820, 2–3. On this lecture see Boylan 1984, 81–8; Edmonds 1991.
3. W. Buckland 1820, 3.
4. Conybeare quoted much of Buckland's lecture in his own popularization (Conybeare and Phillips 1822, li–lviii). On Conybeare see *ODNB; DNBS.*

faint image [. . . .] these surely will be admitted to be objects of sufficient magnitude and grandeur, to create an adequate interest to engage us in their investigation.[5]

Here Buckland was feeding on the vogue for pictorial and textual representations of the Deluge and other apocalypses.[6] He then quotes Cuvier's rapturous prophecy about the mind of man recovering the sublime narrative of geology, and adds, "It is surely gratifying to behold Science, compelling the primeval mountains of the Globe to unfold the hidden records of their origin".[7] This imagery, in which science not only produces spectacular effects but is itself a spectacle of intellectual power, would become paramount in the public rhetoric of the "gentlemen of science" in the 1830s.[8]

Buckland presents several viewpoints from which geology might satisfy the human "appetite" for knowledge, most of them by now familiar. It leads its devotee into "sublime scenery" whose grandeur they enhance by "the magnificence of the speculations which they associate with it"; it gratifies antiquarian tastes by unearthing "the Antiquities of the Globe itself [. . .] the monuments and medals of its remoter eras"; it provides the satisfaction of discovering "the order and harmony of nature" in apparent "disorder and confusion".[9] More specifically, the economically convenient arrangement of strata (exposing coal seams, for instance) reveals "the finger of an Omnipotent Architect" providing for his people's "daily wants" from the earliest eras of the earth's history: this is one of several permutations of natural theology in this lecture, and this multi-denominational discourse must be not be confused with the quite separate matter of geology's support for one particular sacred text.[10] This last argument forms Buckland's rhetorical climax in the form of the geological evidence for the Deluge. At this stage in his career, Buckland's need to unite his two professional worlds—Anglican learning and the new geology—gave him a special interest in this controversial point of encounter between human history and Cuvier's geology.[11] Here the geologist played more than a merely ancillary role in the interpretation of sacred history:

5. W. Buckland 1820, 5.
6. On this topic see Rupke 1983a; Paley 1986; F. Stafford 1994; Paley 1999; Freeman 2004, 163–89. The vogue for apocalypse is discussed in more detail in chapter 7.
7. W. Buckland 1820, 5–6.
8. Morrell and Thackray 1981, 159.
9. W. Buckland 1820, 5–6.
10. W. Buckland 1820, 12. This distinction has been blurred by some scholars (e.g. C. Cohen 2002, 126–7), resulting in an overestimation of the role of Genesis in determining Buckland's procedure.
11. Sommer 2004, 55–8.

the grand fact of *an universal deluge* [. . .] is proved on grounds so decisive and incontrovertible, that, had we never heard of such an event from Scripture, or any other authority, Geology of itself must have called in the assistance of some such catastrophe, to explain the phenomena of diluvian action [. . . .][12]

Like Cuvier, Buckland seems to hint that, in such cases, the geologist's text of "natural monuments" has a hermeneutic authority equal to that of Genesis itself. But no such challenge is posed within his rhetorical framework. His inaugural lecture was designed to placate.

Buckland's published lecture bore the subtitle *The Connexion of Geology with Religion Explained.* Accordingly, its second half is taken up with theological matters, ending with an analysis of four current means by which the Mosaic account of Creation might be harmonized with the findings of geology.[13] The whole section is riddled with quotations from authorities whose piety and orthodoxy is, by implication, established: Chalmers, Sumner, Francis Bacon, Samuel Horsley. But the first authority quoted in this section is Cuvier, "one who deservedly ranks in the very first class of natural observers". As with Jameson, Cuvier's observations on the latest "great and sudden revolution" are made to exemplify geology's harmony with Scripture.[14] In this setting Cuvier, a determinedly secular savant, became a kind of honorary Anglican—a development with which Byron would soon make mischief. Buckland then proposed to conduct detailed research into "diluvial phenomena", understood as the physical evidence of a geologically violent and recent Deluge. This is what he subsequently did, carrying out cave explorations in Britain and the Continent and analysing the bones they contained. That project bore significant fruit in 1821–2, and will be discussed below.

Buckland's inaugural lecture was a model of dignity and restraint. As such it was uncharacteristic of him. It served its purpose, winning him the authority he needed; but thereafter, the key to his effective communication of geology lay not in how it was authorized from above, but in how it appeared to the eager crowds of genteel students who attended his lectures. When assessing the impact of early-nineteenth-century science lectures, we rely heavily on the circumstantial evidence of anecdotes, mostly recorded during a later period when historical geology was a popular and well-established science. In Buckland's case, however, there is enough evidence (some datable to the early 1820s) to shed light on his procedures. Here is not the place to repeat all the usual anecdotes,

---

12. W. Buckland 1820, 23.
13. W. Buckland 1820, 22–38.
14. W. Buckland 1820, 24.

which can be found elsewhere.[15] But this much is clear: Buckland was a born showman. Even the young John Henry Newman, who would later emerge as a major opponent of the vogue for scientific "wonder", informed his mother in 1821 that the geology lectures were "most entertaining, and open[ed] an amazing field to imagination and to poetry."[16]

Newman's words encapsulate a crucial aspect of the importance of Buckland's geology to the educated upper and middle classes: it broadened one's vision. The lectures had to be entertaining, or nobody would attend them: natural science was optional in the Oxford curriculum, and there were no examinations. Buckland had to work hard to persuade career-minded students to attend, and he badly needed the money from their fees to pay for his geologizing.[17] One of his triumphs (anticipated by Jameson) was his practice of lecturing outside, in a quarry or on a hill, in order to demonstrate points of stratigraphy *in situ*. Many students had come to Oxford from country seats, hunting and living the outdoor life, and were now expected to sit inside lecture-halls and become serious men of learning. This eccentric professor delighted them by lecturing in the fresh air and galloping off with an ammonite around his neck.[18]

One of the chief means by which Buckland's geology opened an "amazing field to imagination" was his visual aids, many of which still survive.[19] It is difficult to tell precisely how Buckland employed these materials, but besides the anecdotal evidence, two portraits of Buckland lecturing have come down to us. One of these (Fig. 2.1) shows him lecturing in Oxford's Ashmolean Museum.[20] Visual aids are jumbled around. Buckland is seen holding up an ammonite for the audience to scrutinize; elsewhere he is said to have passed specimens round for this purpose.[21] The visceral thrill of handling and viewing these relics gave force and immediacy to the lessons Buckland wished to convey. His own rooms in Corpus Christi College functioned as a crowded cabinet of curiosities, as celebrated by his friends in verse: "Here see the wrecks of beasts and fishes, / With broken saucers, cups, and dishes".[22] In the lecture-hall, charts presented the landscape in three aspects: in its present visible form (landscape engravings), in its structure (stratigraphic sections), and surveying the structure of the whole of England (geological maps). Engrav-

15. For example in Gordon 1894 and Cadbury 2000.

16. Gornall *et al*. 1961–77, I, 109. Newman's lecture-notes, and those of other students, are reproduced in Boylan 1984, 531–653.

17. Boylan 1984, 75.

18. Boylan 1984, 75–6. On Buckland's lecturing techniques generally see *ibid.*, 103–6.

19. Boylan 1984, 104–5; 1997, 368–9. On the use of visual aids in natural-history lectures see A. Secord 2002, 40–8.

20. Edmonds and Douglas 1975–6. The present Ashmolean occupies a different site.

21. See Daubeny 1869, 85.

22. Daubeny 1869, 81.

ings of fossils could stand in for actual specimens. On the back wall in Fig. 2.1 hangs a picture of the Peale mammoth, its tusks here upturned; about to fall off the table is an engraving of an ichthyosaur, perhaps one of Mary Anning's.

All the visual aids so far mentioned depict geological evidence of Buckland's own day: strata, rocks and fossils, the skeletal ruins of once-living landscapes. But Buckland went further, attempting to bring his audience into immediate visual contact with the former worlds themselves. He introduced into his lectures not only reconstructions of entire skeletons, but also restorations of the living creatures. In 1822 his research on fossil hyaenas allowed him to provide a ground-breaking restoration of a fossil habitat-group, which in turn gave rise to the first known pictorial restoration of a fossil animal in its environment. This breakthrough in the imagining of deep time will be discussed in more detail below. What Martin Rudwick has called "scenes from deep time" developed more seriously in the 1830s with the rise of saurian restoration, initially focusing on scenes from ancient Lyme Regis (ichthyosaurs, plesiosaurs, pterodac-

*Figure 2.1.* Lithograph from 1823 showing Buckland lecturing in the old Ashmolean Museum to an audience of senior members of Oxford University. Oxford University Museum of Natural History, Buckland Papers (Drawings).

*Figure 2.2.* Henry De la Beche's restoration of "a more ancient Dorset", *Duria antiquior* (1830), showing ichthyosaurs, plesiosaurs, and pterodactyles pursuing prey and defecating with alacrity. This is the lithographed version, sold for the benefit of the impoverished Anning family for £2 10s (= *c.* £440 or $870 today). Buckland used this engraving as a "syllabus" for his lectures, distributing copies of it to his students.

tyles) and the Sussex Weald (Iguanodon, Megalosaurus, Hylaeosaurus).[23] The first of these, the geologist Henry De la Beche's watercolour sea-scene *Duria antiquior,* was published as a lithograph in 1830 (Fig. 2.2) and displayed by Buckland in his lectures.[24]

Almost all available pictorial restorations found their way into Buckland's lectures over the ensuing years.[25] Some of these are well known to historians, such as De la Beche's caricature *Awful Changes* (see Fig. 4.3), which depicts "Professor Ichthyosaurus" lecturing on fossil man with the visual aid of a human skull. Besides the published lithograph, Buckland also possessed another version whose much larger dimensions suggest that it was used in his lectures.[26] Others are less well known: the three

23. Rudwick 1992.

24. On Buckland's use of this picture see Gordon 1894, 116–17; Boylan 1984, 440; 1997, 368–9. On the picture itself see McCartney 1977, 44–7; Rudwick 1992; Norman 2000.

25. All the pictures mentioned in this paragraph are held in the Buckland Drawings file in the Oxford University Museum of Natural History.

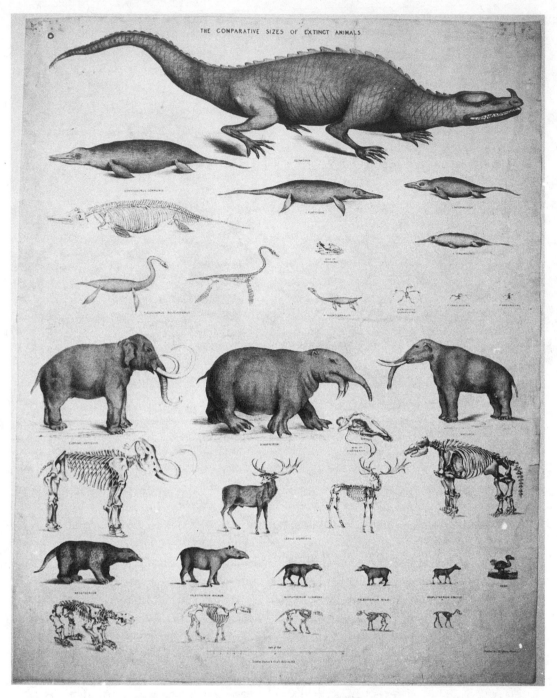

*Figure 2.3*. Buckland's wall-chart showing "The Comparative Sizes of Extinct Animals" (500 × 630 mm), printed not long after 1835 by the educational booksellers Darton & Clark. The Iguanodon at the top is represented as being 100 feet long. Oxford University Museum of Natural History, Buckland Papers (Drawings).

*Figure 2.4.* Two sepia drawings, perhaps by Buckland. The top one represents a Carboniferous landscape; the bottom one represents the Lias, the period depicted in *Duria antiquior* (Fig. 2.2). Oxford University Museum of Natural History, Buckland Papers (Drawings).

reproduced in Figs. 2.3 and 2.4 are published here for the first time. Fig. 2.3 is a large chart depicting the comparative sizes of extinct animals, apparently modelled on a similar chart of living animals dating from 1835. Fig. 2.4 shows two sepia drawings of landscapes in different geological periods. All three have small holes at the upper corners, suggesting that they were mounted as visual aids in lectures to help Buckland's audience picture the past.

Buckland enlivened his lectures with an eccentric sense of humour. He was renowned for his disconcerting swerves between the sublime and the ridiculous, a tension characteristic of Regency show-culture. This mixture of registers served two purposes for Buckland. It enabled him to set a certain distance between himself and his more daring speculations before committing himself in sober prose—as Marianne Sommer has put it, "allowing him to retreat when retreat was found necessary".[27] But it also created theatrical effect, emphasizing the extraordinary nature of the facts he was relating. Buckland brought extinct animals to life in ways that were unorthodox, and hence memorable. According to his student Charles Lyell, Buckland on occasion became a visual aid himself: he "would keep his audience in roars of laughter, as he imitated what he thought to be the movements of the Iguanodon or Megatherium, or, seizing the ends of his clerical coat-tails, would leap about to show how the Pterodactyl flew".[28] According to another former student, Henry Acland, Buckland had other means of bringing fossil specimens to life than merely pointing to posters:

> He paced like a Franciscan Preacher up and down behind a long show-case, up two steps [. . . .] He had in his hand a huge hyena's skull. He suddenly dashed down the steps—rushed, skull in hand, at the first undergraduate on the front bench—and shouted, "What rules the world?" The youth, terrified, threw himself against the next back seat, and answered not a word. He rushed then on me, pointing the hyena full in my face—"What rules the world?" "Haven't an idea," I said. "The stomach, sir," he cried (again mounting his rostrum), "rules the world. The great ones eat the less, and the less the lesser still."[29]

Buckland liked to eat whatever animals he could get his hands on, and his dinner parties were voyages into unknown regions of culinary experience, impressing the rule of the stomach on his hapless guests. Some

26. Rudwick 1975, 537 n. 13.
27. Sommer 2004, 73.
28. Undated account quoted in D. Allen 1976, 63.
29. Gordon 1894, 31.

of his more earnest acquaintances found his theatrical antics "vulgar": Charles Darwin suspected that Buckland was "incited more by a craving for notoriety, which sometimes made him act like a buffoon, than by a love of science".[30] Modern historians have been hardly less forgiving, comparing Buckland's antics unfavourably with the imposing spectacle of Isambard Kingdom Brunel's engineering feats: "In Buckland's hands knowledge was buffoonery; in Brunel's it was domination."[31] For Buckland, buffoonery was a form of power in itself, helping him to build up a reputation and hence attract an audience, to establish complicity with them, and to hold their attention. A family friend recorded that Buckland "feels very nervous in addressing large assemblies till he has once made them laugh, and then he is entirely at his ease."[32]

Buckland's lectures inspired light-hearted verses by students and dons, who circulated these in manuscript or in privately printed broadsheets. These ephemeral poems are richly revealing of how genteel listeners responded to earth history as Buckland told it, and they form a major source for the rest of this chapter. For this reason some introductory remarks on the corpus are called for. Versifying in Latin or English was a common practice among Classically educated gentlemen, and was also beginning to flourish in the mixed society of Regency drawing-rooms. Occasional verses on science appeared in a variety of contexts: in private diaries, in manuscripts intended for circulation, and in albums or commonplace books. This is an extremely fugitive corpus, mostly unpublished, and confident statements on its development cannot yet be made. Some specific loci may be identified: an early and influential example was the body of parodic and celebratory verses surrounding Buckland in Oxford.[33] These poems rarely strayed outside the genteel milieu, and their frequent reliance on the comic potential of fossil turds made them unsuitable for ladies; but Buckland himself recycled some poems to add spice to his own lectures and letters.[34] A second, more public, and more polite group of verses surrounded Gideon Mantell's promotion of geology in Brighton in the 1830s: some appeared in local newspapers to advertise his museum.[35] Later still, younger members of the Geological

---

30. C. Darwin 1958, 102. Compare Murchison's remarks quoted in Rudwick 1985, 170.

31. Morrell and Thackray 1981, 159.

32. Gordon 1894, 27–8.

33. Some examples: [Conybeare] 1822; Daubeny 1869, 69–73, 90–1, 119–22; Magdalene College, Oxford, MS 377, I, 489–91; II, 151–3; Devon Record Office, Exeter, 138 M/F 709, 711, 722, and 725 (original manuscripts now in the collection of Roderick Gordon).

34. Oxford University Museum of Natural History, Buckland lecture notes, "Deluge" 2/3; Edmonds and Douglas 1975–6, 150.

35. Some examples: Magdalen College, Oxford, MS 377, I, 157–9; Alexander Turnbull Library, Wellington, New Zealand, MS-1956; Daubeny 1869, 123–6; G. Richardson 1838, 6–7, 222, 289–95; Curwen 1940, 134. Caricatures such as Fig. 10.5 were part of the same movement.

Survey (founded in 1835) imitated the Geological Society grandees with a cult of the field and hearty annual dinners punctuated by singing. Romantic nature-poetry and rollicking ballads were the result.[36]

These verses are central to a full appreciation of how the science was popularized. They are often quoted today as anecdotal evidence for the science's vogue or for the popularity of individuals like Buckland,[37] but only recently have they become the subjects of serious analysis. Marianne Sommer has shown, in an important article on British cave geology, how Buckland and his colleagues used humorous verses and other informal genres to develop a chivalric, romantic self-image as "knights of the hammer", heroes with prophetic or magical powers over nature.[38] More recently Sommer has shown that Buckland, when presenting his more daring theories *viva voce,* used humour both to reinforce the social cohesion of his gentlemanly audience and to test daring geohistorical scenarios without public commitment.[39] Her account of Buckland's humorous strategies can, I believe, be applied more widely to the whole genre of occasional verses: like caricatures, these were circulated privately for specific audiences rather than placed firmly in the public realm.[40] In this chapter, following on from Sommer's work, I will show how literary techniques which would become central to later public geology— geologist as necromancer, dream-vision frameworks, time travel, resurrection of extinct animals—assumed humorous form in occasional verses well before making a confident public appearance in the sober prose of science books.

These techniques were not invented by the versifying geologists. They drew on eighteenth-century models, which included works that had fallen into official disfavour, such as "theories of the earth" and Erasmus Darwin's speculative botanical verses. The old stories soon found their way into the new science. Philip Shuttleworth's "Specimen of a Geological Lecture, by Professor Buckland" (*c.* 1820–2) is a good example of this continuity and also indicates how Buckland brought the science to life in his lectures. Shuttleworth found it appropriate to parody his style with a sublime sequential narrative clearly modelled on the opening lines of Darwin's *Temple of Nature* (1803):

36. Some examples: Daubeny 1869, 99, 106–16, 158–76; Geikie 1895, 142, 175–6. See Porter 1978a, 825–9.

37. For instance, Gillispie 1951, 98; Cadbury 2000, 58; Rupke 1983b, 71–4, 222–5.

38. Sommer 2003. On magical elements in these occasional verses see also Klaver 1997, 157; R. O'Connor 2003c, 86–96. On the knightly image see Porter 1978a, 819–20.

39. Sommer 2004, 60, 72–4.

40. On caricatures and the testing-out of geological scenarios see Rudwick 1992, 56–7; on scientific caricature as a vehicle of group cohesion see Paradis 1997, 170. See also Janet Browne's study of science and undergraduate humour (1992).

'Twas silence all, and solitude; the sun,
If sun there were, yet rose and set to none,
Till fiercer grown the elemental strife,
Astonished tadpoles wriggled into life;[41] [. . . .]
Now mammoths range, where yet in silence deep
Unborn Ohio's hoarded waters sleep.[42]

This distinctly unbiblical narrative, teasingly reminiscent of discredited evolutionary speculations, was a far cry in both mode and content from Buckland's cautious inaugural lecture. Such grand narratives were not yet suitable for public dissemination. Before the 1830s, Buckland indulged in speculation only when every member of his audience could be vouched for as a gentleman.[43]

These were sensitive times for a visionary science, as can be seen by the example of a slightly earlier series of Darwinian imitations. Between 1818 and 1820 a consortium led by the London bookseller Longman had published a series of mineralogical verses in the form of court scenes and knightly legends about the rocks—Baron Basalt, Lady Serpentine, and so on—in heroic couplets, a rhyming form typical of eighteenth-century narrative poetry. The first volume, John Scafe's *King Coal's Levee,* conveyed the physical characteristics and relationships of the rock types in anthropomorphic terms, rather as Darwin had done for flowers (Fig. 2.5). Later editions of *King Coal* incorporated geological notes by Buckland and Conybeare. These poems were seen by geologists as diverse as J. W. von Goethe and Benjamin Silliman as didactically valuable.[44] In 1819, however, *King Coal* was appropriated by Yorkshire and Lancashire radicals as a satire on the Prince Regent: it contained a Peterloo-like passage where the tyrannical king incurs Giant Gravel's wrath by violently suppressing a plebeian intrusion by the Pebbles. The author and his learned assistants were accused of fomenting revolt, and, though that case came to nothing, the lesson was clear.[45] For now, Buckland and his circle kept their poetry to themselves.

As Buckland's friend Philip Duncan indicated in his "Picture of a Professor's Rooms in C.C.C., Oxford" (1821), only the select few could be admitted to the mysteries of geology, symbolized by Buckland's messy room:

41. Compare E. Darwin 1803, 3 (canto I, lines 3–4). On Darwin's poem see Heringman 2004, 214–15.
42. Daubeny 1869, 84–7. Compare the parody of Sedgwick's lecture discussed in Browne 1992, 181–7.
43. On the social composition of his Oxford audiences see Boylan 1984, 267–304.
44. Rupke 1983b, 223. On Scafe see *ODNB*.
45. On this controversy see [Scafe] 1820b, 60–7; Rupke 1983b, 223–5.

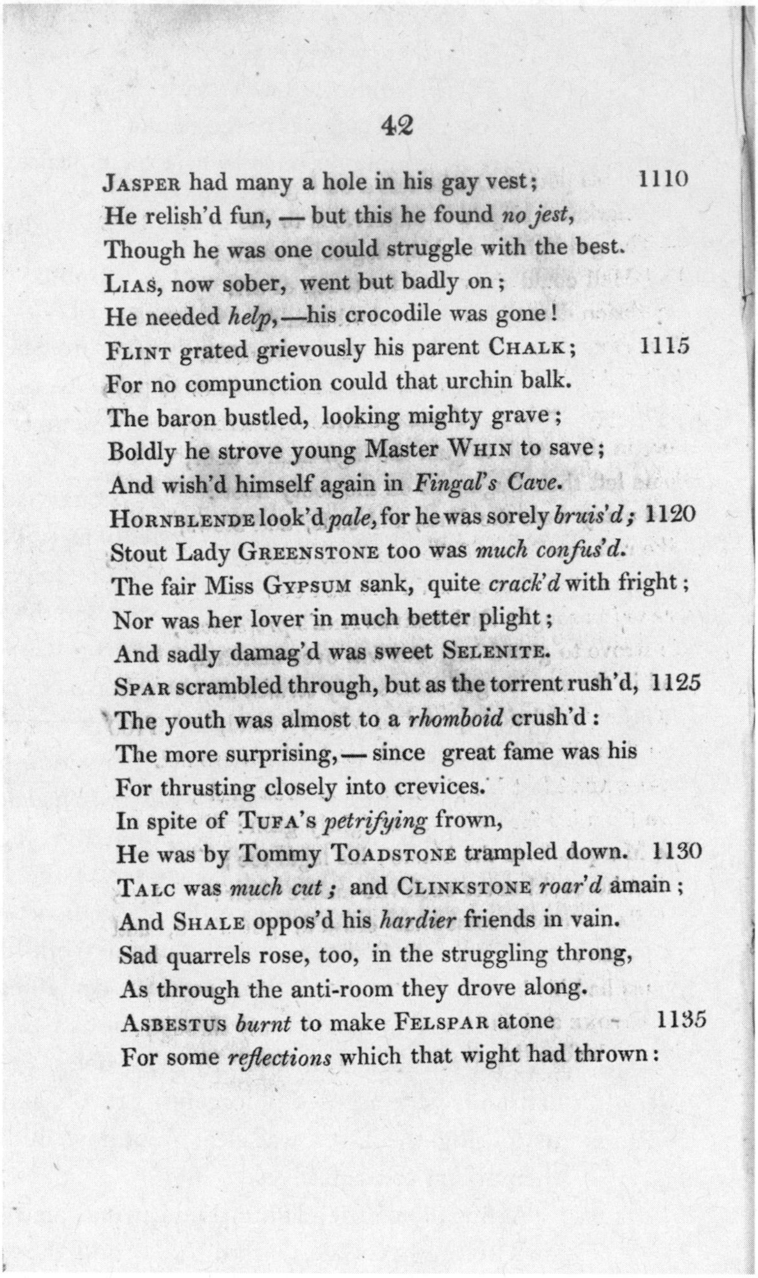

42

JASPER had many a hole in his gay vest;      1110
He relish'd fun, — but this he found *no jest*,
Though he was one could struggle with the best.
LIAS, now sober, went but badly on;
He needed *help*,—his crocodile was gone!
FLINT grated grievously his parent CHALK;   1115
For no compunction could that urchin balk.
The baron bustled, looking mighty grave;
Boldly he strove young Master WHIN to save;
And wish'd himself again in *Fingal's Cave.*
HORNBLENDE look'd *pale*, for he was sorely *bruis'd;* 1120
Stout Lady GREENSTONE too was *much confus'd.*
The fair Miss GYPSUM sank, quite *crack'd* with fright;
Nor was her lover in much better plight;
And sadly damag'd was sweet SELENITE.
SPAR scrambled through, but as the torrent rush'd, 1125
The youth was almost to a *rhomboid* crush'd:
The more surprising,—since great fame was his
For thrusting closely into crevices.
In spite of TUFA's *petrifying* frown,
He was by Tommy TOADSTONE trampled down.  1130
TALC was *much cut;* and CLINKSTONE *roar'd* amain;
And SHALE oppos'd his *hardier* friends in vain.
Sad quarrels rose, too, in the struggling throng,
As through the anti-room they drove along.
ASBESTUS *burnt* to make FELSPAR atone     1135
For some *reflections* which that wight had thrown:

*Figure 2.5.* Geological riot, featuring personified minerals, bringing John Scafe's poem *King Coal's Levee* to a rousing conclusion. [Scafe] 1820a, 42.

Away, ye ignorant and vain!
Away, ye faithless and profane!
Jesters and dainty dandies fly hence,
But enter thou, dear son of science!

Buckland himself appears as a "contemplative" antiquarian "sage", "his eye in a fine frenzy rolling",

whilst he doth utter
Strange sentences that seem to say
I see it all as clear as day;
I see the mighty waters rush,
And down the solid barriers push!
I see the pebbles pebbles chasing,
And scooping out of many a basin;
I see the dreadful dislocation,
And gradual stratification.

Like a poet or prophet, Duncan's Buckland appears as a visionary genius inhabiting two worlds at once, witnessing the Deluge and thus resolving "the strata's strange confusion" in his mind's eye, while simultaneously "strolling" with his "bread and butter" around his room.[46]

In the year following the composition of Duncan's poem, such private expressions of confidence in the new science took on a more authoritative air. Geology came to be presented as having the power to control the deep past, while the geologist was portrayed as able not merely to "see", but to go voyaging into, those landscapes. The catalyst was Buckland's cave research. In terms of the imaginative perception of former worlds, Buckland's analysis of the fossils in a recently-discovered Yorkshire cave was nothing less than a milestone.

### Entering the Hyaenas' Den

In December 1821 Buckland visited Kirkdale Cave, near Kirby Moorside, to study its assemblage of fossil bones (Fig. 2.6). This was on the prompting of Cuvier's assistant Joseph Pentland, who wanted him to find some more material for Cuvier to work his magic on. As it happened, it was Buckland who stole the show.[47]

The assemblage was not in itself unusual: a jumble of fossil hyaena bones, mingled with those of other species (elephant, mouse, duck, hippopotamus, and so on), partially encased in stalagmite and coated in mud. Received wisdom would have it that the more exotic animals had drifted northwards from tropical climes during the Deluge; but Buckland suspected that these animals had lived in Britain until the Deluge brought extinction upon them.[48] His gastronomic approach to fossil detective-work enabled him to sustain his startling deductions: going where no geologist had thought to go before, he established that the stony balls lit-

---

46. Daubeny 1869, 81–3. On the geologist as visionary see Klaver 1997, 157–60; for a more detailed consideration see Sommer 2003, 185–97.

47. For fuller analysis see Boylan 1984; Rudwick 2005b, 622–38. See also Knell 2000, 171–6.

48. Page 1963, 108–9; Rupke 1983b, 31–3.

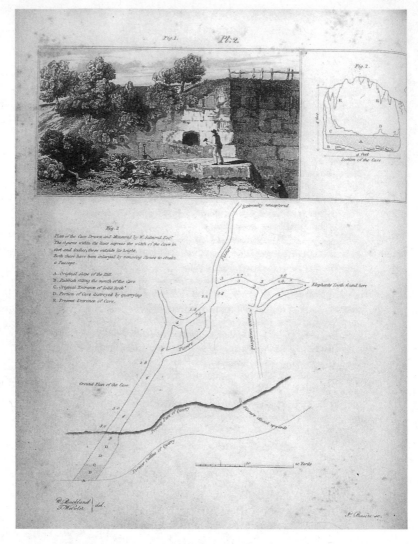

*Figure 2.6.* Kirkdale Cave: a plate in Buckland's *Reliquiae Diluvianae* showing a sketch of the cave opening, a cross-section, and a ground plan. W. Buckland 1823, plate 2.

tering the cave were fossilized hyaena droppings. The bones themselves showed marks of scratching, rubbing, and biting. For Buckland, the Kirkdale assemblage provided "one of the most complete and satisfactory chains of consistent circumstantial evidence I have ever met with".[49] He had already said that he hoped to find there "the wreck of the hyaenas' larder" rather than just more evidence for the action of the Deluge.[50] In February 1822 the Royal Society was regaled with the gory details of

49. W. Buckland 1822, 171.
50. Letter to Jane Talbot, 26 November 1821, quoted in Rupke 1983b, 33.

what the hyaenas ate (which included each other), how they ate it, and how they disported themselves in the antediluvian Kirkdale Cave.

Buckland's paper was published and quoted in periodicals across Europe, to widespread acclaim.[51] This revelation seemed to outdo the wizard Cuvier at his own game. Cuvier now had a string of resurrected fossils to his credit, and in one volume (1822) of the new edition of his *Ossemens fossiles* he had daringly indulged in speculations concerning the animals' feeding habits and behaviour, even providing outlines of four animals' appearances when living.[52] But Cuvier's restorations existed in an environmental limbo. For the first time in the history of the science, Buckland had restored an antediluvian habitat, establishing that the hyaenas had lived and died, over a long period of time, in the country where their remains were found. Kirkdale Cave was a time-capsule in which "a faithful record [. . .] of a past age of the world" was "sealed up".[53]

The imaginative importance of Buckland's hyaena den theory needs to be stressed, first and foremost because scenes from deep time are now too commonplace to be surprising. Buckland himself is still too often dismissed as a blustering parson engaged in the hopeless task of "reconciling" science and religion (even sacrificing his scientific integrity to do so), rather than as a pioneer of new and distinctly unbiblical ways of thinking about earth history.[54] Retrospectively, however, 1822 appears as a defining moment in the history of geology's imaginative impact, and we need to recover something of the mingled shock and pleasure the Kirkdale hyaenas occasioned. Most commentators found something uncanny about Buckland's achievement: it was greeted with rapturous applause, bafflement, or sarcasm, but rarely indifference. Geologists had prided themselves on the physical exertion and capacity for travel which their science required; now Buckland himself had ventured not only into Kirkdale Cave, but into the deep past, where he had stumbled on one of Davy's prehuman "past ages" in all its living detail. In innumerable scientific odes and rhapsodies from the previous century, the personified imagination had followed the astronomers by penetrating the vast tracts of outer space[55]—tracts whose ever-widening extent was indicated by that evocative phrase "the plurality of worlds", which John Sumner

51. Rupke 1983b, 35–6.

52. Cuvier 1821–4, III, plate 66, reproduced and discussed in Rudwick 1992, 32–7.

53. Davy 1827, 49. On the imaginative significance of this restoration see Rudwick 1992, 38–40; 2005b, 637–8. On Buckland's visionary palaeontology more generally see Boylan 1984, 431–43.

54. Cadbury 2000; Freeman 2004, 181–4. For the "sacrifice of integrity" argument see P. Martin 1982, 170 (a tendentious distortion of Page 1969, 263).

55. Nicolson 1963; W. Jones 1966.

had invoked as a possible defence for the existence of geological "former worlds".[56] Buckland's voyage into deep time now realized Sumner's astronomical analogy, bringing one of many distant "former worlds" closely into view—rather too closely for delicacy.

To recapture the novelty of this discovery, some specific responses to Buckland's discovery will now be examined. Because this research was so widely publicized and excited so much comment, it gives us a unique snapshot of how different groups represented and responded to the new science's historical claims. It opens up a vivid picture of popularization in practice at one of the most politically sensitive periods in the history of geology. In the remainder of this chapter we shall examine various oral, textual, and pictorial responses: a speech before the Royal Society, an anonymous letter to the *Gentleman's Magazine,* Buckland's published monograph on cave fossils, and various privately-circulated pictorial and poetical accounts. We begin with a speech made by Humphry Davy, president of the Royal Society.

In November 1822 Buckland was awarded the Royal Society's Copley Medal, the first geologist to be honoured so highly. Davy's speech began by rejoicing that Buckland had helped geology catch up with astronomy, and went on to identify this new achievement in terms of his own vision of the progress of civilization. His language was that of conquest and pursuit, qualities which marked a science in its maturity:

> by these inquiries, a distinct epoch has, as it were, been established in the history of the revolutions of our globe: a point fixed, from which our researches may be pursued through the immensity of ages, and the records of animated nature, as it were, carried back to the time of the creation.[57]

Davy next mentioned how "gratifying" it was that "the progress of science" proved the biblical Deluge "beyond all doubt". Such exegetical reflections were not typical of Davy, but these were sensitive times. Having thus covered himself, and adding a jibe at Lamarckian transmutationism for good measure, he asserted that God had intended "the laws of nature" to be discovered by man's own "labour and industry". Authorized from above, the man of science was "left to exert these god-like faculties, by which reason ultimately approaches, in its results, to inspiration."[58] Davy was of course speaking of "inspiration" as a manifestation of "genius"; but, because he had just been advocating the freedom of scien-

---

56. Sumner 1816, I, 285.

57. Davy 1827, 51. On the language of conquest in geology see J. Secord 1982 and (for its application to cave geology) Shortland 1994, 39–42, and Sommer 2003, 193–4.

58. Davy 1827, 51–3.

tific inquiry and contrasting it with the moral authority of the "sacred history", his use of the term "inspiration" hinted at the problem of the Bible's literal inerrancy.

Perhaps unwittingly, Davy here pointed up the challenge which Buckland's findings posed to biblical literalists holding to a young-earth cosmology. First, Buckland's work showed (far more vividly than a stratigrapher could have done) that the Deluge could not have formed the bulk of the strata in the earth's crust. But there was a subtler challenge implicit in Buckland's description. His identification of "a distinct epoch" of antediluvian history, in which wild hyaenas and other animals roamed England, laid claim to a narrative domain traditionally governed by Genesis and *Paradise Lost*. It was not the monstrosity of the hyaenas that proved problematic: after all, in 1824 Buckland's literalist opponent William Eastmead published an account of the Kirkdale fauna in which the hyaenas appeared even more "savage and [. . .] frightful" than in Buckland's description.[59] The difference was that Eastmead's hyaenas were postdiluvian. Even if Buckland's hyaenas were made out to be contemporary with the antediluvian human population in the Fertile Crescent, the very existence of this alternative, unscriptural, literally bestial narrative (revealed, in Davy's terms, by the "god-like [. . .] inspiration" of "philosophy") threatened the imaginative hegemony of Genesis.[60] In retrospect, Buckland's description emerges as a crucial step on the way to the full-scale assault on biblical literalism which he and his colleagues would later conduct. These scenes from deep time may have aligned themselves with biblical, anthropomorphic narrative patterns, for instance presenting the cannibalistic hyaenas in typological terms as "fratricidal" Cain-figures, deservedly destroyed by the Deluge;[61] yet they expanded the popular view of the world before the flood well beyond its human actors and biblical scenery. Cuvier's historical geology may have been welcomed into establishment culture to support Moses against his radical detractors, but it was quick to strain at the leash.

On the evocative word "inspiration", Davy paused in order to hand the medal to Buckland. Davy then elaborated on the specific uses of geology, as in his introductory lecture of 1805. This time he placed particular

59. Eastmead 1824, 25, discussed in Sommer 2003, 186.

60. On the literalist reaction to the hyaena-den theory see Rupke 1983b, 42–7; Sommer 2004, 59. Sommer (2004, 59–60) has suggested that Buckland was unwilling to date the arrival of humans in Britain to the antediluvian period because of the argument from design, according to which humans were not created until the earth was ready for them: "placing humankind within [. . .] [a] Britain populated by gigantic mammals and cannibalistic carnivores would have been [. . .] heretical". However, this in turn raises the question of how human coexistence with Asian tigers or Egyptian crocodiles was to be explained, and I have not found any evidence that this was a problem even for literalists.

61. Daubeny 1869, 72–3, 119–20.

emphasis on the new vistas opened up by Cuvier's historical geology. In so doing he aligned the science clearly with the antiquarian spectacle to be seen in museums at the time:

> in the history of the past changes of the globe, what a sublime subject is there for the exercise of the imagination!
>
> If we look with wonder upon the great remains of human works, such as the columns of Palmyra, broken in the midst of the desert, the temples of Pæstum, beautiful in the decay of twenty centuries, or the mutilated fragments of Greek sculpture, in the Acropolis of Athens, or in our own Museum, as proofs of the genius of artists, and power and riches of nations now past away; with how much deeper a feeling of admiration must we consider those grand monuments of nature, which mark the revolutions of the globe [. . . .][62]

This passage is effectively a descant on Cuvier's theme. Eloquently expressive of both human progress and the humbling grandeur of earth history, and emanating from an author of impeccable philosophical credentials, it soon became one of the most quoted passages in the literature, doing public service in the epigraphs and chapter conclusions of geological books and museum guidebooks.[63]

On a more private level, Buckland's friends composed and circulated occasional verses to celebrate his discovery. The twofold response in 1822 of his friend Conybeare is particularly interesting. First, Conybeare composed a poem entitled "The Hyaena's Den at Kirkdale near Kirby Moorside in Yorkshire, discovered A.D. 1821". This translated Buckland's already colourful account of his findings into grotesque doggerel, presenting him as one who can "spy" on "what was done ere the birth-day of time".[64] The hyaenas themselves are portrayed as unnaturally powerful, their monstrosity enhanced by their alignment with the novelties of modern technology (in this case an early form of pressure cooker):

> Their teeth had the temper of steel,
> Skulls & dry bones they swallowed with zest, or
> Mammoth tusks they dispatch'd at a meal,
> And their guts were like Pappin's digester.

62. Davy 1827, 54–5.
63. Mantell 1834b, 2; 1836b, verso of title-page; 1844, x; 1847, x.
64. These quotations are taken from the published broadsheet, [Conybeare] 1822; I am grateful to Martin Rudwick for lending me his copy. The poem is also printed, with slight variations, by Daubeny (1869, 92–4) and Rudwick (1992, 40–3).

Conybeare mocks his friend's ambition to eat his way through the animal kingdom: the poem's Buckland-persona exclaims that "potted venison" is a delicacy far inferior to "Hyaenas' bones potted in mud". Even without Buckland's own oddities, the poem's portrait of these swarming "monsters" crunching up bones and rats, defecating, "scratching their hide", and even "slic[ing] up each other" fits awkwardly with the concluding reflections on the "Mystic Cavern" with its "chasms sublime". This disjunction extends the mixed modality of Buckland's own rhetorical style, as well as the "rude harmony" Sommer identifies as characteristic of the "romantic cave".[65] When Buckland told his colleagues at the Geological Society a story about how the "half starved" hyaenas used to leave their cave and "help themselves" to water-rats by the nearby lake, his former pupil Lyell was baffled, commenting that "Buckland in his usual style enlarged on the marvel with such a strange mixture of the humorous and the serious, that we cd. none of us discern how far he believed himself what he said".[66] Buckland's antediluvian anecdote was founded on a comparison of the Kirkdale fossils with the known habits of modern hyaenas, but his "usual style" was to play up the drama of the story, ensuring that it became a "marvel" in his fellow geologists' eyes.

Conybeare's poem was printed as a broadsheet, and copies were distributed among selected gentlemen both in Britain and on the Continent. Buckland certainly relished the joke: in 1822 he sent a copy to the Yorkshire clergyman Francis Wrangham, apologizing that his paper would not be printed for some months and offering Conybeare's doggerel as "a poetical Version" of the paper.[67] What made this broadsheet special, however, was that the poem accompanied a skilful lithograph of "Mr. BUCKLAND peeping into the Hyænas' den, to see what they are about, &c." (Fig. 2.7), as another commentator put it.[68] This was one of the very earliest geological restorations of a true scene from deep time, in that the animals were drawn as they would appear when living, and in their own habitat, rather than merely being outlines with a minimal background. Restoration was an art as well as a science, and the imagina-

---

65. Sommer 2003, 199–200.

66. Lyell to Mantell, 8 February 1822; printed in J. Thackray 2003, 2. On this passage as a typical example of Buckland's blend of fact and jest see Rupke 1983b, 71–2; Sommer 2004, 60.

67. Buckland to Wrangham, 14 March 1822, written onto a copy of Conybeare's broadsheet, itself bound into Wrangham's copy of W. Buckland 1822 (separate offprint), now held in the British Library (1254.k.1.(1.)). The letter reveals that by this time the poem had been leaked to the *Yorkshire Gazette*, though with injunctions to maintain its anonymity. Wrangham was the dedicatee of Eastmead's *Historia Rievallensis* (1824).

68. Gibson 1822 (this article from the *Yorkshire Gazette* was enclosed in Buckland's letter to Wrangham mentioned in n. 67). Rudwick (1992, 38–41) has provided the fullest analysis of Conybeare's drawing.

*Figure 2.7.* Drawing of Buckland entering the antediluvian Kirkdale Cave like a time-traveller, engraved alongside William Conybeare's poem *The Hyaenas' Den* on a printed broadsheet.

tive daring of Conybeare's drawing carried a risk that its content might be deemed overly speculative.[69] For this reason it was circumscribed, like the occasional verses discussed above, by its humorous presentation and its limited circulation among gentlemen. Such images were not displayed to a wider public as serious scientific representations until the 1830s.

Rudwick has stated that Conybeare's cartoon-like drawing was not "translated into a more soberly scientific scene from deep time".[70] However, a closer look at the pictorial evidence to hand suggests that Buckland's Oxford students were in fact shown a soberer version of the drawing, without the comic element added by Buckland's caricatured figure. In 1970 two copies of a previously unpublished portrait of Buckland were analysed by Patrick Boylan (Fig. 2.8). The portrait, dated 1823, shows the

69. The drawing was probably not executed by Conybeare himself (Rudwick 1992, 257 n. 14), but for convenience it is here called "Conybeare's".

70. Rudwick 1992, 40.

Drawn from Nature & on Stone by Geo. Rowe.

Printed by C. Hullmandel

GUL. BUCKLAND B.D. F.R.S. MIN.. ET GEOL.ᴸ OXON PROFF.ᴿ 1823.

*Figure 2.8.* Lithographed portrait of Buckland by George Rowe (1823). Note the hyaena restoration behind his left shoulder and, below this, a reversed version of the sketch in Fig. 2.6.

Professor standing behind a table littered with fossils, holding a hyaena jaw with one hand while pointing at it with the other. Behind him are geological maps and posters, many of which represent his own discoveries. One poster clearly shows two hyaenas in all their furry, reconstructed glory, gnawing bones in a cave. As Boylan has noted, the poster has clear affinities with Conybeare's drawing.[71]

This portrait of Buckland raises two important points. First, all the other animals depicted in the portrait are in a fossil state. The hyaenas are the only ones restored in the flesh, even though it would not have been hard to restore the mammoth's outlines because an entire mammoth had been found frozen into the permafrost in Siberia two decades earlier. This distinction confirms that Buckland's discovery of an ancient habitat was of crucial importance for the depiction of the ancient earth. Second, this seems to me to be a portrait of a man engaged in demonstrating something to an audience, rather than (as Boylan has suggested) an "ikon-like" image showing the man of science "surrounded by symbolic representations of his triumphs". The latter interpretation would indeed make this portrait "uncharacteristic of the period";[72] but it has clear affinities with the contemporary engraving shown in Fig. 2.1. In both pictures Buckland's pose is identical (though holding different fossils); the table is at the same height; some of the same fossils and pictures can be seen in the room; and, if we allow for the accidental reversal of images,[73] the arrangement of the posters on wooden boards leaning against the wall beneath a suspended map is exactly the same. Buckland's mouth is slightly open: he is in full flow, giving one of his lectures.[74] According to Rudwick, the first "soberly scientific" scene from deep time was the lithograph of De la Beche's *Duria antiquior* (1830), which Buckland used as a visual aid.[75] Figure 2.8, however, suggests that this was predated in the 1820s by a serious large-scale restoration of the Kirkdale hyaenas.

In fact, this now-lost restoration reinforces rather than alters the underlying historical model proposed by Rudwick for the development of these scenes among gentlemanly geologists, with their wider popularizations coming after a process of internal dissemination and testing.[76] Buckland's Oxford lectures played their part in this earlier phase, functioning in part as an extension of the private-circulation network around

71. Boylan 1970, 351.

72. Boylan 1970, 353.

73. Lithographs had to be drawn in mirror image, so this kind of mistake was common.

74. This interpretation seems implicit in the observations made in Edmonds and Douglas 1975–6, 142.

75. Rudwick 1992, 44.

76. Rudwick 1992, 233–5.

which occasional verses and jokes moved. Members of this audience, like those in the Royal or Geological Societies, were all vouched for as respectable gentlemen.[77] Even if this portrait of Buckland was originally intended for the frontispiece of a projected geological treatise,[78] Rudwick's "caution hypothesis" would be upheld by the fact that the hyaena poster, if reproduced in this context, would be too small and marginal to attract much attention.[79] But whatever its purpose and transmission may have been, one thing is clear: Buckland is, yet again, depicted showing off, and a Kirkdale hyaena provides the focal point of his display. With this in mind we return to Conybeare's drawing, which dramatizes the moment of discovery required for this display to take place.

If Conybeare's poem can be seen as a translation of Buckland's academic hyaena paper into jog-trot verse, his drawing depicts the discovery, complete with the time-travelling figure of Buckland, crawling into the cave bearing the candle of Enlightenment science.[80] In Rudwick's words, Buckland is seen "penetrating the epistemic barrier between the human world and the pre-human, and looking perhaps as surprised to see the hyaenas as they are to see him." This is rather an understatement. Buckland's hair is standing on end, and so is that of the hyaena in the foreground, which arches its back menacingly. Mouths are gaping all round: the hyaenas closest to Buckland are snarling, while Buckland's jaw has dropped (whether in delight or fear is impossible to tell). Notwithstanding the picture's humorous mode, it conveys a strong sense of hostility between man and beast, in keeping with the traditional image of the cave as a site for primal conflict.

In its deliberate use of fantasy—the fantasy of time travel—Conybeare's broadsheet flags up the uncanny nature of Buckland's achievement, restoring what no human eye could have beheld. At this stage in the science's development, the idea of viewing such a scene was so new and counter-intuitive that Buckland and others seem to have felt that the best way of representing this leap of imagination among themselves was to translate it into a literal journey in time, jamming together prehuman

77. Several of those in the Oxford audience in Fig. 2.1 were members of the Geological Society anyway (Edmonds and Douglas 1975–6, 154–65).

78. As suggested by Boylan (1970, 353).

79. Compare Buckland's miniature and marginal restoration of the pterodactyle in his Bridgewater Treatise (1836, II, plate 22, fig. P), taken from another equally marginal German restoration of 1831 and reproduced as Fig. 9.1 below. On this marginality see Rudwick 1992, 57, 68–9. Compare also Mantell's miniature restoration of ventriculites in his first book (1822, 175).

80. This interpretation of the candle is based on Rudwick's (1992, 39). Shortland (1994, 31) sees the candle rather as "the torch that harks back to our primitive ancestors [. . .] conjur[ing] up [. . .] the era of Neanderthal man, a habitual troglodyte who [. . .] used the scourge of fire to clear his abode of potential foes." Neanderthal man, however, was not described until 1857. Sommer (2003, 195) has suggested that the candle also represents "the flag of the conqueror".

*Figure 2.9*. Resurrecting extinct bears in Gailenreuth Cave (and making them perform). Detail from a sketch in Oxford University Museum of Natural History, Buckland Papers.

past and human present with comical incongruity. This is confirmed by the fact that the earliest known pictorial restorations of extinct cave-bears and pterodactyles as living animals—both, again, associated with Buckland's work—used similar devices. Figure 2.9 was drawn, probably by Buckland, sometime after his trip to a bone-cave at Gailenreuth in Bavaria in 1816: it represents an encounter between a top-hatted man and the extinct cave-bears whose presence Buckland had deduced from the bones littering the cave. The man's gesture appears to combine the menagerie showman with the conjuror: it is as if he has resurrected the bears from the bones seen on the cave floor and now shows his power over them by making them dance for him or sit down quietly.[81] (This recalls the pet bear named Tiglath-pileser which Buckland acquired later in his career and dressed in a cap and gown.) Another version of the time-travel fantasy is seen in Fig. 2.10, a painting by the Lyme Regis

81. This picture has been reproduced and discussed in Rudwick 2005b, 607–8.

*Figure 2.10.* Painting by George Howman (1829, now in Lyme Regis Museum) showing a dragon flying by night over a romantic seascape. An inscription on the back states, "By the Revd George E Howman from Dr Buckland's account of a flying Dragon found at Lyme Regis, supposed to be noctivagous". This last word means "wandering by night", as Buckland described the pterodactyle in his 1829 paper (W. Buckland 1835, 218).

clergyman George Howman, inspired by Buckland's 1829 Geological Society paper on a pterodactyle found at Lyme.[82] No human figures are seen, but the ruined castle and storm-tossed boat indicate that this pterodactyle—if we may call it such[83]—has somehow flown into the present day. Its portrayal as a dragon, complete with pointy tail, enhances the scene's fantastic nature, reminiscent of the *Arabian Nights*; but it also indicates how difficult it was for people in the early nineteenth century to imagine these new monsters without drawing on old iconographies. The pterodactyle is not represented *like* a dragon, but *as* a dragon.

Both pictures present extinct animals travelling forward in time into the present day. Conybeare's drawing, by contrast, presents the geolo-

82. W. Buckland 1835. On Howman's painting see Fowles 1984, 27; Torrens 1995, 266.

83. If so, this appears to be the first British pictorial restoration of a pterodactyle. The earliest known restorations outside Britain were a pair of distinctly mammal-like sketches made by Jean Hermann of Strasbourg in 1800 (Taquet and Padian 2005).

gist travelling *back* in time, and in the poem Buckland gazes upon the distant past like a visionary. As Sommer has shown, Conybeare's lines on the "Mystic Cavern" form part of a strategy by which Buckland and other geologists asserted their own authority over the earth's secrets by drawing on contemporary mythic and literary associations of caves.[84] As for the Bible, the significance of Buckland's glimpse into a lost world is double-edged. Rudwick, taking his cue from Conybeare's twelfth and thirteenth stanzas, has commented that "Buckland's reconstruction fills in what the bare textual records of the world before the Deluge hardly begin to suggest".[85] Indeed, in his subsequent volume, *Reliquiæ Diluvianæ* ("Diluvial Relics", 1823), Buckland marketed his cave researches as historical evidence for the biblical Deluge. But his world of hyaenas does not so much fill in the biblical account as provide a starting-point (Davy's "distinct epoch") for an alternative Creation-narrative which would soon be shrugging off the historicity of the biblical Deluge as being of limited geological significance.[86] Geology was beginning to flex its muscles.

Many biblical literalists resented the triumphalism with which earth history was being wrested from their grasp. A pseudonymous letter to the popular and relatively conservative London periodical *The Gentleman's Magazine* by a native of Kirby Moorside ("Kirkdaliensis") may also reflect annoyance at this metropolitan bigwig plundering the provinces for his own glory.[87] Before we examine this letter, however, it is worth glancing at two previous mentions of the cave in the *Gentleman's Magazine,* since this will show us the overall context in which its readers first viewed the discovery.

The editors first heard of the cave in 1822. In February, while Buckland was presenting his discoveries to the learned of London, readers of the *Gentleman's Magazine* were learning of an "Antient Cave" discovered in Yorkshire, in which animal bones had been found. The likely explanations were that the bones were "Antediluvian" and had been washed there by the flood, or that the animals had lived in the cave after the flood "if they ever existed in this island".[88] This notice appeared in the "Antiquarian Researches" section, and anyone reading the magazine sequentially would have just finished a longer notice about a barrow in Nettleton, with speculations about "early British Antiquity". After read-

---

84. [Conybeare] 1822, final verse; Sommer 2003, especially 196–7.

85. Rudwick 1992, 39.

86. Nevertheless, at this point Buckland still maintained that the "geological deluge" (indicated by the physical evidence) was the same event as the biblical Deluge. See Rudwick 2005b, 609–38.

87. "Kirkdaliensis" 1822.

88. Anon. 1822c.

ing about Kirkdale Cave, he or she would then have read about an "Egyptian Mummy" and some Roman ruins discovered in North Africa.

In April, part of Buckland's paper on the Kirkdale hyaenas (as printed in the *Annals of Philosophy*) was extracted and printed in the *Gentleman's Magazine*, again under "Antiquarian Researches". Again, a sequential reading reveals the way in which many readers first met the new geology. Having perused a notice of "Egyptian Antiquities in the British Museum", our reader would come to an article entitled "Antediluvian Cave" (note the change from the vaguer "Antient"), which promised to give a "minute and interesting detail" from Buckland's paper: "It gives a curious account of an antediluvian den of hyænas".[89] The word "curious" implies that Buckland's account is intriguing and worth reading, rather than intending any veiled sarcasm: this was an article designed to keep the polite reader abreast of the latest philosophy, rather than a reaction or opinion. This reader now learns how, in Buckland's view, the evidence of the hyaena den showed that "no important or general physical changes" had affected it since. The quotation from Buckland's paper ends without editorial comment, and the reader moves on to read about another "Antient Barrow" and an "Ancient Seal" (of the heraldic variety).

These two accounts, then, present the Kirkdale hyaena as an "extinct species", a more furry and less cultured analogue of the ancient Britons, Romans, and Egyptians. In both issues, information on the cave is surrounded by accounts of man-made underground chambers whose historical secrets, like Kirkdale's, were being plundered and interpreted by intrepid antiquaries. Although these accounts are dryly narrated, we can see how the popular resonances of human antiquity (Ossianic and Oriental respectively) came to be associated with "antiquated beasts".[90] The hyaenas' position within these accounts reflects the placing of fossils in museums like William Bullock's: his 1812 guidebook contains a section entitled "Miscellaneous Articles" in which mammoth bones and Egyptian mummies constitute the chief attractions.[91] These associations were natural enough, given geology's antiquarian roots, and were already proving useful to its popularizers.

"Kirkdaliensis" was moved to write his sarcastic letter on 15 May 1822, and it was printed in the *Gentleman's Magazine* in June, taking up nearly four pages. His main objection was to the imaginative liberties taken by this upstart southerner. His letter begins by recalling the "*luminous* account of [. . .] the *dark* 'Antediluvian Cave'": insisting that "it is very far

89. Anon. 1822d.
90. The phrase is Conybeare's, taken from a poem printed in Daubeny 1869, 79.
91. Bullock 1812, 129–34.

from my intention to set up my little spark of knowledge against the blazing splendour of his", he ironically invokes the Enlightenment trope of
scientific knowledge as illumination. He reasserts the scriptural basis of
true historical knowledge against Buckland's "antiquarian" speculations,
invoking the then-standard dating of the Deluge to 2349 BC. Against
this established authority he presents the "extremely curious" evidence
of "DILUVIAN MUD" with which Buckland had argued that the hyaenas
were antediluvian. "Kirkdaliensis" uses the term "MUD" three times in a
single sentence, each time in full capitals, presumably to discredit Buckland's grand conclusions by drawing attention to the unglamorous material on which they are founded. A page later he descends even lower,
commenting that the fossil faeces are not "a whit less curious, and must
doubtless afford a high treat, and perhaps *relish,* to the real lover of antiquity".[92] This innuendo plays on the same level as Conybeare's suggestion
that Buckland enjoyed eating "hyaenas' bones potted in mud",[93] though
the two writers used low humour with very different intentions.

"Kirkdaliensis" further attacks Buckland by ridiculing his visionary
powers:[94]

> Incredulous persons might here be tempted to inquire, how this profound
> Antiquary knows what changes took place in these bones *before the Flood,*
> that is, "while the den was inhabited," [. . . .] Perhaps Mr. Buckland, like
> many of our brethren of the isles of North Britain, may have possessed the
> gift of *second sight,* in a remarkably acute manner; and possibly, ere long,
> the world may be favoured with some more of his speculations; or as we
> may say, "visions, having his eyes open," [ . . . ][95]

The power claimed by the geologist is invoked with ironic intent, just as
the Geological Society elite had defined themselves against "theories of
the earth" by ironically praising their poetic qualities. The conclusion of
the letter suggests that "Kirkdaliensis" had heard the discovery applauded
in terms similar to Davy's, even if he had not heard Davy's own words:[96]
Buckland was seen by many as an inspired genius, and "Kirkdaliensis"
aimed to undercut such vague intimations of supernatural powers. Here

92. "Kirkdaliensis" 1822, 491–2.
93. [Conybeare] 1822.
94. As noted by Sommer (2003, 189–90).
95. "Kirkdaliensis" 1822, 492.
96. The letter ends by parodying the language of progress and genius: "we have had
a secret laid open to our view in this discovery, which for above 4000 years past has been
concealed from 'mortal ken;' [. . . .] In the mean time hope [. . .] will doubtless keep alive in the
minds of *Philosophers* the expectation of having wonders hereafter revealed, which may make
air balloons, steam-boats, gas-lights, and other wonders of this enlightened age in which we
live, appear like mole-hills compared with the Grampian-hills" ("Kirkdaliensis" 1822, 493–4).

he invokes the familiar "primitive" figure of the supposedly clairvoyant Hebridean peasant, signalling connotations of credulity, obscurantism, and fraudulence which could be made to apply as much to the Enlightenment image of the "sage of science" as to the priestly or magical authorities this image was designed to displace. Strikingly, he then invites Buckland to indulge in further "visions" and "present us with a correct picture of this curious spotted animal, as the same presented itself to his 'mind's eye,' when he wrote this elegant illustration"—a wish that turned out to be amply granted by Conybeare's drawing and the related poster.[97]

"Kirkdaliensis" then attempts to tar Buckland's discovery with the brush of commercial sensationalism. He proposes that a *"Bazaar"* be opened for selling such "curiosities" as *"Antediluvian Album Græcum"* [98] and *"Diluvian mud"*:

> And as Mr. Belzoni's curiosities are advertised to be very soon sold, your Correspondent is of opinion, that the owner of the Egyptian Hall [. . . .] [should] open such a Bazaar there, and your Correspondent, who lives very near the Kirkdale Cave, will readily become his country Agent.[99]

This reference needs some explanation. In 1815 the itinerant showman and circus strongman Giovanni Battista Belzoni had become involved in excavations of ancient Egyptian sites such as the Valley of the Kings and the Lost City of Berenice. Some of the resulting plunder had been displayed in Bullock's Egyptian Hall (see Fig. 1.2) in May 1821, once the fashionable London publisher John Murray had successfully promoted Belzoni's memoir on the excavations.[100] The sensation caused by these exhibits lasted well beyond the obligatory season, drawing crowds for over a year. In June 1822, as "Kirkdaliensis" mentions, the relics were to be auctioned off, many going to the British Museum. But one aspect of Belzoni's display had created a particular stir: he had installed replicas of part of the tomb of Seti I, into which the visitor could walk as if it were a more comfortable version of the real thing. This display was designed to re-create Belzoni's first impressions as he entered the tomb. It offers tempting parallels with Conybeare's caricature, the first scene from deep time, which also depicts the immediate reaction of an antiquary-*cum*-showman on crawling into a newly-discovered tomb (albeit a tomb

97. "Kirkdaliensis" 1822, 492. On the relation between the concept of second sight and the "mind's eye" see Larrissy 1999; on the Celtic connotations see Sims-Williams 1986. On the "sage of science" generally see Knight 1967. Summer (2003, 182–4 and 189–90) has commented on the relations between magic and "enlightenment" in Regency cave science.

98. Dung.

99. "Kirkdaliensis" 1822, 493.

100. Altick 1978, 243–6; Leask 2002, 128–56.

of antediluvian hyaenas). Whether intentionally or not, both Conybeare and "Kirkdaliensis" associated Buckland's discovery with the same exhibition of antiquarian time travel. Both men drew out the spectacle's vulgarity, which allowed Conybeare to explore the science's imaginative possibilities but which, for "Kirkdaliensis", only confirmed the disreputable nature of these trespassers upon biblical (and Yorkshire) territory.

In a final irony Murray, having succeeded with Belzoni's Egyptian memoir, published Buckland's new book on cave fossils in 1823, *Reliquiæ Diluvianæ*. Like Belzoni's book, this handsome treatise united learned antiquarianism with the pleasures of sublime topography. It sold well. Its yoking of Genesis and fossils discomfited some geologists, but even Buckland's most outspoken opponent admitted that it had "greatly contributed to render the science of geology popular" and supported "the authority of revelation".[101] Indeed, the challenge to a literal six-day Creation implied by its contents was distinctly underplayed by the surrounding matter (such as the impressively orthodox-sounding title). Geology's cultural status was still precarious. Before we dismiss the letter of "Kirkdaliensis" with retrospective wisdom, we should remember that it was printed in a widely circulating magazine and read by thousands, whereas Conybeare's statement was privately printed and circulated among a chosen circle of savants. Davy's speech was delivered before a similarly select audience and not published until 1827. The implications of Buckland's research for cherished views of Scripture were more likely to damage him than the literalists if aired publicly—and would in all likelihood be used against him by radicals.

Here Byron's verse-drama *Cain* served as an object lesson.[102] Published by a reluctant Murray in December 1821, *Cain* was a sophisticated biblical "problem play"—subtitled *A Mystery*—about how Adam's elder son came to kill his brother Abel. In it Byron remorselessly anatomized the concept of rebellion against established order. At the drama's heart is a pair of scenes in which Lucifer, using the language of Enlightenment rationalism, tempts Cain to rebel. Lucifer teaches Cain that the earth is immeasurably ancient, cosmically irrelevant, and involved in a cycle of divine destruction, extinction, and replenishment. He takes him on a cosmic voyage into "the Abyss of Space", where Cain watches the earth dwindle to nothingness. By this interstellar route Lucifer conducts him into deep time or "Hades" (Fig. 2.11), where phantoms of former worlds with their extinct pre-Adamites, mammoths, and leviathans overwhelm Cain with wonder,

101. Fleming 1825–6, 208.
102. On Byron see Franklin 2000; *ODNB*. The most reliable editions of Byron's poems are Peter Cochran's online texts at http://www.internationalbyronsociety.org, but for bibliographic convenience I cite McGann and Weller's printed editions (Byron 1980–93 and the much more accessible Byron 1986).

LUCIFER.

Enter!

CAIN.

                 Can I return?

LUCIFER.

Return! be sure: how else should death be peopled?
Its present realm is thin to what it will be,
Through thee and thine.

CAIN.

             The clouds still open wide
And wider, and make widening circles round us.

LUCIFER.

Advance!

CAIN.

        And thou!

LUCIFER.

          Fear not—without me thou
Couldst not have gone beyond thy world. On! on!

                [*They disappear through the clouds.*

## SCENE II.

*Hades.*

*Enter* LUCIFER *and* CAIN.

CAIN.

How silent and how vast are these dim worlds!
For they seem more than one, and yet more peopled
Than the huge brilliant luminous orbs which swung

*Figure 2.11.* Lucifer leads Cain into Hades, the realm of the geological past: the beginning of Act II, Scene ii of Byron's verse-drama *Cain*. Byron 1821, 384.

then sadness, then anger at God's cruelty.[103] These didactic visions were explicitly based on Cuvier's geology, as Byron noted in a mischievously pious preface. Several commentators accordingly saw *Cain* as an unwelcome form of geological popularization. Byron's friend Tom Moore, the Irish poet, regretted that Byron had promoted the potentially "desolating" (but as yet little-known) catastrophe-theories of Cuvier "in poetry which every one reads".[104]

By demonizing the new science as a gift from Lucifer, Byron confronted the imaginative challenge it posed to biblical literalism. If you base your faith on the literal truth of Genesis, he seemed to suggest, modern science will wreck it for you. But the play created such a furore that its subtleties were ignored or flattened out by many readers. Outraged clerics and delighted radicals saw Lucifer as a cipher for Byron's opinions, preaching the liberating power of reason in opposition to "priestcraft".[105] This reading completely ignored Lucifer's evident bad faith, not to mention the dénouement in which Cain, reduced to a state of abject depression by Lucifer's cosmic showmanship, gets into a fight with Abel and accidentally kills him. Nevertheless, the radical journalist Richard Carlile remarketed Byron's drama as cosmology for the people, reprinting it in a cheap pirate edition and marshalling it alongside explicitly didactic works by Percy Shelley, Elihu Palmer, and George Toulmin. Moore's fears were confirmed. *Cain* spurred radicals in Carlile's circle to start wielding Cuvier's historical geology as a weapon against the Church (for instance in Carlile's newspaper *The Republican*), rather than relying solely on the increasingly outdated eternalist cosmologies of Hutton, Toulmin, and Palmer.[106]

Meanwhile, Buckland was engaged in the delicate operation of trying to persuade the English establishment that this new science was not harmful to the Christian faith, but could positively support it.[107] His Kirkdale research invited a still bolder self-image for the science. But *Cain* set the cat among the theological pigeons, and one suspects that, in the resulting flurry, Buckland was particularly feathered. His response was characteristically offbeat. In order to refine and reinforce geology's self-presentation as a safe "dark art", he added his own satirical verse "an-

---

103. *Cain* II.i and II.ii (Byron 1986, 901–21).

104. Marchand 1973–94, IX, 103–4; Dowden 1964, II, 620.

105. The same interpretation is common today. See Goldstein 1975; for a subtler reading see Schock 1995.

106. Carlile 1822a; 1822b; 1823, 407. On Carlile's other scientific interests see Goldstein 1975; Cooter 1984, 201–23. I have so far found no evidence for other radical writers using Cuvier's geology to this end before 1821.

107. Rupke 1983b; Sommer 2004.

tidote" to the growing number of hostile reviews of *Cain*. In its close engagement with well-known poems of the day, Buckland's response stands out from the writings so far examined. Here, too, Kirkdale Cave served as a focal point for the new self-confidence with which geologists presented their science to each other, if not (yet) to the wider public.

## Wizards and Radicals

Buckland's cave research rejuvenated the image of geologist as wizard.[108] One aspect lent itself especially well to characterization as a dark art. The study of fossil turds or coprolites, pioneered by Buckland, yielded valuable information about the diets of extinct animals. William Wollaston, writing to Buckland to confirm that the gritty nodules in the hyaena caves were indeed droppings, added that "though such matters may be instructive and therefore to a certain degree interesting, it may [be] as well for you and me not to have the reputation of too frequently and too minutely examining faecal products".[109] Philip Duncan took the opportunity to recast his earlier poem about the "dear son of science" quoted above. This time the novice is shown that geology's exalted status rests on very unglamorous material.

> Approach approach ingenuous youth
> And learn this fundamental truth
> The noble science of Geology
> Is bottomed firmly on Coprology
> For ever be Hyaena's blest
> Who left us the convincing test [. . . .][110]

In the same letter Duncan also suggested that "Coprologia" should be set as the subject for the university's Latin Verse Prize ("Ars Geologica" was in fact chosen as the subject for 1823).[111] Tellingly, Duncan scribbled on the back that Buckland was not to reveal these secrets to "Mrs B." Lavatory humour notwithstanding, the image of the geologist as the uncanny possessor of mysterious, potentially rather grim secrets reinforced the

---

108. Contrasting assessments of this image are offered by Shortland 1994 on the one hand and Sommer 2003 and 2004 on the other. On the image's seventeenth- and eighteenth-century ancestry see F. Stafford 1994, 34–41.

109. Wollaston to Buckland, 24 June 1822; quoted in Rupke 1983b, 33.

110. Oxford University Museum of Natural History, Buckland Papers, Miscellaneous MSS./1. This poem was not printed by Daubeny.

111. Rupke 1983b, 223.

notion that geology was an elite science, not to be tampered with by the "profane"—nor, in these delicate matters, by women either.[112]

This earthy dimension of cave geology chimed in with the mythic resonances of caving in this period. As Michael Shortland has put it, "the deeper one goes into the bowels of the earth, the closer one gets to something forbidden and threatening, alluring yet repulsive."[113] Oddly, however, Shortland has tried to set up a disjunction between the "poetic" aspect of caves celebrated by "the Romantics", and the "coarse" mining-related practice of cave geology exemplified by Buckland. Shortland's claim that geology should be "disengage[d] [. . .] from Romanticism as a source and context" is founded on a popular misunderstanding of who and what the English "Romantics" were.[114] This so-called movement (invented at the end of the nineteenth century)[115] is often represented in literary encyclopaedias by six very different canonical poets of the period 1780–1830: William Blake, William Wordsworth, Samuel Taylor Coleridge, Lord Byron, John Keats, and Percy Shelley. Shortland's definition of Romanticism excludes political engagement, scientific thought, the sexualization of nature, masculinity, physical exertion, and individual experience. All of the six poets just mentioned, in their different ways, strongly embraced these concerns and rejected the watered-down pantheism Shortland attributes to them. There was no such thing as a homogeneous, anti-scientific "Romantic attitude to nature":[116] all these men engaged enthusiastically with the sciences of their time, including geology (which Shortland claims had no appeal for Wordsworth, Shelley, or Coleridge).[117] Turning then to the "earthy" science of geology, Shortland uses the example of Buckland to suggest that geologists' experience of caves had nothing in common with "the manipulated delights offered to the Romantics" or "Romantic effusions to sublimity and sensation".[118]

However, the following passage from Buckland's treatise *Reliquiæ Diluvianæ* tells against this view:

> This cave is one of the most remarkable I have ever seen, for the beauty of its roof, and perfection of its stalagmite [. . . .] presenting the varied fea-

112. "Mrs B." (née Mary Morland) was an accomplished geologist in her own right (Kölbl-Ebert 1997; Burek 2001; *ODNB; DNBS*).

113. Shortland 1994, 13–14.

114. Shortland 1994, 5.

115. St Clair 2004, 212–13 n. 10.

116. Shortland 1994, 39.

117. Shortland 1994, 19. On these figures' interest in geology see Levere 1981, 167–71; Wyatt 1995; Leask 1998b; R. O'Connor 1999; Heringman 2004. On other interlockings of science and Romanticism see Cunningham and Jardine 1990.

118. Shortland 1994, 36. Shortland's conclusion has been disputed along similar lines by Sommer (2003) and more briefly by Heringman (2004, 26 n. 58).

tures and irregular undulations of large and beautiful cascades, suddenly congealed into a mass of transparent alabaster [. . . .] The roof also of the main chamber, as well as of its side aisles, is in all parts broken into, and clustered over with irregularly grotesque forms of exquisite beauty, rivalling the richest combinations of the most complicated gothic fretwork, and far surpassing them in the wild and irregular varieties in which its masses descend, like inverted pinnacles, to meet the icy lake of stalagmite that covers the floor.[119]

Buckland's indulgence in this kind of language for the benefit of genteel readers belies the notion that his geology was the opposite of "romantic" and points up the danger of using popular dichotomies and literary-critical labels as a cultural-historical shorthand. Even if we retain the label, "Romanticism" was clearly no distillation of pure idealism. There is, of course, a discrepancy between the conventional sublimity of an enchanted cavern and the image of Buckland eagerly gathering dung with his bare hands; but the British public had an appetite for the curious and the sensational. In this age of sensibility, caves appealed to early-nineteenth-century tourists because they united the "grotesque" and the "exquisite", the material and the spiritual.[120] Geology's appeal cannot be put down to one side or the other: geological fieldwork was, in Rudwick's words, "loaded with sentiments that united elements of romanticism [. . .] with those of robust, manly Christianity and the gentleman's love of the countryside",[121] to which one might add (here following Shortland) a delight in transgressing the boundaries of genteel respectability. Several anecdotes suggest that Buckland and his friends, secure in their social status, relished the doubt which their grubby pursuits cast over their gentility in the eyes of humbler onlookers.[122]

The geologists' enthusiasm for this heterogeneous "cult of sensibility" certainly contributed to their developing self-image as necromancers. As Shortland has rightly noted, they drew on a long-established image of the cave as a site of "spiritual confrontation" charged with "classical monstrous associations" and "able, through the category of the sublime, to unfetter the overcivilized mind and thrust it forth on wild imaginings".[123] This complex chain of associations was not confined to the traditional "Romantic period":[124] it can be traced, in substantially similar forms,

119. W. Buckland 1823, 127–8, describing Forster's Höhle.
120. Sommer 2003, 199–200.
121. Rudwick 1985, 40–1.
122. See Gordon 1894 and Shortland 1994.
123. Shortland 1994, 13–14; Sommer 2003, 178–82.
124. In present-day literary-historical accounts, the limits of this period vary from 1789–1824 at its narrowest to 1770–1850 at its widest.

back to the early eighteenth century (and arguably beyond). The cult of sensibility which developed in the 1750s drew strength from antiquarian writings on the Old North and the Celtic West (including Macpherson's Ossian). Caves became a standard trope for "spiritual conflict" and "wild imaginings" such as were later recast by Shelley and Keats.[125]

Thomas Gray's ode *The Descent of Odin*, first published in 1768, was an early example. The Norse god Odin, mediated through Paul-Henri Mallet's *Introduction à l'histoire de Danemarc* (1755) and various later works, was the necromancer *par excellence*. Here he makes a voyage down to Hel, the underworld, where he summons a dead prophetess from her grave to question her about the future. Gray's ode is a free adaptation of an English translation of the seventeenth-century Danish scholar Bartholin's Latin translation of *Vegtamskviða* ("The Wayfarer's Song"), a mediaeval Icelandic poem based ultimately on Norse mythology.[126] By the beginning of the nineteenth century, *The Descent of Odin* was extremely well known and had been reprinted many times.[127] In view of the material discussed above, it is easy to see how a geologist might make the link between Gray's depiction of the Norse necromancer and Buckland's summoning of hyaenas from beyond the grave. One piece of occasional verse made just this link.

*The Professor's Descent* was written in 1822. The only surviving manuscript is in Buckland's own spidery hand, and internal evidence suggests that he composed it himself rather than just copying it down.[128] Formally speaking, it is a close parody of *The Descent of Odin*, taking the "high" theme of the Norse gods and applying Gray's form and metre (and many lines verbatim) to the "low" subject of geological investigation, specifically Buckland's visit to Kirkdale Cave:

> Uprose the King of Rocks with speed
> And saddled strait his War-bro' steed:

125. Romantic cave-poems are discussed more fully in Sommer 2003. On the cult of northern antiquity see Wawn 2000, from which the term "the Old North" is here lifted (see especially pp. 30–3). Wawn's usage covers Scandinavia and various perceptions of an ancient "Scandinavian Britain". Before the 1830s, however, the poetic perception of Scandinavian and Celtic antiquities in Britain often overlapped, as seen in Gray's "Odes" (which adapted Norse, Welsh, and Orcadian traditions). For wide-ranging reflections on northern otherworlds as sites of spiritual conflict see Davidson 2005.

126. Gray's poem is edited in Lonsdale 1969, 220–8. *Vegtamskviða* is translated in Larrington 1996, 243–5. On Gray and the Old North see Wawn 2000, 27–30.

127. On the reception of Gray's Norse odes see Gray 1973, unpaginated introduction; Clunies Ross 1998, 105–66.

128. The most compelling evidence for Buckland's authorship is the presence of crossings-out and corrections in his handwriting. The manuscript (of which the Devon Record Office, Exeter, holds a microfilm, 138 M/F711), is in the collection of Roderick Gordon, to whom I am grateful for allowing me to print extracts here. The full poem is printed in R. O'Connor 2006a.

To the Yorkshire steep he rode
The Old Hyæna's drear abode.
Him the Dogs of Darkness spied
Their shaggy throats they opened wide [. . . .]

Buckland takes the place of Odin as the hero of an otherworld voyage, battling monsters and demons.[129] But, although the poem is a parody of Gray's ode, the object of its satire (quite a different matter) was Byron's *Cain*, which for Buckland represented a direct encroachment of "mad" radicalism upon his hard-won scientific and theological territory. Buckland was not alone in responding poetically to *Cain*. Many hostile responses were written in verse: it was widely felt that the poetic power of *Cain* spread a moral poison which required a poetic "antidote".[130]

In Buckland's poem the Professor enters the cave and, on speaking the magic words ("the verse that vocalizes stone"), has a vision of the antediluvian world or "Lord Byron's Hell & Chaos". Lucifer appears and, thinking his visitor an ally, reveals that he is brewing a poisoned drink to drive radical science writers into further acts of madness:

A drink to madden Byron's brain,
To nonsense madder still than Cain;
To fire mad Shelly's impious pride
To final crisis, suicide.
This quaff'd in vulgar Carlisle's alehouse
Shall quickly urge him to the gallows.

To which the Professor retorts, "D—— their souls with all my heart!" before asking Lucifer to unveil further secrets of the strata. Realizing that this questioner is a geologist and therefore a "Foe of hell", Lucifer goes back to bed, grumbling that he will not be outwitted again.

In his study of Buckland, Nicolaas Rupke has briefly mentioned *The Professor's Descent* as a piece intended for "amusement", citing it as evidence that geologists "were proudly aware of the literary use of their work".[131] This interpretation seems to dampen the poem's fire somewhat. The aggressive jocularity of the Professor's retort suggests that the poem also had a tactical dimension. In writing this poem, Buckland was fiercely discrediting a set of writings which threatened the status he and his colleagues had begun to claim for geology—a position they had

---

129. On cave geology and the quest-romance see Sommer 2003.
130. Anon. 1822a; 1822b; 1822e; Battine 1822; Adams 1823; Wilkinson 1824. On these reviews see Chew 1924; Steffan 1968.
131. Rupke 1983b, 76; 1983a, 38.

not yet consolidated, and which was therefore vulnerable. In the light of Buckland's stealthy campaign to win geology greater independence from the clerically oriented disciplines which had acted as its academic patrons, the radically anticlerical science of Byron's Lucifer (and Carlile's Byron) needed to be shown up as "nonsense". This poem apparently circulated only in Buckland's circle: it seems to have been aimed at bolstering the self-image of geology, rather than effecting any direct change in its public projection.

But why use Gray's ode in this context—why, in particular, model this satire formally on Gray rather than on its chief target, Byron? It may be that Byron's later styles—exemplified by *Don Juan* on the one hand and *Cain* on the other—were too difficult to parody without implying admiration. Buckland's use of an older form was not unique: other hostile reviewers of *Cain* who responded in verse tended to employ older satiric forms such as (mock-)heroic couplets. Perhaps Buckland felt that his purpose was best served by the insistent rhythms of Gray's iambic tetrameters, which often slid into pounding, incantatory trochees.[132] Perhaps, too, the landscape of the Old North was felt to be more familiar geological territory than the unspecified cosmic spaces of Byron's drama. Bakewell, Playfair, and several eighteenth-century geological writers had manipulated the subjective connotations of the Old North to render their books more appealing. Moreover, during the late eighteenth and early nineteenth centuries, British pioneers of Icelandic geology (John Thomas Stanley, George Mackenzie, Henry Holland) also spearheaded a new wave of enthusiasm for Norse literature in Britain.[133] Imaginatively speaking, the worlds of Scandinavian antiquity and geology were never far apart: as the layout of the *Gentleman's Magazine* demonstrates, geology was just another branch of "antiquities". It is not in itself surprising that Buckland should employ these associations in his poem.

But none of these factors explains why Buckland chose *The Descent of Odin*. In view of his satiric design—to show that geology was not "demonic" or anticlerical—it would surely seem inappropriate to cast himself as hero in the role of Odin, that fearsome death-god of a heathen people usually known for performing hideous atrocities on monks, nuns, and each other at the drop of a helmet. Only a few months before, Walter Scott's historical novel *The Pirate* (1821) had been published, selling in large numbers and profoundly affecting the way people viewed

132. A tetrameter is a line constructed from four pairs of syllables. Iambic tetrameters place the stress on the second syllable of each pair ("Their shaggy throats they opened wide"); trochees place the stress on the first syllable of each pair ("Foe of Hell I know thee now").

133. Playfair 1802, 485; Bakewell 1813, 21–2. On British geologists in Iceland see Wawn 2000, 41–59.

the Old North before the canonical saga-translations of the 1840s and 1850s. *The Pirate* conveyed Scott's vivid enthusiasm for Scandinavian antiquities, but the swashbuckling, bloodthirsty "Viking ethos" came in for sustained moral condemnation (much more so than the "Highland ethos" in Scott's first historical novel, *Waverley*). Indeed, Scott's representation of a modern-day Viking, the corsair Clement Mertoun, owes not a little to Byron's very own verse-romance *The Corsair* (1814), whose harsh individualism had been mocked a few years earlier in two verse-satires by Buckland's Oxford colleague and friend Charles Daubeny.[134]

Buckland's choice of Odin seems less strange when we examine the English poetic tradition which had grown up around this figure. He had arrived in early eighteenth-century England in the hybrid uniform of neoclassical Gothic. At first, readers were shocked, and some regretted Gray's abandonment of sense for sensibility, of country churchyards for heathen rituals.[135] However, the new aesthetic soon became familiar, and Odin's image began to mellow. As a hero of Regency odes he was still uncanny, yet distinctly decorous. Unlikely as it may seem, in Regency England Odin could even be viewed as a primitive Protestant culture-hero, thanks in part to a scholarly euhemeristic tradition according to which the "real" Odin, far from being a god, had been a powerful military leader at war with a tyrannical Rome.[136] Two Odin poems written in 1827 by the poet laureate, Robert Southey ("The Race of Odin" and "The Death of Odin"), are typical in representing Odin as a vigorous and noble warrior. His magical powers were merely another aspect of his might, and his love of battle was more Homeric than demonic. This Odin was more suitable for Buckland's purposes.

These observations are supported by the iconography of Odin in eighteenth-century English book-illustration. One image merits particular attention, for it was used to illustrate Gray's *Descent of Odin* in editions of his verse from 1776 onwards (Fig. 2.12). It illustrates the moment in the poem when Odin summons the dead prophetess from underground by tracing the "Runic rhyme" on a rock in front of him. Odin is shown in neoclassical splendour, his Gothic wildness discreetly signalled by the shock of curly hair protruding from beneath his clean helmet, neatly tied in a knot. His gesture is confident, his physique virile. The surrounding scenery is wild and barren, with beetling cliff and dark mountains. It has been an arduous journey: he has come on horseback, and his "coal-black

---

134. On *The Pirate* and its reception see Wawn 2000, 60–88. On Scott's career see *ODNB*. See Daubeny's commonplace book: Magdalen College, Oxford, MS 377, I, 146–8.

135. Clunies Ross 1998, 110.

136. Wawn 2000, 188.

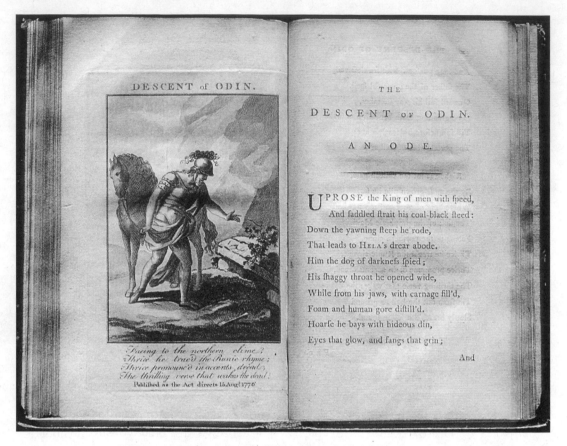

*Figure 2.12.* Odin tracing the "Runic Rhyme" in a John Murray reprint of Gray's *Descent of Odin,* on which Buckland modelled *The Professor's Descent.* Gray 1786, 119 and facing plate.

steed" stands nearby.[137] What more fitting image for the gentlemanly geologist as heroic "knight of the hammer"?[138]

Now look again at the picture, this time with Buckland in mind. A hero stands at the mouth of a cave, gesturing confidently towards a well-lit rock face on which strange runic symbols are seen. This rock face is particularly striking, because it is clearly stratified. Seen from this viewpoint, the runes take on a new meaning—in fact, an old meaning which dates back well into the eighteenth century, but which Buckland and oth-

137. Odin's horse Sleipnir was traditionally held to have eight legs, as Gray noted in his commonplace book (Lonsdale 1969, 223 n.); the picture sacrifices accuracy to decorum.

138. Thor would fit this epithet more closely, but in the early 1800s he was not a particularly prestigious member of the Norse pantheon. During the 1840s his reputation grew, however, as witness the pseudo-mediaeval title-page of the (recently lost) 1850 manuscript volume containing the songs and ballads sung at Geological Survey dinners, "Ye Recorde Boke off ye Royale Hammereres, off whyche Anciente Ordere Tooballcane and Thorr were erlie Knyghtes" (Geikie 1895, 142).

ers were developing into the idea of strata as pages of earth history.[139] The lines beneath the picture describe Odin having just "trac'd" the runes on the rock, but both the word "trac'd" and the figure's gesture can be given another meaning outside the context of the poem itself: the figure can be seen as interpreting rather than writing the runes, "tracing" them as an antiquary traces the history of an ancient people. The lines also describe the "Runic rhyme" as having the power to "wake the dead"; similarly, Buckland's decipherment of the book of nature allows him to "resurrect" the bones of extinct creatures. Accordingly, the weathered bones of the long-dead prophetess lie on the floor at the mouth of the cave, soon to be awakened by our "knight" (like the bears' bones in Fig. 2.9). These geological resonances are solely in the mind of the viewer; but had Buckland been one such viewer, it seems unlikely that he would have missed them. Gray was one of his favourite poets; he may have encountered this picture during his reading, prompting him to employ *The Descent of Odin* as a basis for his own poem.

Whether Buckland knew this picture or not, his need for a heroic model was well answered by Gray's Odin, with his propensity for travel, his vigorous physicality, his supernatural authority, and his mastery of a secret language. These characteristics all pass to the Professor. His dialogue with Byron's Lucifer fits the common poetic trope of spiritual warfare in caves, a site in which (as Sommer has shown) geologists staged their contests with poets—and Moses himself—for cultural authority.[140] Buckland was only just beginning to embark on a long struggle against widespread biblical literalism. At stake was the authority to tell (part of) the story of Creation, and *Reliquiæ Diluvianæ* represented Buckland's public consolidation of the territory claimed at Kirkdale. In *The Professor's Descent*, the struggle between Byron's Lucifer and the Odinic Professor enacts a contest over the same patch of imaginative ground, but Buckland is fighting on the opposite frontier—that of the radical materialists, to whom the poem's savage and cannibalistic hyaenas are compared.

Buckland had to fight the cause of historical geology against both extremes: while retaining its respectability, geology had to provide a sublime frisson of adventure and magic. He therefore presented a hybrid character for his hero, a geological "parson" modelled on the heathen knight/god Odin and exemplifying the tension between rational neoclassicism and irrational sublimity.[141] The Professor's uncanny supernat-

---

139. On fossils and minerals as "lithic writing" or "natural runes" see B. Stafford 1984, 293–314.

140. Sommer 2003, 196–7.

141. Compare Wawn's judgement (2000, 29): "The neo-classical and the sublime co-exist in a creative tension."

ural powers are repeatedly pitted against forces of chaos and darkness (the "Radical" hyaenas, the vision of "Lord Byron's Hell", and Lucifer himself), and the villains are outwitted. Lucifer comes across as a particularly feeble character, partly because most of the advantages he is given in Byron's *Cain* are denied him here and given to the Professor instead. It is the Professor who takes the initiative, compelling Lucifer to reveal secrets; the Professor, rather than Lucifer, is a "Traveller" who can pierce the bounds of space and time, summoning up Byron's former worlds; and the Professor is the one portrayed as "fearless" and Promethean in his transgressive boldness. In short, the Professor possesses all the supernatural powers one would usually associate with "Satanic" figures, but (and this is the point) he is a "Foe of hell", because he is a true geologist. To make this point clear, Buckland adds a footnote: "(i.e. not dealing with the Devil tho' some have their doubts)". There is a world of difference, Buckland seems to insist, between the imaginative "Hell & Chaos" of Byron's "mad" and self-destructive vision, and the sensational but ultimately rational wonders of geology as taught by those qualified to pronounce on it.

*Cain,* and the upsurge of radical geology-writing in 1822, may well have prompted some geologists to abandon all ideas of making their science appeal to the public, and once again to avoid taking imaginative risks. Buckland did not react in this way, and *The Professor's Descent* may have been meant to show his fellow-geologists that their science could remain inviolate without sacrificing its attention-grabbing potential. It also demonstrates just how closely antiquarianism and geology were entwined in the early 1820s. Finally, it leaves no room for doubt over the importance of literary culture and poetic sensibility to the rise of geology. We have seen Buckland engaged in a complex literary and metaphorical debate in order to define the territorial claims of historical geology over the picturing of life before man. Far from being a mere *jeu d'esprit,* this poem reveals Buckland wielding his sense of humour as an instrument of aggressive exclusion, ensuring that the cohesion of his chosen group (the genteel geologists) is not threatened by radicals' deployment of Cuvier.[142]

The local effect of Buckland's poem is unknown, but geology's cultural self-confidence continued to increase, apparently undeterred by the *Cain* controversy. In his inaugural lecture, Buckland had promised that geology would serve "a subordinate ministry" in the temple of the humanities; but by the mid-1820s he was lecturing on the evidential superi-

142. On Buckland's humour as an instrument of exclusion in another context see Sommer 2004, 66.

ority of fossils over texts in reconstructing the past.[143] More and more occasional verses were written in which geologists were presented with a supernatural authority rivalling that of Old Testament prophets.[144] Conybeare penned an "Ode to a Professor's Hammer" in the late 1820s which celebrated recent fossil restorations in apocalyptic vein:

Hail to the hammer of science profound! [. . . .]

Beneath the storm of its thundering blows
    Bending, and opening, and staggering, and reeling,
Mountains reluctant their story disclose,
    The secret of millions of ages revealing.

The fossil dead that so long have slept,
And seen world after world into ruin swept
        Start at the sound
        Of its fearful rebound.[145]

Geology here presides over a secular Apocalypse, replacing the Last Trump with hammer-blows. The aggressive instrument of "science" forces nature to give up her most precious treasure, a "story" which contains "the secret of millions of ages". This is an image of supreme confidence in the power of science, echoing Cuvier's exhortation to "burst the limits of time" and dramatizing the boldest claim in Buckland's inaugural lecture: "it is surely gratifying to behold Science, compelling the primeval mountains of the Globe to unfold the hidden records of their origin".[146]

Such unashamed subversion of biblical language underlines the challenge posed by the new science to the old Creation-narrative. We have moved a long way from Sumner's cautious approval of Cuvier. Only a few years after Sumner's book came out, clerics like Buckland and Conybeare were revelling in the rhetoric of Enlightenment science, embodying Cuvier's "antiquary of a new order".[147] The hidden records were beginning to be deciphered, and the monuments restored. Of course, the image of the geologist as antiquary was far from being Cuvier's inven-

143. Here Buckland was reviving an older rhetoric: see Rappaport 1982, 27–31.
144. Sommer 2003, 197, 201.
145. Daubeny 1869, 78–9.
146. Cuvier 1813, 4; W. Buckland 1820, 5. Compare Cuvier's explicit allusion to the Last Trump and Ezekiel's valley of dry bones (Cuvier 1812, III, 3–4). On the persistence of this imagery in British geology see Shortland 1994, 39–40; Sommer 2003, 193–201.
147. Cuvier 1813, 1.

tion. Fossils had long been associated with antiquaries, and this label was bound to attach itself to anyone collecting old bones and inspecting ancient caves. This association is borne out in early-nineteenth-century journals and museums, where fossils jostled against Egyptian mummies for the public's attention. Images of "bringing to light" or "penetrating the darkness" of "a hoar antiquity" had been associated with the discovery of ancient monuments long before geologists began to use such language.

Consequently, when some of the geologists' bolder claims began to reach a wider public, they chimed in with pre-existing expectations that geologists should have stories to tell about the distant past. This narrative dimension was not the primary concern of most Geological Society members, but it dominated the science's constitution in the genteel and clerical setting of Oxford University, thanks partly to the peculiar institutional role geology was brought in to play there, and partly to Buckland's talent for showmanship. With the help of colleagues like Conybeare, Buckland became the pre-eminent British advocate of Cuvier's vision of geology. In this capacity he transformed the science's public profile and helped establish it as "a branch of the science of Archæology".[148] His eccentric sense of humour and vivid lecturing style played an important part in this strategy: the personality cult surrounding him spilled over into enthusiasm for the science, resulting in a melting-pot of self-confident rhetorics of display and authority. A decade later, the same techniques would be flourished before a wider public: under Buckland's guidance, Oxford would launch the BAAS's deliberate cultivation of spectacle.[149]

Buckland's identification of an antediluvian hyaena's den at Kirkdale was crucial to this process. Here Cuvier's hints at time travel ("bursting the limits of time") actually took pictorial form, resulting in the first-ever graphic restoration of a former geological era. When Humphry Davy celebrated Buckland's hyaena research in 1822 as the conquest of "a distinct epoch" in the "immensity of ages",[150] he was not merely indulging in hyperbole, but expressing a real sense of scientific breakthrough. The rest of antediluvian history remained a vast, chaotic *terra incognita* as well as a *terra incognitorum*, a savage land awaiting intellectual colonization. The old cartographic motto "Here be dragons" was to prove only too appropriate for the fossil researches Buckland's work now helped stimulate.

148. Francis 1839, 16. See also Torrens 1998.
149. Morrell and Thackray 1981, 158–9.
150. Davy 1827, 51.

# Lizards and Literalists

3

The 1820s saw a marked rise not only in the number of geological investigations taking place, but also in the number of large English fossil vertebrates identified. Besides the ancient mammals of North America, Siberia, and the Paris Basin, an older, more alien world than that of the hyaenas was coming into focus. In the updated edition of his *Ossemens fossiles,* Georges Cuvier took his readers back "to another age of the world [. . .] when the only creatures which walked the earth were cold-blooded reptiles".[1] The evidence for what Gideon Mantell would soon christen the "Age of Reptiles" was drawn from the impressive array of recently collected English saurian fossils:[2] ichthyosaurs, plesiosaurs, and other marine reptiles from the West Country and Yorkshire; Iguanodon bones from Sussex; a Megalosaurus from Oxfordshire. With this fossil repertoire, the pictorial possibilities of deep time expanded. Fragments

1. Cuvier 1821–4, V part ii, 10: "à un autre âge du monde [. . .] où la terre n'étoit encore parcourue que par des reptiles à sang froid".
2. Mantell 1831.

from the new story of Creation could now be narrated with more confidence and patriotic pride.[3]

This confidence is evident in the increasing freedom with which British writers on earth history exploited the spectacular potential of fossils to stage the world before man. In the first part of this chapter I will chart this movement by examining the work of two geologists of the new school, both (in their different ways) peripheral to the Geological Society of London: Gideon Mantell and Robert Bakewell.[4] In the late 1820s they produced revised versions of older treatises and in both cases, a new tone can be detected when we compare their previous work. But spectacle and confidence were not confined to the new geology. In the second and third parts of this chapter we shall see how the same rhetoric was seized on by several biblical-literalist writers in the late 1820s to popularize their own geologies, much to the dismay of the Geological Society's leading lights. Men like Lyell, were, I suggest, spurred on by these literalist writings to step up their publicity, competing for public assent and attention. In this literary battlefield over virgin territory, spectacular rhetoric was wielded by both sides. Even so, not all participants saw themselves as combatants. The stirring idea of a clear-cut struggle between old-earth and young-earth geologies, so dear to the hearts of popular writers today, emerged only in the 1830s—and even then, not in the minds of everyone who sought to promote either version of the science.

## A New Fossil Repertoire

Of all the names associated with the unearthing of the "Age of Reptiles", none is more famous than Gideon Mantell, the "discoverer" of the Iguanodon. Present-day dinosaur books routinely begin with an anecdote about a fossil tooth allegedly picked up by Mantell's wife on the roadside in 1822 while her husband was visiting a patient (see Fig. 3.1).[5] Despite opposition from other men of science, Mantell identified this tooth as belonging to a gargantuan herbivorous reptile. He was vindicated by Cuvier, and the animal was christened "Iguanodon". The latter part of the story fits easily onto the heroic narrative template which has served publishers so well ever since *Longitude:* the history of British science now appears to be littered with struggling provincial geniuses who doggedly

3. On the expanding repertoire see Delair and Sarjeant 1975; Taylor 1997; Rudwick 2008 in press. On the patriotic dimension see Torrens 1995, 266–7.
4. On Mantell see Dean 1999; *ODNB*. Mantell's quest for elite status was critically examined by Wennerbom (1999).
5. It has so far proved impossible to verify the precise timing and location of the tooth discovery and the role played by Mary Ann Mantell (Dean 1999; Cleevely and Chapman 2000).

persisted in the face of upper-class bigotry and eventually managed to change the face of the world as we know it.[6] Mantell is a journalist's gift: a frustrated country surgeon and fossil collector with strong scientific ambitions, powerful enemies, a melancholy life-story and a diary packed with wounded remarks. Heroic story-patterns are bound to suggest themselves when we are examining a hierarchical scientific culture whose self-appointed leaders relied heavily on the fieldwork of provincial investigators to build up their own authoritative and well-publicized syntheses. The heroic-biography mode may seriously distort our picture of "the establishment" (scientific or otherwise), yet it remains true that when men like Mantell wanted to gain scientific recognition beyond the merely local level, they did have to struggle. Such status required money and time; Mantell had neither, and on both counts he spent well beyond his means.

For Mantell, writing books was no luxury; it was part of a high-risk strategy for making his name among the great and good.[7] In this chapter we examine his first two books, both published in London by Lupton Relfe: *Fossils of the South Downs* (1822) and its sequel, *Illustrations of the Geology of Sussex* (1826).[8] These were handsome quarto volumes, lavishly illustrated (see Figs. 0.1, 3.1, and 3.2) and with prices to match: the first cost £3 3s (roughly equivalent to £550 or $1,090 today), while the second cost £2 15s (roughly £480 or $950). They bore French and Latin epigraphs, broad margins, and a widely spaced, leisurely typography. All this combined to suggest a high social, as much as scientific, status. They were clearly written with half an eye on the gentlemanly specialists, and half on the collectors who owned Parkinson's *Organic Remains*. In this one respect Mantell's strategy paid off: the volumes earned him much respect among his intended readerships. However, they were marketed poorly and sold abysmally. The first incurred a massive loss, while the second sold fewer than fifty copies.[9]

Mantell's books became more financially viable in 1833, when his third treatise was published by the prestigious London firm Longman. The man who persuaded Longman to take Mantell on was Bakewell, the geologist and surveyor whom Mantell had recently befriended. The successive editions of Bakewell's *Introduction to Geology* provide a useful point of comparison with Mantell's work, and this chapter will compare the first

6. Mantell takes on this role in Cadbury 2000 and McGowan 2001. On storytelling in popular history of science see Fyfe and Smith 2003.

7. Wennerbom 1999.

8. The title-page records "1827" but it was published in December 1826 (Dean 1999, 90). A second imprint was made around 1829 (Cleevely and Chapman 2000).

9. Mantell 1851, 226; Dean 1999, 50–1, 95–6.

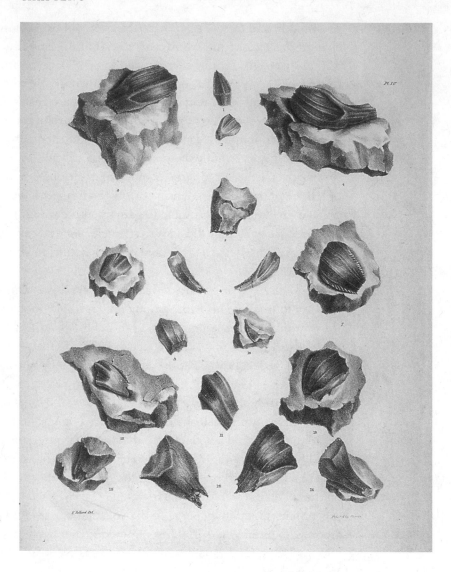

*Figure 3.1.* Iguanodon teeth. Mantell 1827, plate 4.

edition of 1813 (published by J. Harding) with Longman's third edition of 1828.[10] Bakewell, too, was a scientific outsider courting the attention of gentlemen, but his books, unlike Mantell's, were initially targeted at landowners with more practical interests. These were the men for whom Bakewell himself acted as a freelance surveyor. This slant is clear in the epigraph to the first edition, quoting the president of the Board of Agriculture: "A knowledge of our subterranean wealth would be the means of furnishing greater opulence to the country than the acquisition of the mines of Mexico and Peru."[11] But Bakewell increasingly appealed to his

10. The second edition of 1815 is textually almost identical to the first, which has been discussed in chapter 1.
11. Bakewell 1813, title-page.

readers' cultivated interest in landscape aesthetics and travel; successive editions gave more space to this aspect of geology, and in 1828 the utilitarian epigraph was dropped.

These were less ostentatious books than Mantell's: all editions of Bakewell's *Introduction* were fat octavos, though still expensive in middle-class terms (14*s* in 1813 = *c*. £120 or $240 today). The text was packed in thickly, with narrow margins and few plates. Both editions sold very well by Regency standards, and Mantell's books soon followed suit: his 1833 treatise was an upmarket octavo, and during the 1830s and 1840s his publications became increasingly miniature (ultimately appearing as duodecimos) as they attracted more middle-class readers.[12] Bakewell's octavos, however, grew fatter and fatter, each new edition proudly claiming to be "Greatly" or "Considerably Enlarged".[13] Mantell's and Bakewell's books belonged to different genres not only in terms of style and readership, but also in their content. Mantell's primary focus was upon past life-forms, revealed within a context of local antiquarianism; Bakewell sought to provide a useful guide to geology more generally, aiming at utility rather than aesthetics—except where he wished to render the science entertaining. On the latter occasions his aims harmonized closely with Mantell's, and both men used similar rhetorical tropes when it came to depicting the wonders of the past.

Comparing each author's second book with his first, we can gain a vivid sense of just how fast the science was changing in the 1820s, and how its public profile was being raised and developed. The preface to Mantell's 1826 volume emphasizes the "rapid [. . .] advancement of geological science" since 1822,[14] while the subtitle of Bakewell's 1828 volume reflects the same sense of change: as Fig. 3.3 shows, this text is advertised as having been "Entirely Recomposed", and instead of "Comprising" merely "the Elements of the Science" (as it had in 1813), it now comprised "the Elements of the Science in its Present Advanced State, and All the Recent Discoveries".[15] It was almost two hundred pages longer than the first edition.

*Gideon Mantell: From Facts to Romance*

The growing self-confidence of geology in general and Mantell in particular is first of all reflected in the differing amount of space taken up

12. The exception is the large quarto *Pictorial Atlas of Fossil Remains* (Mantell 1850), but this was an annotated reprint of plates from works by Parkinson and Artis.

13. Bakewell 1828, 1833, and 1838, title-pages.

14. Mantell 1827, v.

15. Bakewell 1813 and 1828, title-pages.

by the Genesis question in each book.[16] This problem is addressed cautiously and at length in the 1822 volume, in the form of an anonymous letter to Mantell by "a clergyman of the established church". His "Essay on the Mosaic account of the Creation" refers repeatedly to "Jameson's Cuvier". Mantell had it printed as a preface, intending to "silence the idle clamours that have been raised against geological speculations, from their supposed tendency to scepticism."[17] His 1826 volume, by contrast, makes no reference to this matter. For Mantell, the discovery of new facts about ancient reptiles (above all his beloved Iguanodon) was at once a justification for this new-found confidence in the science and his chief reason for writing the second volume. We now compare those passages in his 1822 and 1826 volumes in which he departed from the restrained enumerative mode characteristic of scientific monographs in order to project a more overt sense of wonder.

*Fossils of the South Downs,* though by far the longer work, contains perhaps only two such passages. Neither is in the author's own words, again reflecting a certain reticence: both are quotations from more established authorities. One of these occupies a crucial position within the text, being the book's concluding passage. Mantell's "Concluding Observations" on the changes of the earth's surface end on a devotional note, with a dual appeal to external authority. First comes a paraphrase of a remark made by the distinguished naturalist and medical writer John Ayrton Paris, who had suggested that geology's grandeur derived from its indications of the Creator's ordering wisdom:

> to discover order and intelligence in scenes of apparent wildness and confusion, is the pleasing task of the geological enquirer, who recognizes in the changes which are continually taking place on the surface of the globe, a series of awful but necessary operations, by which the harmony, beauty, and integrity of the universe are maintained and perpetuated; and which must be regarded [. . .] as wise provisions of the Supreme Cause, to ensure that circle of changes so essential to animal and vegetable existence.[18]

Like many English Nonconformists at this time, Mantell's personal attitude towards natural theology was rather more emotive and instinctive than rationalistic in the Paleyan tradition[19]—another possible reason for leaving the preliminary section on Moses and Genesis as a letter from a clergyman, instead of tying himself down with his own rationalization.

16. On developing relations between science and religion in the 1820s see Corsi 1988, 49–60.

17. Mantell 1822, vii–ix. The letter occupies pp. 1–13 and was written by Henry Hoper (Dean 1999, 152 n. 4).

18. Mantell 1822, 304–5.

19. See Astore 2001.

This doxological attitude is reflected in the mode by which the passage (and, with it, the book) concludes, a quotation from the final "Hymn" in James Thomson's *Seasons*. Mantell invites his readers to join in this concluding paean, where the sentiments endorsed by Paris are cast into poetic form:

> Mysterious round! what skill, what force divine
> Deep felt in these appear! ——
> *   *   *   *   *   *   *   *   *   *
> Were every falt'ring tongue of man,
> Almighty Father! silent in thy praise,
> Thy works themselves would raise a general voice,
> Even in the depths of solitary wilds,
> By human foot untrod, proclaim thy power.[20]

For Mantell, the significance of fossils went beyond their individual appearance; throughout his life he was at pains to communicate these intimations of a divine system of which the philosopher is vouchsafed glimpses.

On the other hand, Mantell elsewhere hinted at a more sensational form of display, concentrating on particular scenes as much as cosmic schemes. Just as parts of Thomson's *Seasons* reveal the poet's awkward sense that many things in nature were disturbing, even appalling, and hard to fit into an ordered pattern, so Mantell's successive treatises revelled more and more in the uncouth appearances of former creatures. In 1822, he had identified the pleasure experienced by the geologist as the observation of order in apparent "wildness and confusion"; his later treatises would capitalize on that "wildness" by emphasizing the incredible size and strangeness of the Iguanodon and its companions. His 1822 volume contains only one example of this rhetoric, at the end of a section on the saurian fossils found in the Tilgate Forest quarries. After describing the vestiges of "an animal of the lizard tribe, of gigantic magnitude", Mantell's imagination seems to be fired, and, borrowing the voice of his French master, he exclaims:

> Reflecting upon these extraordinary facts, may we not inquire with the illustrious Cuvier, *"At what period was it, and under what circumstances, that turtles and gigantic crocodiles lived in our climate, and were shaded by forests of palms, and arborescent ferns?"* [21]

20. Mantell 1822, 305.
21. Mantell 1822, 57.

The italics bring Mantell's observations on the Tilgate fossils to a suitably dramatic conclusion: the question is left hanging over the reader, encouraging them to ponder the mystery of a tropical England.

These hints at antediluvian spectacle became more pronounced in Mantell's second treatise, *Illustrations of the Geology of Sussex*. Devotional ecstasies are nowhere to be found in this book: all its heightened language is concentrated in a chapter on the Tilgate saurians, now greatly extended. This time Cuvier's stirring question is reworked in Mantell's own words so that it opens the chapter, inviting the reader to envisage a tropical landscape before revealing that Sussex once looked like this:

> Before we enter upon the description of the fossils of Tilgate Forest, let us for a moment consider what would be the nature of an estuary, formed by a mighty river flowing, in a tropical climate, over sandstone rocks and argillaceous strata, through a country clothed with palms, arborescent ferns, and the usual vegetable productions of equinoctial regions, and inhabited by turtles, crocodiles, and other amphibious reptiles?[22]

Admittedly, this passage does not yet display that mastery of evocative prose for which Mantell would become celebrated: the bit about "the usual vegetable productions" does not quite trip off the tongue. But the rhetoric of display is evident in his first words: "Before we enter" inscribes Mantell as a guide, accompanying the reader into an imaginary museum. Indeed, he recycled this passage in the printed guidebooks to his own museum.[23]

Mantell's enthusiasm takes on its most characteristic form when he indicates the dimensions of the Iguanodon. Strict analogy with the iguana, the only way of estimating the size when the specimens were so fragmentary, produced an approximate length of seventy to a hundred feet (see plate 6).[24] Most of the Tilgate chapter is taken up with dispassionate enumeration, but the Iguanodon stretches this sober language to breaking point. The words "gigantic", "enormous", and "monstrous", strictly unnecessary when exact figures are always given, appear nine times in three pages.[25] His description of a fragment of a thigh-bone soars rapidly into rhetorical showmanship, coupling enumeration with exclamation: "no less than 23 inches in circumference!" The fragment from which this calculation is drawn is spectacularly depicted, in its actual size, in a foldout plate (Fig. 3.2).

22. Mantell 1827, 51.
23. Mantell 1834b, 12; 1836b, 10. For the 1829 version see Dean 1999, 97.
24. Fig. 2.3 and Plate 6, both executed in the 1830s, represent 100-foot Iguanodons.
25. Mantell 1827, 76–8.

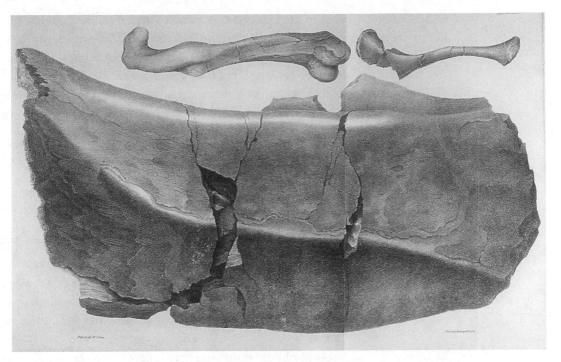

*Figure 3.2.* Fragment of an Iguanodon's thighbone. The smaller bones above were understood to be the thighbone and collarbone of a Megalosaurus. Mantell 1827, plate 18.

Directing us to this plate, Mantell invites us to clothe it with flesh in our mind's eye: "Were it clothed with muscles and integuments of suitable proportions, where is the living animal, with a thigh that could rival this extremity of a lizard of the primitive ages of the world?"[26] As in Ashe's *Memoirs of Mammoth* (the first fossil book Mantell owned),[27] the emotive appeal of vital statistics shines through the exposition. So strange was the idea of a vast herbivorous reptile in 1826 that Mantell borrowed a vocabulary hitherto reserved for mammoths.

We need to register this strangeness: the Iguanodon was a monster according to the strict usage of the term, seemingly composed from parts of known animals. (The only other extinct land saurian then known, the Megalosaurus, was less extraordinary: conceived as a carnivorous lizard, it was pictured as a kind of crocodile and was also thought to be smaller than the Iguanodon, which counted for a great deal.) Cuvier's expressions of amazement, which Mantell quoted in the original French, serve to introduce the Iguanodon in Mantell's book. He first quoted from the 1824 volume of *Ossemens fossiles* to introduce a new creature "even more extraordinary than all those we know of". He next quoted proudly from

26. Mantell 1827, 77.
27. Dean 1999, 53.

Cuvier's letter to him of 20 June 1824, in which Cuvier acknowledged him to be correct in identifying the creature as reptilian: "Might we not have here a new animal, a herbivorous reptile?"[28] Later Mantell observes that some of the bones are "so large that they appear more like the bones of mammoths or elephants, than of reptiles", quoting Cuvier's admission that he had originally taken them for those of a large hippopotamus because of their size.[29] So when Mantell sought to elicit public wonder at the Iguanodon in his new book, he appropriated the showmanlike rhetoric hitherto reserved for fossil mammals. The "Age of Reptiles" had only just been identified as a distinct chronological entity. As a new scene in the pageant of life on earth, it had to follow existing theatrical conventions.

This new mammoth for the 1820s—older, bigger, and stranger than the fossil elephant—is then described in language reminiscent of Byron's *Cain*: "the Iguanodon was one of the most gigantic reptiles of the ancient world; and a colossus in comparison to the pigmy alligators and crocodiles that now inhabit the globe." Its size alone is so surprising, claims Mantell, that "our readers might well exclaim that the realities of Geology far exceed the fictions of romance."[30] This topos was to become a hallmark of subsequent popularization, and here it provides a rhetorical springboard for the book's conclusion. Mantell at last attempts to answer Cuvier's enthralling question which he had excerpted and italicized in 1822.[31] Until the final paragraph, however, the last two pages comprise only a tentative description of the landscape there evoked:

> We cannot leave this subject, without offering a few general remarks on the probable condition of the country, through which the waters flowed that deposited the strata of the Tilgate Forest; and on the nature of its animal and vegetable productions. Whether it were an island, or a continent, may not be determined [. . . .][32]

In later publications Mantell would learn to omit such statements of uncertainty, emphasizing in a more showmanlike manner only those parts of the landscape which he could delineate with confidence. But presently the reptiles return and, with them, Mantell's descriptive confidence. Here the "fact-or-fiction" topos functions ostensibly as an apology for apparent extravagance, but also serves to advertise the science:

28. Mantell 1827, 71, 72 ("encore plus extraordinaire que tous ceux dont nous avons connoissance", "N'aurions-nous pas ici un animal nouveau, un reptile herbivore?"). On the controversy over these teeth see Dean 1999, 71–85.

29. Mantell 1827, 77.

30. Mantell 1827, 78. Compare *Cain* II.ii.67–74 (Byron 1986, 909).

31. Mantell 1822, 57.

32. Mantell 1827, 82.

If we attempt to pourtray the animals of this ancient country, our description will possess more of the character of a romance, than of a legitimate deduction from established facts. Turtles, of various kinds, must have been seen on the banks of its rivers or lakes, and groups of enormous crocodiles basking in the fens and shallows.

The gigantic *Megalosaurus,* and yet more gigantic *Iguanodon,* to whom the groves of palms and arborescent ferns would be mere beds of reeds, must have been of such a prodigious magnitude, that the existing animal creation presents us with no fit objects of comparison.

At this point, and only at this point, Mantell addresses the reader directly, as if in a lecture-hall. He commands the reader to summon the animal up before their mind's eye:

Imagine an animal of the lizard tribe, three or four times as large as the largest crocodile; having jaws, with teeth equal in size to the incisors of the rhinoceros; and crested with horns; such a creature must have been the Iguanodon! Nor were the inhabitants of the waters much less wonderful; witness the Plesiosaurus, which only required wings to be a flying dragon; the fishes resembling *Siluri, Balistæ, &c. &c.*[33]

The exclamation mark intensifies the visual rhetoric: in the second sentence we move from "imagining" to actually "witnessing". The Iguanodon's resemblance to the dragons of romance is hinted at by Mantell's excited speculation about what the plesiosaur might look like, were it to have wings. But this passionate paragraph ends rather anticlimactically: "&c. &c." suggests a loss of theatrical nerve (or perhaps a looming publisher's deadline). The book's remaining paragraph continues the retreat from full-blooded showmanship with cautious references to other authorities.

Yet Mantell was bold to incorporate such a detailed scene. A single sentence from the previous book had expanded to fill a whole subsection, now occupying the book's final paragraphs. This was the first time the British reading public had been treated to a detailed verbal restoration of a specific prehuman landscape. Even Buckland's most vivid published descriptions of monsters up to that point occur in his anecdotal descriptions of the observed habits of present-day hyaenas.[34] But Mantell made an "attempt to pourtray the animals of this ancient country"; to do so was indeed to venture into the realms of "romance". His vision of the

33. Mantell 1827, 83–4. Note the inconsistency, common for this period, in the italicization of the name "Iguanodon" in this passage.

34. W. Buckland 1823, 21–4, 37–8.

Iguanodon would continue to define public perceptions of that creature for several decades,[35] and the great lizard would remain the jewel in the crown of English vertebrate palaeontology for over a century.

*Robert Bakewell: From Curiosity to Astonishment*

The 1828 edition of Bakewell's *Introduction to Geology* was mainly devoted to stratigraphy and geological processes rather than fossils, yet it too reflects the increasing need for spectacular display in 1820s earth science. Bakewell was not a palaeontologist, so he relied in these matters on other writers—notably his friend Mantell, whose more dramatic passages on the Iguanodon and its country he quoted at length in the new edition.[36] In chapter 1 we saw how, in the first edition, Bakewell had already painted a visionary verbal "retrospect" of the deep past, as part of the book's conclusion. For purposes of comparison I quote it in full again:

> What various reflections crowd upon the mind, if we carry back our thoughts to the time when the whole surface of the globe was agitated by tumultuous and conflicting elements; or to the succeeding intervals of repose, when all was one vast solitude; and again to a subsequent period, when the deep silence of nature was broken by the bellowings of the great mastodon and the mammoth, who stalked the lords of the creation, and perished in the last grand revolution of the globe before the formation of man!

There had followed a brief account of the organic remains supporting such "reflections", then a brief discussion of hypotheses about the earth's planetary origins. Then came a caution against undue speculation, but Bakewell defended geological speculation in principle for its ability to awaken "curiosity" and lift mankind above the brutes by "exalt[ing] his hopes 'beyond this visible diurnal sphere.'"[37] With that the main text had ended.

The most obvious change in the new version of the "retrospect" concerns the period represented by the secondary strata, originally described as "one vast solitude". In 1828 this period has become a rather lively time "when enormous crocodilian animals scoured the surface of the deep, or darted through the air for their prey".[38] This change points up, with pe-

---

35. It continued to be influential even after the Crystal Palace Gardens monster-sculptures (Fig. 7.10, Plate 7) popularized a radically different interpretation from the 1850s onwards.

36. Bakewell 1828, 282–4.

37. Bakewell 1813, 326–7, 327–30.

38. Bakewell 1828, 481.

culiar clarity, the transformation of imagined deep time brought about by the reptiles unearthed in the intervening years. Other significant alterations and additions surround this passage. The defence of geological speculation has been made more elaborate, moving away from the moral-religious emphasis of the 1813 version: Bakewell now stresses intellectual pleasure, developing the notion that speculation "excite[s] our curiosity" to answer a new question, *What advantage can be derived from the study of Geology?*[39] This question is introduced by a passage from James Beattie's much-loved poem *The Minstrel* (1771–4), whose untutored boy-hero learns philosophy, piety, and the spirit of true poetry from his observations of nature. Like later authors of children's "introductions to geology" who employed the very same passage, Bakewell asks his readers to be similarly receptive to geology's moral implications, at the same time as commanding their childlike wonder at the phenomena described:

> it is not a mere fiction of fancy, that
>
> > "Earthquakes have raised to heaven the lowly vale,
> > And gulphs the mountains mighty mass entomb'd,
> > And where the Atlantic rolls, wide continents have bloom'd."
> >
> > *Beattie.*
>
> Yet whatever proofs we may have of such changes, they are so remote from our present experience [. . .] that we cannot avoid regarding them as more proper subjects for the poet, than for the sober records of the natural historian.[40]

Bakewell, like Mantell, readily admits the challenge which geology poses to the imagination. Also like Mantell, however, he manipulates the apparent wildness and "fiction" of geology's facts and speculations. He invests them with a romantic intellectual "excitement" under whose influence the study of nature reveals "its most attractive charms" and stimulates "moral impressions and pleasures".[41]

Geology's appeal to the questing, speculative imagination is thus defended anew. This appeal is then embodied in the vivid retrospect, via this new link-passage:

39. Bakewell 1828, 479.

40. Bakewell 1828, 479. Later uses include Hack 1835, 113; Mantell 1836a, 16; Miller 1847b, 109; 1859, 129. Hack's book, in novelized "conversation" form, explicitly constructs the young student (and hence her young reader) as a modern type of Beattie's Edwin: see Hack 1835, xvi, 13–14.

41. Bakewell 1828, 480–1. In the fourth edition of 1833, Bakewell would add Sedgwick's eloquent philosophical defence of geology's benefits for mankind's "imagination", "feelings", and "active intellectual powers" (Bakewell 1833, 537).

> Geology discovers to us proofs of the awful revolutions which have in former ages changed the surface of the globe, and overwhelmed all its inhabitants: it reveals to us the forms of strange and unknown animals, and unfolds the might and skill of creative energy, displayed in the ancient world: indeed, there is no science which presents objects that so powerfully excite our admiration and astonishment. We are led almost irresistibly to speculate on the past and future condition of our planet, and on man, its present inhabitant. What various reflections crowd upon the mind, if we carry back our thoughts [. . . .][42]

For Bakewell in 1828, geology had usurped astronomy's traditional place at the pinnacle of scientific sublimity: it was now the high road to "astonishment". Bakewell followed the growing trend in science writing for recasting amazement as religious sentiment—or teaching natural theology by clothing it as sensationalism. In Bakewell's opinion, geology owed its newly exalted position to two aspects in particular: causal processes ("awful revolutions", with particular reference to their role in species extinction) and the fossil monsters themselves ('strange and unknown animals").

The changes Bakewell made to his 1828 edition illuminate the sharpened sense of history which the study of fossils was now conferring on geology. The new chapter on organic remains ("petrifactions") begins with the analogy of nature as a book. This device goes back to the Renaissance, but the new understanding of fossils in the 1820s now lends the analogy the air of a sudden revelation, a new "discovery":

> If it had been predicted a century ago, that a volume would be discovered, containing the natural history of the earliest inhabitants of the globe, which flourished and perished before the creation of man, with distinct impressions of the forms of animals no longer existing on the earth,—what curiosity would have been excited to see this wonderful volume; how anxiously would Philosophers have waited for the discovery! But this volume is now discovered; it is the volume of Nature, rich with the spoils of primeval ages [. . . .][43]

We have already seen this Baconian analogy in Cuvier's *Theory,* again with reference to fossils. It would soon be taken up by Lyell, whose *Principles of Geology* would enshrine it in its most sophisticated form. In Bakewell's hands it reveals the additional curiosity-value which geology had earned since the first edition appeared in 1813.

---

42. Bakewell 1828, 481.
43. Bakewell 1828, 27.

To some extent, Bakewell could now take the reader's curiosity for granted. The new edition opened with the statement that "There are perhaps few persons possessed of much curiosity in early life, to whom the following question has not frequently presented itself—*What is the world made of?*"[44] This initial interest established, Bakewell can then go on to suggest that geology offers even richer and stranger rewards to those who also ask about the history of the world as well as its composition. The enthusiasm with which Bakewell wrote about this aspect of geology is all the more striking given that fossils were not his primary interest, and that he relied largely upon outside authorities in these sections. The new prestige of fossil studies is also reflected in the frontispieces of these three editions. The first (1813) depicts a traveller sitting amongst the toothy rocks of Cader Idris: geology was, by implication, a healthy outdoor activity. The second (1815), while its text is identical, bears a new frontispiece (this time coloured by hand) depicting a bird's eye view of a geological process, "Salt Formation at Cardona in Spain". But the third edition's frontispiece (1828; Fig. 3.3) displays a single fossil curiosity: it shows a life-size, hand-coloured drawing entitled "The Gigantic Trilobite", a very large invertebrate from Bakewell's own collection. The reader, and prospective buyer in Longman's shop, was drawn in by a blatant appeal to old-fashioned antiquarian curiosity.

So, although Mantell and Bakewell had begun their literary careers targeting two distinct kinds of upper-class reader—the genteel collector and the practical-minded proprietor respectively—they now found themselves using very similar devices to attract their readers. This convergence partly reflects an overlap between these two upper-class groups (proprietors were often collectors too); but the devices Mantell and Bakewell were using point to a growing and well-heeled middle-class readership who were adopting practices associated with the landed gentry, such as fossil collecting and building up libraries. These readers wished to be enlightened and entertained on their own terms. Many of them professed suspicion of popular romance-fiction: these sensational stories were widely read by all classes, both as books and in magazines, but their affordability in the dawning era of cheap print earned romances a "lower-class" label among the upwardly mobile.[45] The changing content of Mantell's and Bakewell's books reflects a wider movement in which scientific writing was steered to meet the new demand for "rational amusement". Fossils offered an obvious route: their veiled revelations of vanished worlds inhabited by fantastic beings could easily be framed

44. Bakewell 1828, 1.

45. On the relation between sensation fiction and the concept of the "popular" see Shiach 1989, 88–100. On the mistrust of fiction see Gallaway 1940; for a representative middle-class evangelical perspective see "Priscus" 1838.

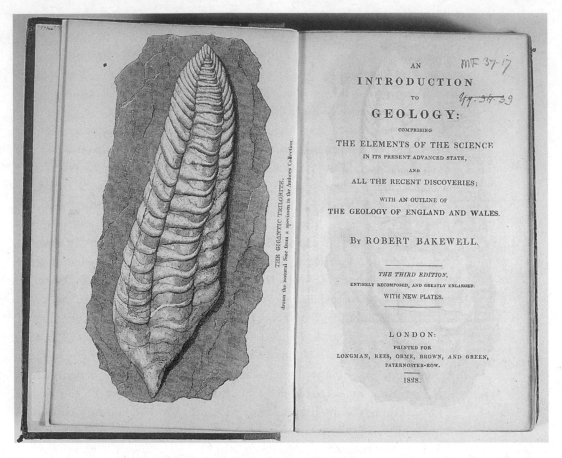

*Figure 3.3.* The third edition of Bakewell's *Introduction to Geology*. Bakewell 1828, frontispiece and title-page.

as rational truths which "far exceed the fictions of romance",[46] providing the thrill of sensational fiction without the stigma of falsehood. This combination of self-improvement with aesthetic allure was reinforced with edifying reflections and quotations from well-known poems, shadowing the rising middle-class vogue for collectable anthologies of verse and prose.

It is no coincidence, then, that when Lyell wanted to promote the work of the Geological Society for the *Quarterly Review* in 1826, he focused almost entirely on the "recent and splendid discoveries" of its fossil "department". The fossil saurians are seen to realize the "fictitious animals of romance" and "the winged dragons of fabulous legends", and quotations from Shakespeare, Ariosto, and Milton support this poetical tone.[47] But these strategies were not restricted to adherents of the new

46. Mantell 1827, 78.
47. [C. Lyell] 1826b, 509, 524, 518, 538.

geology. Fossils could be, and always have been, interpreted in a number of ways. No account of the science's popularization in the 1820s would be complete without considering the influential part played in this process by writers who held that the earth was only a few thousand years old. In what remains of this chapter we shall explore the ways in which biblical-literalist writers employed the new repertoire of ancient monsters to generate enthusiasm among the same readership for their own very different theories of the earth.

### The World without Eden: Literalist Earth Science in the 1820s

Every heroic narrative needs a good cardboard villain. In the received history of nineteenth-century geology, the so-called scriptural geologists have served this purpose well as Bible-wielding opponents of the new science. They are seen lurking balefully behind the scenes, their faces contorted with anger as Lyell, William Smith, or some other chosen hero marches centre stage to transform Western thought for ever. Someone in the audience shouts, "Look behind you!"—the dagger is poised to strike—but just in time our hero wheels around. A single blow of his geological hammer sends the benighted malefactors back to church where, with a sinister rustling of pamphlets, they huddle together and prepare to fight Thomas Huxley in the next act.

This is the picture which emerges from older works on the history of science by pioneering scholars such as Charles Gillispie and Milton Millhauser.[48] As an explanatory model of the relations between geology and biblical scholarship in the early nineteenth century, this kind of Whiggish, oppositional narrative is no longer accepted by most historians of science, especially in the light of recent reassessments of evangelical science.[49] Yet the old model retains its power, for three reasons. First, heroic narratives about scientific discoverers are back in fashion at the popular level. Like snobbish aristocrats, conservative clergymen provide gratifyingly reactionary symbols of the difficulties the hero must overcome in his or her quest for the truth. Second, books like Gillispie's *Genesis and Geology* are still treated as unquestioned authorities by some specialists in other fields who touch on this area—a problem compounded by the current blurring of boundaries between academic and popular publications.[50] For many readers, early-nineteenth-century biblical-literalist writers on earth history are still confined to Gillispie's "lunatic fringe",

---

48. Gillispie 1951; Millhauser 1954. Millhauser's article remains useful as a general overview.

49. Examples of such reassessments include Hilton 1988; Brooke and Cantor 1998; the essays in Livingstone *et al.* 1999; Astore 2001; Fyfe 2004a.

50. For example, Freeman 2004, 85–8, 186–8.

their writings "too absurd to disinter".[51] Third, among historians of science, an overriding interest in the celebrated "pioneers" of British geology has led to a situation in which the only literalists one hears about are those whose voices were most strident against the new science. Even among historians who repudiate the model described above, the relative lack of interest in literalist writers means that the latter tend—understandably—to become flattened out as a purely oppositional phenomenon, a (usually provincial) backdrop of reaction against which the work of the Geological Society and its allies stands out more clearly.[52] In such cases, the old rhetoric of scornful incredulity sometimes slips in through the back door.

Recent years, however, have seen a welcome reassessment of provincial scientific investigators and marginalized scientific practices, and as a result some literalist writers are beginning to attract more interest in their own right.[53] In particular, Terry Mortenson has recently begun the long-overdue process of rehabilitation, questioning the old assumption that early-nineteenth-century literalist writers on earth history were a homogeneous group of scientifically illiterate anti-geologists and analysing their marginalization as a consequence of broader social and intellectual shifts. Mortenson's book *The Great Turning Point* is a mine of information about these neglected figures and the debates in which they engaged. It is, however, unlikely to receive much positive attention among academic historians of science, because it is intended to serve the purposes of a very different community: the young-earth creationist movement.[54] Mortenson's aim has been, as the back cover proclaims, to "open eyes and hearts to the veracity of God's Word", presenting the widespread acceptance of old-earth geology by the Victorian church as (in the subtitle's words) a "Catastrophic Mistake" while displaying contemporary literalist writers as spokesmen for the mainstream view and a shining example to their present-day successors.[55] This purpose is as Whiggish as the secular triumphalism of much popular-science writing, causing Mortenson to overestimate the geological knowledge of several

51. Gillispie 1951, 152.

52. Useful work on the oppositional dimension of the literalists includes Morrell and Thackray 1981, 234–44; Rupke 1983b; Rudwick 1985, 43; J. Moore 1986; Stiling 1999. By contrast, no literalist geological writers are included in the recent *Dictionary of Nineteenth-Century British Scientists,* apart from those deemed to have "contributed" to other sciences (Lightman 2004).

53. For example, Osborne 1998, 44–9, 178–95; Michael Roberts 1998; Knell 2000; Lynch 2002a; 2006. See also Yule 1976, 99–129, 317–34.

54. On the history of this movement see Numbers 1992.

55. Mortenson 2004, back cover and title-page. However, as Mortenson is quick to point out (2004, 237), there is no direct historical link between the early nineteenth-century movement and its present-day counterpart.

literalists and to underestimate the exegetical skills and religious commitments of some old-earth geologists.

On the other hand, some of the counter-claims made by some of Mortenson's opponents are equally overdrawn, particularly the claim that most people around 1800 already assumed that the earth was created over vast ages rather than in six literal days. This claim is designed to push literalism back to the historical fringes. But biblical exegesis has always been a thoroughly contested terrain, and in the late eighteenth century literal and non-literal interpretations of Genesis 1 were equally current. Both had their roots in the branching theological debates of the third, fourth, and fifth centuries AD. The allegorical approaches taken by some of the Church Fathers had opened the possibility for old-earth cosmologies (though in opposition to Aristotelian "eternalism"), but interpretations grounded in a literal reading went back to Eusebius and Jerome, the founders of sacred chronology. This discipline had provided a firm basis for subsequent thinking about earth history in the Christian world, especially after the Protestant Reformation:[56] its practitioners had used the new humanist methods to compare biblical texts with those of other historical traditions in order to pronounce on the shape and meaning of human history. These works provided an authoritative scholarly foundation for the Bible's pre-eminence in structuring world history.[57]

To some extent, these foundations had been eroded in the late eighteenth century by new critical approaches to biblical texts emanating from the German universities,[58] some of which viewed the Creation-account as a mythical narrative with spiritual rather than historical or scientific authority—a move mirrored by the rising authority of philosophers and naturalists in the sphere of antediluvian *history*. The "higher criticism", as the German-led movement became known, was digested by many scholars and philosophers in late-eighteenth-century Britain, but assent was far from unanimous (and demoting the historical authority of Genesis still left open the question of how old the earth was). Furthermore, the doctrine of the Bible's inerrancy—not new in itself—had been framed in a newly clear-cut and conservative sense around the start of the nineteenth century, partly fuelled by counter-revolutionary anxiety.[59]

Claims about what "most people" believed require clear definition. If we are talking about most naturalists engaged in geological fieldwork,

56. Protestants took a new interest in interpreting Genesis 1 literally, but they were far from inventing this practice (*contra* Bowler and Morus 2005, 105–6).
57. On sacred chronology see Grafton 1991, 104–44. On its links with earth history see Rupke 1983b, 52–7; Rudwick 1986; Fuller 2001.
58. Frei 1974; J. Moore 1986, 332–5.
59. On the denominational distribution of biblical literalism see Rosman 1984, 224–7.

then the claim that an old-earth cosmology was widely credited in 1800 stands firm and needs to be emphasized. If, however, we are talking more generally about "the majority of educated Christians"[60]—let alone the uneducated—then the claim begins to look distinctly shaky. In the 1790s John Hunter was warned by a Royal Society colleague against popularizing an old-earth cosmology for fear of offending the "pardonable superstitions" of the literate public; since 1701 Bibles had had the date "4004 BC" printed in the margin alongside the earth's creation in Genesis 1; and schoolteachers and textbooks often reinforced this view.[61] The rush of furious pamphlets which greeted the publication of Byron's fossil-rich verse-drama *Cain* in 1821 confirms that the radical associations of old-earth cosmology remained strong in the minds of political conservatives, and not only among diehard eccentrics.[62]

Such evidence is, of course, only anecdotal: norms of belief can be established only on the basis of a socially comprehensive and theologically nuanced survey of a broad range of literature, from sermons and Bible commentaries to newspapers and diaries, and no such survey has yet been undertaken.[63] Conversely, it would be foolish to assume (with many popular expositions of the history of geology) that just because the date 4004 BC was taught in schools and printed in the margins of Bibles, it was necessarily believed by all. Yet the evidence mentioned points to the currency of young-earth views among a significant cross-section of Georgian and early-Victorian society, not just a handful of reactionaries on the "extreme fringes".[64] After all, if most educated people already believed in an immeasurably ancient earth, it would hardly have been necessary for its popularizers to expend so much energy in hammering this view home. In a country as politically volatile, regionally diverse, and intellectually heterogeneous as early-nineteenth-century Britain, confident claims about what constituted theological or scientific orthodoxy are out of place. Different groups had different orthodoxies.

The heavily politicized controversy surrounding present-day young-earth creationism is not the ideal setting for historical debates about nineteenth-century literalists. One way in which we might grope towards a balanced reassessment of these figures is to shelve (for the present) the

60. Michael Roberts 1998, 254; Lynch 2002b, x; 2006, 132.
61. On the Hunter case see Rudwick 2005b, 126–7. On Bible margins see Fuller 2005. On schoolteachers and their books see [Hawthorne] 1860, 14; Astore 2001, 158. See also K. Lyell 1881, II, 168.
62. For an example from 1845 see J. Secord 2000a, 339.
63. Useful (but diametrically opposed) summaries of the stances taken towards the date of Creation in a small selection of nineteenth-century theological writings are given in Michael Roberts 1998 and Mortenson 2004, 40–7, but in both cases it is unclear on what basis the selection has been made.
64. This phrase is Fyfe's (2004a, 24).

vexed questions of expertise and norms of belief, and concentrate instead on how these writers participated in the literary culture of science and promoted the study of rocks and fossils on their own terms.[65]

First it is worth mopping up some potential confusions of terminology. Nineteenth-century literalist earth-histories are normally referred to as "scriptural geology" or "Mosaic geology". These labels were current in the 1820s: the former comes from George Bugg's *Scriptural Geology* (1826–7), the latter from a book by Granville Penn, and both labels were applied to literalist writers by Buckland and Lyell. The labels are, however, misleading. They tend to bracket all such writers together as a single movement, even a conspiracy (which is precisely how Buckland and Lyell aimed to present them), and they tend to elevate Bugg or Penn as typical representatives. Both implications are untenable. Furthermore, most of the old-earth geologists discussed in this book believed (in some sense) in the inspiration and authority of Scripture, and many took a lively interest in interpreting Genesis; some of them even sought to reclaim the phrase "scriptural geology" by applying it to themselves.[66]

We may not, therefore, draw a clear line between "geology" and "scriptural geology"; using the latter label today only clouds the issue.[67] Confusion also results from recent attempts to reclaim the word "creationism" from today's young-earth movement and reapply it to Victorian figures who believed in "creation rather than evolution":[68] the word "creationism" carries connotations of opposition against evolutionary orthodoxies which do not apply to early-nineteenth-century Britain. Calling men like Buckland and Hugh Miller "creationists" not only reduces the scope of their work,[69] but facilitates the popular myth that they were somehow struggling against the tide. Finally, the term "evangelical" still causes confusion among non-historians. Neither today nor in the nineteenth century has evangelicalism necessarily entailed a literal interpretation of Genesis, or indeed a rejection of evolutionary theories: evangelicals were ranged on both sides of these debates.[70]

Recently, some scholars have turned to the more promising distinction between "geologists and interpreters of Genesis":[71] the former read the Bible in the light of the rocks, the latter read the rocks in the light of

65. A similarly dispassionate approach is sketched by Brooke and Cantor (1998, 57–64) as a test-case for more general analytical procedures.

66. "A Scriptural Geologist" 1839; "Fides" 1839.

67. It becomes especially confusing when Buckland and his like are referred to as "Mosaic" or "Scriptural" geologists (Haber 1959, 204; Hilton 1988, 150; Freeman 2004, 85).

68. Hilton 1988, 150; Lynch 2002a.

69. Dean 1981, 120; Merrill 1989, 237.

70. This has been emphasized by Yule 1976, 328–9; Michael Roberts 1998; Brooke 1999.

71. J. Moore 1986.

the Bible. But here, too, it is practically impossible to draw a dividing-line. Of the exegetical scholars who entered the debate, many favoured an old-earth cosmology. Moreover, many geologists and scholars tended in practice to read both "texts" in the light of each other, rather than simply twisting one to suit the other. Both "texts" were obscure enough to suggest many possible interpretations and controversial enough to encourage strong rhetoric, so a simple, transparent hierarchy of interpretative priority was rare.

Our problems are compounded by the term "literalist", not helped by its widespread confusion with the separate doctrines of biblical authority, divine inspiration, infallibility, and inerrancy. Even if we leave aside the problem of translation, there are many different possible "literal" readings of any given text, especially one as laconic as Genesis 1. What, for instance, does the phrase "In the beginning" really mean? Does it refer to the beginning of all time, or the beginning of earth history, or the beginning of human history? All three could claim to be "plain" or "literal" readings. Determining the correct reading introduced considerations external to the passage concerned, such as other parts of Scripture or evidence from the "book of nature". In the 1830s Buckland himself would tout the view that "the world had existed for millions of years antecedent to the Hebrew account of the creation" as being entirely compatible with "the most literal translations of the Mosaic legends".[72] What united the writers we are now examining was their own particular *variety* of literalism with respect to Genesis 1, according to which the earth was made around six thousand years ago in six solar days. My use of the word "literalist" should therefore be understood in this qualified sense.

These writers were not necessarily hostile to geologists of the new school. In the 1830s, certainly, the rising confidence and popularity of the new geology would force many literalists to take a more consistently oppositional stance. As James Moore has suggested, this opposition was partly clerical, "an urgent bid to maintain socially needful interpretations of the Genesis stories, hitherto dependent upon their own mediation and lately, it seemed, threatened by a band of upstarts [. . .] who proposed to go about their work as if Genesis did not exist."[73] But in the 1820s, many literalist writers were not particularly interested in opposing these upstarts; nor were they all clergymen. Penn was a distinguished literary scholar who had published on Classical and Byzantine literature as well as early Christian writings; Thomas Rodd was a poet and scholar; An-

72. Anon. 1836b, 634. For similar old-earth assertions of "literalism" see "A Scriptural Geologist" 1839, 27 n.; Christmas 1850, 105.

73. J. Moore 1986, 337; see also Rudwick 1986, 312. Moore restricts the term "scriptural geologist" to the anti-geological literalists of the period 1832–1860.

drew Ure was a respected chemist, science lecturer, and political econo-
mist; and James Rennie was a naturalist.[74]

Interpreters of Genesis did not fit easily into two sharply-opposed
camps of "literalists" and "liberals" (these terms were coined much
later), but formed a spectrum of debate. Take, for instance, the pamphlet
by "A Christian Philosopher" entitled *An Outline Sketch of a New Theory
of the Earth* (1824).[75] The pseudonym might tempt us to group this author
among the literalists, as might his frequent recourse to biblical exegesis,
his theory that the Yorkshire bone caves prove the violence of the Deluge,[76] and his emphasis on Genesis as the "inspired" account of a "faithful historian", let alone the uncompromising epigraph from 1 Corinthi-
ans: "The wisdom of this world is foolishness with God". On the other
hand, he also promoted the idea of a prehuman world ruled by enormous
animals and openly declared that Genesis 1 was not "a literal relation
of facts". More controversially, he also suggested that Noah was a black
man and denied the reality of eternal torment.[77]

Speculation about earth history could lead in many directions, and
literalists were not members of a single "party":[78] they would not necessarily have enjoyed each other's company at the dinner table. Penn's *Comparative Estimate of the Mineral and Mosaical Geologies* (1822) was influential,
but other literalists disagreed with some of his theories or recommended
that people should also read the work of his opponents. The second edition of George Young and John Bird's *Geological Survey of the Yorkshire
Coast* (1828) cautiously refers to Penn's book: "We are not prepared to
admit all that Mr. Penn has advanced; but his theoretical views appear to
us, on the whole, much more judicious than those which he combats."[79]
Rennie's anonymous children's book *Conversations on Geology* (1828) goes
to such lengths to recommend Penn's *Comparative Estimate* that bibliog-
raphers and historians used to attribute this book also to Penn[80]—an
unlikely attribution given that the *Conversations* are engagingly written,
whereas Penn's style is as pedantic as George Eliot's Mr. Casaubon, every
sentence weighed down by italics. Yet Rennie also encourages his readers
to read the works of Penn's theoretical opponents for their factual con-

74. For biographical information see *ODNB*. On Ure see also Copeman 1951; *DNBS*. On
Rennie see also *DNBS*. Useful summaries of the literalist arguments of Penn, Bugg, Young,
and Ure are given in Mortenson 2004.

75. The pseudonym is not to be confused with those of William Martin (author of anti-
Newtonian pamphlets and lectures, and brother of the artist John) or Thomas Dick (who used
the sobriquet as the title of his first book).

76. "Christian Philosopher" 1824, 28–9.

77. "Christian Philosopher" 1824, 23–7, 36–7.

78. For this view see Millhauser 1954, 73. On the plurality of literalist attitudes before the
1830s see Rupke 1983b, 48; Michael Roberts 1998, 234, 245.

79. Young and Bird 1828, 356.

80. Gillispie 1951, 264; Millhauser 1954, 71; Oldroyd 1979, 247; Merrill 1989, 84–5.

tent, since "the Mosaic geology is so recently published, that [. . .] it cannot be looked upon as established".[81] Before old-earth geology became popular enough as an alternative Creation-narrative to galvanize many literalists into furious reaction and fixed opinions, literalist accounts often manifested themselves as provisional "theories of the earth" to compete with John Playfair, Jean-André Deluc, and Jameson's Cuvier. To characterize their entire *oeuvre* as a "tedious little accumulation of polemic pamphlets" is (among other things) to ignore the great difference between the "movement" as it appears in the 1820s and its later, more oppositional manifestation.[82] We should take Miller's favourite term for such writers, "anti-geologists", with a large pinch of salt.[83]

There were, of course, some "anti-geologists" before the 1830s. Bugg's *Scriptural Geology* was an influential example, but its aggressive tone was unusual. Bugg's attitude harks back to the counter-revolutionary paranoia of the 1790s and 1810s, under whose influence Erasmus Darwin and James Hutton had been vilified, and around which Buckland had wriggled to establish geology at Oxford. Bugg energetically deplores every aspect of "modern *Geology*" (or simply "*Geology*", typically printed with table-thumping italics). The first volume concludes that "modern *Geology* cannot possibly exist consistently with a fair and literal construction of the *Word of God*", and that it contradicts sound science as well as "the plainest dictates of common sense".[84] The second volume alludes more directly to political fears, concluding:

> Whether, however, *Geologists* have any deep-laid or concealed design against Revelation, or not, the mischievous tendency of this modern Theory is too evident to pass unnoticed. [. . .] I ardently hope, however, that we shall [. . .] value more our Bible and its plain and obvious instruction; [. . .] and *"meddle not with them that are given to change."* [85]

Bugg seems to have been especially concerned about the growing elitism of these *"Geologists"*, imagined here almost as a secret society dedicated to the overthrow of the established order—like those which Edmund Burke had claimed were responsible for the French Revolution. Such imagery would be played on again during the 1830s (after another revolution in France, and violent Reform agitation in Britain). Bugg's aim was to channel the public's attention away from geology and back to the

---

81. [Rennie] 1828, 45, 290–1.
82. Millhauser 1954, 65.
83. For attacks on "anti-geologists" see "Fides" 1839; Miller 1857, 383–422.
84. [Bugg] 1826–7, I, 363. On Bugg see Mortenson 2004, 77–97.
85. [Bugg] 1826–7, II, 355–6.

Bible, whose meaning is available to all. In his treatise, writers who give ground to *"Geology"*, such as Buckland, Cuvier, and Sumner, are castigated mercilessly.

Uncompromising, suspicious, and shrill, *Scriptural Geology* has long served as an emblem of the phenomenon to which it lent its title;[86] but it is not typical of 1820s literalist writings about the earth. Other purposes besides chastising the new geology emerge as more significant for these writers. It is possible to tease apart three of these purposes, although in individual texts they are often combined. First and most obvious was the concern to shore up the Bible's traditional authority in the sphere of earth history. Bugg and Penn shared this aim, as well as a love of italics; but their procedures differed widely, since Bugg appealed only to the divine inspiration of Scripture, whereas Penn ostensibly tested the two "geologies" ("mineral" and "Mosaical") by the external standard of Baconian induction. Penn also displayed a critical attitude to the Genesis text itself, dismissing some passages as later scribal additions and going well beyond a conventionally "literal" interpretation in other particulars.[87] Accordingly, Penn was less dismissive of the new geology. Whereas Bugg denounced Buckland to the skies, Penn took a more courteous tone. But the difference went beyond scholarly etiquette: the two literalists directed their readers' attention in opposite directions. Bugg tried to coerce his readers into averting their eyes from the rocks, whereas Penn aimed to "stimulate" their "curiosity", exploiting the imaginative potential of geology in order to show that Moses was still the best guide to interpreting these "wonderful" and *"amazing"* monuments.[88] For this reason Penn's tome was seen by some as "sublime".[89]

A shorter work, the *Defence of the Veracity of Moses* by "Philobiblos" (the poet Thomas Rodd the elder), also begins with dark references to the French Revolution;[90] but Rodd's enthusiasm for the wonders of geology prevented him from taking an anti-geological stance. His book is divided into three chapters, "Of the Creation", "Of the Deluge", and "Of the Caverns in the Peak of Derby". Rodd's aim was to use these caverns to defend Moses's reliability on the subject of subterranean waters; but his object in visiting the Speedwell Cavern seems to have been more complex. He was, after all, a poet:

86. Millhauser 1954, 71; Haber 1959, 212; D. Allen 1976, 70.

87. Penn 1825, II, 209–43. Penn's use of textual criticism has been misunderstood as simple inconsistency or "suspect" practice (Miller 1857, 405–6; Mortenson 2004, 75; Lynch 2002b, xiii–xiv), but textual-critical tools could be used to support conservative as well as liberal interpretations of the sacred texts. Such ambiguities point up the problematic nature of the concept of "literalism" and call for further examination.

88. Penn 1825, I, 1–2.

89. [Rennie] 1828, 45.

90. [Rodd] 1820, i–ii.

A more frightful Cavern does not exist: by the help of water-rockets it may be seen to advantage. [. . . .] I could not avoid penning a few lines on this place, which I have visited several times with fresh delight and astonishment:—

> Hail to the generous spirit of the brave,
> That fearless rang'd the seas from shore to shore,
> [. . . .]
> No seas I range, no foreign soil I tread,
> But far below the earth pursue my way,
> Through regions inaccessible to day;
> The high-arched cavern tow'rs above my head;
> Whilst in the terrible abyss below
> Rush the wild waters, thundering as they flow.[91]

It should not surprise us that literalist geology so easily embraced imaginative enthusiasm. Thomas Burnet too had felt himself to be interpreting Genesis literally, and his sonorous *Sacred Theory of the Earth* had long served as a model of sublimity for subsequent poets.[92]

The second purpose claimed by literalist writers on earth history was to use geology to illustrate the Genesis accounts of Creation and the Deluge. This aim goes back to Burnet's *Sacred Theory* and was shared by many other early "theories of the earth", which are the ultimate ancestors of works like *A Short Narrative of the Creation and Formation of the Heavens and the Earth* (by "Philo", 1819). Such writings assume (rather than insist) that the reader accepts the Bible as primary narrative authority, but use science to deepen the reader's imaginative grasp of the events narrated in Genesis. Their appeal to antiquarian amazement recalls the parallel (and ultimately competing) efforts of the new school of geology:

> All of us must be anxious to know with certainty something, at least, of the creation and formation of the planet upon which we live and move; yet how few of us who possess an authentic history of the process, have given to it that consideration which a subject so grand deserves! Who can contemplate the process without amazement! Who can review and reflect upon it without acknowledging the grandeur and power of the Almighty [. . .]?[93]

Taken out of context, the only significant difference between this appeal of "Philo" and analogous exhortations by contemporary old-earth

91. [Rodd] 1820, 104–5.
92. Nicolson 1963.
93. "Philo" 1819, 3.

writers such as Bakewell is the location of that "authentic history".[94] For the former, it is in the book of Genesis; for the latter, in the strata. For both groups, unearthing the narrative excited a thrill of discovery. Like Burnet, these authors played on their readers' curiosity and love of spectacle.

A vital context for this enterprise was to be found in the art galleries and theatres of the 1810s and 1820s, where artists like John Martin reconstructed biblical catastrophes in the light of the latest historical and scientific researches.[95] Martin's approach was painstakingly antiquarian, aimed at reproducing the events as precisely as possible. He was particularly skilled at suggesting vast, almost infinite distances which dwarfed the human figures involved. With *The Fall of Babylon* (see Fig. 7.21) and *Belshazzar's Feast* (plate 8) in 1819–21, Martin had become one of the best-known artists of the day, especially in North America and France, where the adjective *martinien* was coined to describe his characteristic style.[96] He commanded massive audiences, often at commercial locations such as Bullock's Egyptian Hall. His mezzotint engravings brought him to an even wider public, and he was commissioned to provide illustrations for books such as Milton's *Paradise Lost* (see Figs. 7.18 and 10.4).

*The Deluge* (1826, Fig. 3.4) is a typical example of Martin's procedure, illustrating Genesis 7 with meticulous attention to detail. Like earlier theorists of the earth, he used astronomical speculations to shed light on this event, linking the Deluge with the conjunction of the sun, the moon, and a comet, just about discernible in the brighter patch of sky in the background.[97] Cuvier himself allegedly visited Martin's studio and expressed his approval of the last-mentioned detail (although Martin's early work owed nothing to Cuvier).[98] In the 1830s, Martin would find himself taking a step into deep time, restoring England's ancient saurians for Mantell and other geologists;[99] but British apocalyptic spectacle in the 1820s was firmly rooted in a literal interpretation of the Bible, with little or no input from the new school of geology. These pictures reinforced the Bible's narrative authority in the most impressive way imaginable.[100]

Apocalyptic spectacle proved a useful resource for those literalists who wrote earth history with a view to promoting the science among a wider public. In these texts, the controversy surrounding the age of the

94. See above, p. 130.

95. On Martin see Todd 1946; Balston 1947; Feaver 1975; Paley 1986; *ODNB*.

96. See also Figs. 7.11 and 7.20. On Martin's French reputation see Seznec 1964.

97. On this painting see Matteson 1981; Rudwick 1992, 22–4; below, pp. 289–90, 303–4.

98. Rudwick 1992, 256 n. 16. On Cuvier's alleged visit see Matteson 1981, 222–3; on Martin's influence on geologists see Pointon 1979.

99. Figs. 7.6, 7.8, and 10.3.

100. Poetic and iconographic aspects of apocalyptic spectacle are discussed in more depth in chapter 7.

Figure 3.4. *The Deluge* (1828), mezzotint engraving by John Martin, based on his painting of 1826. Prints like this were expensive, with proof copies typically selling at £2 10*s* 6*d* (= *c*. £440 or $875 today), but they enabled Martin to reach a wide audience. British Museum Prints & Drawings, Mm.10.3.

earth became less prominent: the authors deployed geology's wonders to attract the uninitiated to a new and exciting science. In their view, of course, the former worlds opened up by Cuvier's and Mantell's gigantic fossils were not prehuman, but antediluvian; yet they were no less "wonderful" for all that. The antediluvian world outside Eden and the Fertile Crescent was as thrillingly alien for these writers as the world before Eden was for Cuvier. The popularity which geology won by the end of the 1820s was due in no small measure to such writings, and we now examine three of them in more detail.

## Three Literalist Treatises

These three books were all written by Scotsmen and published in England. Formally and stylistically, however, they are quite different, and their authors would certainly have been surprised to find themselves occupying a single literary category here purely because of their geohistorical conservatism. *A New System of Geology* (1829), written by the chemist Andrew Ure and published in London, was a fat octavo of over six hundred pages which proved popular as a compendium of facts.[101] Its self-confessed "novelty" consisted in its arrangement of these facts into a single philosophical argument, intended "to draw forth the accordances of Science and Revelation".[102] Unlike Penn and Bugg, however, Ure also aimed to impart useful knowledge about geology for "general perusal", and it is this aim which dominates the treatise rather than a need to defend Moses at every turn. Ure was a Fellow of the Royal Society and a member of the Geological Society, and he lectured to working men on physical science at Glasgow's Andersonian Institution. His book was designed to render geology, especially fossils, "attractive to the English reader".[103] It succeeded well enough to be excerpted by later popularizers, despite a thunderous put-down by Adam Sedgwick in 1830 (which itself testifies to the volume's success).[104]

*A Geological Survey of the Yorkshire Coast* (1822) was written by two fossil collectors, the clergyman George Young and the artist John Bird.[105] It was aimed at a wealthy and genteel audience of fossil collectors, the same class of people who bought Parkinson's *Organic Remains* and Mantell's early treatises. This book was published in Whitby, where Young was a Presbyterian minister. It is a quarto, lavishly illustrated by engravings of fossils. Young was a prime mover in the Whitby fossil market, which gave him additional reasons for promoting local geology, besides his duty as a literalist minister to counter Buckland's hyaena-den theory. An earlier volume of Young's, his *History of Whitby* (1817), had also included an essay on geology by Bird which apparently stimulated local interest in fossil collecting.[106] The new book was also admired, even by old-earth geologists like Sedgwick, and went into a second, enlarged edition in 1828.

A much smaller book, the anonymous duodecimo *Conversations on Geology* (1828), was probably written by the natural-history lecturer James Rennie and was published in London by Samuel Maunder.[107] This book was modestly illustrated, but with some engravings done in colour. It takes the form of conversations between an omniscient mother and her

---

101. See Anon. 1831, 80, which faults Ure for not being comprehensive enough but claims that "as an introduction to geology [. . .] it will be found highly useful".

102. Ure 1829, lv.

103. Ure 1829, viii.

104. Hack 1835, 166, 214, 270; Blewitt 1832, 108; Sedgwick 1826–33a, 207–10. On Sedgwick see Clark and Hughes 1890; *ODNB; DNBS*.

105. On Young see Osborne 1998; Knell 2000, 137, 142–3; *ODNB*.

106. Knell 2000, 44.

107. This attribution was made in Anon. 1829d and confirmed by John Thackray. It was later noted in print by J. Secord (2000b, 381). As discussed above (p. 139), this book has since often been attributed to Granville Penn because of the prominence of his name on the title-page (Fig. 3.7).

two well-behaved and inquisitive children, Edward and Christina. Edward is characterized chiefly by his curiosity and sceptical attitude towards received wisdom, while Christina is anxious to retain her romantic fondness for landscape and asks fewer questions. The book was aimed at securing geology a niche as a science suitable for children, and equally appealing to boys and girls. Modelled on Jane Marcet's extremely popular *Conversations on Chemistry* (1806),[108] it claimed to be the first treatise on geology written specifically for children, and was popular enough to be reprinted twice.[109]

All three books were written in the late 1820s, and each reflects the extended fossil repertoire that geology had gained by that time. They also reveal a permeable boundary between literalist and non-literalist geologies. Rennie and especially Ure made positive use of the work of their theoretical opponents. Rennie encouraged his readers not to confine their reading to literalist ("Mosaic") geologists, and his motherly persona Mrs. R. recommends a heterogeneous group of geologists as "the most instructive, though not always the most interesting authors": Saussure, Humboldt, Webster, Macculloch, Conybeare, Brongniart, Von Buch, "and many others". These writers are distinguished above all by the fact that they "describe what they have examined in nature, sometimes, though not always, unbiassed by theory"—this last qualification being the only criticism voiced.[110] On the other hand, Buckland is singled out for moderate criticism for his "singular" and ultimately "fabulous" hyaena den theory, which few literalists of the 1820s could avoid confronting.[111]

Ure openly admitted his debt to the gentlemanly elite. He acknowledged the "inestimable" *Outlines of the Geology of England and Wales* (1822) by William Conybeare and William Phillips as the main source for his own account of the secondary strata.[112] The *Outlines'* avoidance of theoretical issues allowed Ure to yoke its stratigraphy to his literalist cosmology: Gillispie called this coupling "forced",[113] but such individualistic syntheses were common at a time when most geologists were lying low on theoretical questions. More striking was Ure's enthusiasm for the fos-

108. Marcet 2004; for an acknowledgement see [Rennie] 1828, 4. Marcet's book had reached an eleventh edition by 1828; for discussion see Myers 1997 and Fyfe 2004b. On science for women and children see Gates 1998; Fyfe 2000.

109. See [Rennie] 1828, 4: "Geology is still a new science, and has not yet produced any popular writer". Earlier examples, however, include [Scafe] 1820a; 1820b; Mawe 1821; [Lowry] 1822. Rennie's third edition was published in London by J. W. Southgate in 1840. The second edition of 1828 was identified by John Thackray (handwritten note in Thackray's copy of this edition, property of Maggs Bros., London).

110. [Rennie] 1828, 290–1.

111. [Rennie] 1828, 333–6. See also Rupke 1983b, 42–50.

112. Ure 1829, vii; compare [Rennie] 1828, vii–xi, 290.

113. Gillispie 1951, 194.

sil researches of Cuvier, Buckland, and Mantell. His preface claims that he has dwelt on these for their imaginative value:

> The author has likewise diligently availed himself of the ample means accumulated in the *Ossemens Fossiles* of Baron Cuvier, the Philosophical and Geological Transactions, &c. of enlivening the dark catacombs of the earth, by interspersing among his descriptions of its mineral planes, an account of their ancient tenants. By transferring to his pages, systematic exemplars of the analytical science displayed by the great naturalist of France, in restoring antediluvian zoology, he expects to make them peculiarly attractive to the English reader.[114]

In other words, Cuvierian restorations of these beasts would help the book sell. The ease with which fossils fitted literalist paradigms is borne out by Ure's effortless allusions to the work of old-earth geologists. A section on "sea lizards" opens by paraphrasing the equivalent section of Cuvier's *Ossemens fossiles,* and later quotes the most dramatically pictorial part of Conybeare's 1824 Geological Society paper on plesiosaurs, likening their behaviour to that of swans.[115] Reading the section on the "Marvellous Iguanodon of Mantell" on its own, isolated from Ure's theoretical framework, one could be forgiven for thinking Ure a follower of the new geology in all its particulars. One would never guess that the "ancient epoch" to which he alludes was, in his view, a period in human history—particularly as the section concludes with a eulogy on Cuvier, praising his "soundness of inference" and "general enlargement of thought".[116]

The imaginative possibilities opened up by Cuvier were used as enthusiastically by Ure as they had been by Bakewell, Buckland, and Mantell. Like these men, Ure elicits wonder with a rhetoric of spectacular display, romanticizing this "ancient empire of the dead" with the help of attractive illustrations (Fig. 3.5). In the Introduction, Cuvier himself guides Ure and the reader through the catacombs:

> In accompanying him [Cuvier] through the dark cemeteries of the earth, a mysterious gleam from the primeval world penetrates our soul, and solemnly awakens its deepest faculties. We seem to walk among new orders of beings, endowed with extraordinary forms [. . . .] These all speak of a world unlike our own, the fashion of which has long passed away.[117]

114. Ure 1829, viii.
115. Ure 1829, 226–7, 242–3. See p. 330 n. 20 below.
116. Ure 1829, 243–7.
117. Ure 1829, liii.

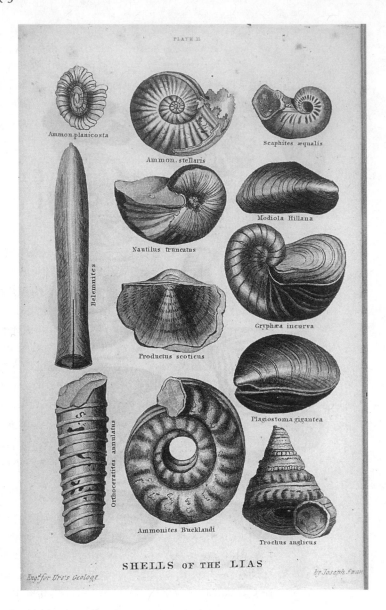

PLATE II.

Ammon.planicosta

Ammon. stellaris

Scaphites æqualis

Belemnites

Nautilus truncatus

Modiola Hillana

Productus scoticus

Gryphæa incurva

Orthoceratites annulatus

Plagiostoma gigantea

Ammonites Bucklandi

Trochus anglicus

SHELLS OF THE LIAS

Engᵈ for Ure's Geology          by Joseph Swan

*Figure 3.5.* One of the handsome plates in Andrew Ure's *New System of Geology*, showing fossil shells of the Lias. Ure 1829, plate 2.

This "world" is, if anything, even more alien than Mantell's, since here Ure relaxes his grasp of comparative anatomy and presents a veritable gallery of monsters. The "crocodiles furnished with fins, but no feet" and "lizards of whale-like dimensions" are familiar enough as ichthyosaurs and Iguanodon; but the "sloth [. . .] of the gigantic stature of a rhinoceros" (the ground-dwelling giant Megatherium) was apparently capable of imitating its arboreal cousins, "suspending itself [. . .] from trees of colossal growth" on its "enormous arms and claws". His saurian quadrupeds "bearing wings on their toes"—a slightly too literal translation of the word "pterodactyle"—are more reminiscent of Bosch's *Garden of Delights* than Cuvier's fossil menagerie.[118] This passage aims to catch readers' curiosity and tempt them to read on. The main body of the book is more restrained, but sensationalism persists in subtitles and running headers: "Two Monstrous Sea Lizards", "Enormous Megalosaurus of Buckland", "Crocodiles Crushed to Death".[119]

The crushed crocodiles call for some explanation. They bring us back to the question of how Ure fitted fossils into his literalist chronology. According to Ure, such fossils testify to the massive convulsions with which God had periodically afflicted the earth as punishment for the sins of Cain's "apostate brood", that is, antediluvian humans. These catastrophes formed the strata, whose deposition was extraordinarily violent, crushing animals to death where they stood (hence the damaged state of many fossils). The most recent cataclysm was the biblical Deluge, as the chapter-heading "Animal Remains, or Ruins of the Deluge" indicates.[120] All fossil species were part of a "primeval stock" destroyed in the Flood, after which the face of the earth was "revivified" in a new burst of creative activity, preparing for the spread of human civilization.[121] This apocalyptic vision is reminiscent of Martin's *Deluge*, first published as a mezzotint engraving the year before Ure's treatise appeared. As in Martin's biblical scenes, there is a clear moral, channelling the thrill of spectacle into religious reflection:

> But that world, the victim of sin, will not have perished in vain, if its mighty ruins serve to rouse its living observers from their slumberous existence, if they lead them to meditate seriously on the origin and end of terrestrial things, and [. . .] the works and ways of Providence.[122]

A similar rhetorical movement from apocalypse past towards apocalypse future was employed by other literalists in the 1820s. In 1822 Penn recommended geology for its imaginative potential, arguing against recent claims to the effect that the evidence of geological convulsions was *"of less immediate importance to us,* than the events which raised the *obelisk* and the *pyramid,* the *temple* and the *tower"* (claims which themselves confirm the extent to which geology and antiquarianism were being viewed as potentially competing branches of the same endeavour).[123] Sallying in

118. Ure 1829, liii.
119. Ure 1829, 226, 221, 259.
120. Ure 1829, 348–9, 498, vi, 501.
121. Ure 1829, 500–4.
122. Ure 1829, liii–liv.
123. Penn 1825, I, 268.

on behalf of "the *spires of primitive granite*", Penn asserted in his characteristic style that human antiquities "tell us only of that which is *gone by
for ever*". But geological monuments reveal past revolutions, "leading the
thoughts by an indissoluble chain from that which '*was*' and '*is*,' to that
which '*is to come*,' [. . . .] to *the revolution which still impends*", so that "the
thoughts will naturally travel *forward*, in contemplation of *another earth* [.
. .] not figurative or allegorical, but real and habitable".[124] Aesthetic pleasure merges with religious sentiment as the strata disclose the shape of
salvation history. Penn placed this argument in his book as a final salvo
against those who claimed geology was unimportant; Ure positioned his
own apocalyptic reflections at the end of his Introduction, inviting readers to venture further and preparing them to react with the right kind of
awe, which he later topped up with a melodramatic passage describing
the ruins of the Deluge.[125]

Sacred history is less prominent in the books by Young and Rennie,
who preferred to restrict their religious reflections to natural theology.
Buckland's inaugural lecture was still, for many, the classic statement
of this position. Despite his reservations about Buckland's "fabulous"
hyaena-den theory,[126] Rennie quoted Buckland's inaugural lecture at
length when introducing the argument from design: "On this subject
we cannot do better than make a few extracts from the observations of
Professor Buckland".[127] Four pages of Rennie's seven-page preface are entirely taken up with defending this argument, and, in a pattern which is
now becoming rather familiar, his strategy is distinctively compilatory.
From the moment that the subject is raised until the end of the introduction, Rennie himself hardly contributes anything to the text. Buckland
is given the lion's share of the argument, after which its significance is
transmuted into poetry. Rennie invites his readers to sing along with
eight lines from Robert Montgomery's recently-published devotional
epic *The Omnipresence of the Deity* (1828):

May we not, then, well exclaim in the words of a modern poet,

"There is a voiceless eloquence on earth,
Telling of Him who gave her wonders birth; [. . .]"

and so on.[128] With these lines Rennie's preface abruptly ends. The technique of ending with a pious peroration soaring into verse recalls Mantell's *Fossils of the South Downs*.

124. Penn 1825, I, 269–71.
125. Ure 1829, 505, quoted and discussed below, p. 321.
126. [Rennie] 1828, 336.
127. [Rennie] 1828, vi.
128. [Rennie] 1828, xi.

Young and Bird ended their own chapter on Yorkshire organic remains in a similar vein of aesthetic piety, using the increasingly common protest that they "cannot" end the chapter "without" making the spectacular nature of this science explicit in some form of peroration:

> We cannot close this part of our Work, without expressing our astonishment, at the rich variety and beauty of the animal remains discovered in our strata [. . . .] Whether we survey the largest, or the most minute [. . .] we see matter for the highest admiration, and the warmest praise. "Great and marvellous are thy works, Lord God Almighty!" [129]

This quotation from the apocalypse sequence in the biblical Book of Revelation (15.3) performs an analogous function to Rennie's use of Montgomery: all are apparently invited to sing with the heavenly choir. In their Introduction, too, Young and Bird defend their science's religious value at length, and to reinforce the point they break into quotation from two poetical books of the Bible, Psalms and Job. The quotation from the Psalms casts a holy aura around intellectual pleasure:

> while they [geological researches] enlarge the bounds of our knowledge, and present a wide field for intellectual employment and innocent pleasure, they may serve to conduct us to that glorious Being, "who by his strength setteth fast the mountains [. . . .]" [130]

By 1828, both natural theology and the argument from intellectual pleasure had found their way for the first time into Bakewell's *Introduction to Geology*. Young and Bird went on to defend the speculative imagination in the Bakewellian manner: "if her [imagination's] wanderings prove a stimulus to exertion, producing a larger accumulation of facts [. . .] they will forward, rather than retard, the progress of real knowledge".[131] And this knowledge, far from "savouring of presumption", forms "part of the homage due to the Creator, whose infinite perfections are more and more illustrated and displayed, in proportion as his works are explored by the light of science."[132] These words "illustrated" and "displayed" suggest the distinctively visual nature of geology's appeal in the hands of its literalist apologists, no less than with its old-earth proponents.

However, one significant difference between the two schools of geology begins to emerge here. Literalist geohistory, with its definite, knowable time-scale, could not partake of the aesthetics of the infinite. Worlds

129. Young and Bird 1822, 278; 1828, 310.
130. Young and Bird 1822, 2; 1828, 2.
131. Young and Bird 1822, 2–3; 1828, 2–3.
132. Young and Bird 1822, 4–5; 1828, 4–5.

inhabited before man made for a less anthropocentric cosmos, and this is reflected in the way the two groups conceptualized the aesthetic status of their science. Its relation to astronomy, the sublime science *par excellence*, provides a useful touchstone. Cuvier, Buckland, Davy, and others had used the analogy with Newton and astronomy to predict that geology would enable the mind to travel as far in time as astronomy propelled it in space. Geology could thus partake of an aesthetic of wonder by which unlimited immensity (spatial or temporal) bewildered the mind into a temporary and pleasurable loss of rationality. Aesthetically, geology was claimed as the equal of astronomy (or, for an astronomer like John Herschel, almost its equal).[133] Such rapturous bewilderment was denied to the literalists, for whom earthly time was always calculable by referring to Scripture or the chronologers. Accordingly, the analogy with astronomy was weaker, and the stars retained their aesthetic superiority. Young and Bird therefore began their Introduction by appealing to their readers' utilitarian and parochial instincts to draw their attention from the stars down to the mud:

> [Geology], though less attractive than the pursuits of the astronomer, has an equal claim to our attention. There is something, indeed, peculiarly fascinating in the contemplation of those bright orbs which bespangle the sky; and the imagination more willingly attends the understanding, in soaring through the regions of immeasurable space [. . .] than in viewing the features [. . .] of the earth on which we tread [. . . .] Yet this science, fewer as its attractions are, is by no means uninteresting, or unimportant. While we study the laws which govern the remotest planets, shall we remain ignorant of the planet which we inhabit?[134]

The new school of geology was already beginning to make use of the attractions of the "immeasurable": not holding to a bounded time-scale, they could quite simply create a bigger spectacle. In the 1830s this is what they would do, much to some literalists' annoyance. Size could be made to matter: in the late 1820s the French novelist Balzac was already musing on Cuvier's "sublime" ability to multiply temporal distances by adding noughts to the number of years which had elapsed in deep time.[135]

Rennie took a different approach. Within the confines of literalism he turned geology's lack of ultimate sublimity into a positive virtue. Towards the beginning of *Conversations on Geology*, Edward calls astronomy "sublime" because "it raises our thoughts above the earth and its little

---

133. Herschel 1830, 287; Miller 1835, 48; Francis 1839, 2–4.
134. Young and Bird 1822, 1; 1828, 1 (replacing "fewer" with "scanty").
135. Balzac 1964, 25.

scene of change and bustle, and leads the mind to contemplate the starry universe and the infinity of space";[136] but his mother has a ready answer. Geology's more limited scope makes it more suitable for human study:

> I think Geology is, perhaps, better fitted for our limited comprehensions than astronomy; [. . .] it is easier to bring the mind to rest on the comparative littleness of the earth at its creation, than to let our thoughts travel abroad through the boundless fields of infinite space. When we descend to the earth, we feel ourselves more at home; we are not so overpowered by sublimity as in the contemplation of astronomy; we can think more calml[y] and reason more at ease; and we can trace the finger of God more visibly, perhaps, because more nearly.[137]

This passage, however, forms part of a larger discussion over geology's aesthetic status in relation to other sciences. Edward's favourite science is astronomy because it is "sublime" and adventurous, while Christina loves botany because it is "beautiful":[138] these two established sciences had become clearly gendered as educational subjects for children. To make geology attractive to both boys and girls, Rennie situates it within a gender-neutral space between overpowering sublimity and small-scale beauty, devoting much didactic energy to showing how geology is "romantic".[139]

Both Christina and Edward are puzzled by their mother's use of the term "romantic" to describe a science, although Christina reacts with stereotypically gendered enthusiasm:

> I am sure I shall like it, for I delight in romances; and whenever I hear the word, I think of the Happy Valley in "Rasselas," Robinson Crusoe's Island, or the Enchanted Gardens of Armida; but I always thought there were no romances in philosophy.[140]

To explain the odd mixture of "romance" and "philosophy" which geology supposedly represents, Mrs. R. now embarks on a detailed defence of its imaginative importance. But, as we have already seen, "imagination" was a double-edged concept when applied to scientific writing,

---

136. [Rennie] 1828, 7.

137. [Rennie] 1828, 8–9.

138. [Rennie] 1828, 6–7.

139. Geology's aesthetic status was commonly sited in this "picturesque" or "romantic" middle position between astronomy and botany: see Anon. 1821b, 430–1. On gender and scientific dialogues see Myers 1989.

140. [Rennie] 1828, 5. Rasselas and Crusoe recur as part of the constitution of geology's "romantic" appeal among the non-literalist geological writings of the 1830s and 1840s.

potentially both positive and negative. In what follows we see Rennie guiding his readers through these troubled straits to reach some stable definition, in his own terms, of geology's "romantic" appeal. Theories of the earth are discredited, on the one hand, as "pretty romances"; on the other hand, Burnet's theory is vindicated on aesthetic grounds, albeit only as poetry and not as serious philosophy. For when Christina scoffs at the notion of a perfectly smooth globe as tedious and unpoetical in the extreme,[141] her mother tartly retorts, "On the contrary, his book is highly poetical, and is now read on that account alone."[142]

However, when vindicating "true" geology (i.e. "Mosaic geology"), Mrs. R. is entirely positive about its "romance", appealing both to Christina's love of scenery and Edward's dreams of distant stars:

> I [. . .] call Geology romantic, because it not only leads us to travel among the wildest scenery of nature, but carries the imagination back to the birth and infancy of our little planet, and follows the history of deluges and hurricanes and earthquakes [. . . .] Would you not think it romantic to travel [. . .] among mountains and valleys, where the tempests have bared and shattered the hardest rocks [. . .] ? And would you not think it romantic to dream about the young world emerging from darkness, and rejoicing in the first dawn of created light? To think of the building of mountains, the hollowing out of valleys, and the gathering together of the great waters of the ocean? And will it not be romantic to discover the traces of the ancient world before the time of Noah, in every hill and valley which you examine?[143]

The verbs are revealing: geology "leads" and "carries" us, so that we "dream about", "discover", and "travel among" stark landscapes and sublime historical processes. Scientific inquiry becomes a form of time travel. Strange as it may seem, the writer whom Mrs. R. recommends for this purpose is Penn, whose system "will give you much more sublime views of the creation than are to be found even in the inspired poem of Milton; and that is saying a great deal."[144] It certainly is saying a great deal, as is the later suggestion that Burke, the best-known eighteenth-century theorist of the sublime, would have called Penn "the 'architect of ruin'" (a comment which elicits another eager squeak from Christina).[145] Such comparisons may seem excessive: for all his qualities, Penn was

141. Compare Miller 1859, 192–3.
142. [Rennie] 1828, 11–12.
143. [Rennie] 1828, 7–8.
144. [Rennie] 1828, 45.
145. [Rennie] 1828, 295.

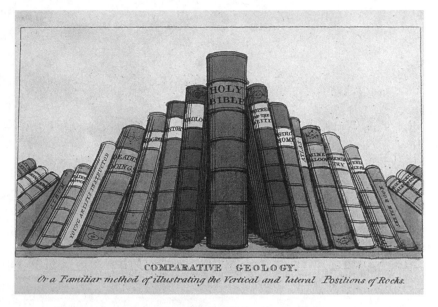

COMPARATIVE GEOLOGY.
Or a Familiar method of illustrating the Vertical and lateral Positions of Rocks.

*Figure 3.6.* The strata represented as a bookshelf. [Rennie] 1828, plate opposite p. 40.

surely one of the dullest writers of his generation. But the discrepancy only serves to underline the extent to which the subject-matter of geology was already seen as sublime and poetical in itself, independently of literary style. For Rennie, Penn was no learned pedant, but a poet with the power to call up Miltonic views of earth history.

These points are all made at the beginning of Rennie's book, in order to induce the reader to read on. The rest of the book maintains this aesthetic appeal with, for example, scattered quotations from poetry: "Conversation Fifth" opens with a twenty-five-line quotation from Akenside's *Pleasures of Imagination,* here quoted by Edward (much to his mother's admiration) to prove to the wilting Christina that scientific knowledge increases rather than reduces one's enjoyment of rainbows.[146] Although most of the book is devoted to evaluating theories and expounding facts, its conversational style bears out the preface's promise to render the science "inviting".[147] Particularly engaging are the twelve pictures. One of these (Fig. 3.6) represents the arrangement of strata in the crust as a pile of books leaning towards the centre.[148] At first glance this emblematic image seems to embody the Enlightenment analogy of the "book of na-

146. [Rennie] 1828, 77–8.

147. [Rennie] 1828, v.

148. [Rennie] 1828, plate facing p. 40. In the third edition the books have vanished from the equivalent plate, which instead shows a coloured stratigraphic section of the Brocken mountain in Germany, with the strata leaning in towards the centre as in the earlier picture ([Rennie] 1840, plate facing p. 33).

156
157

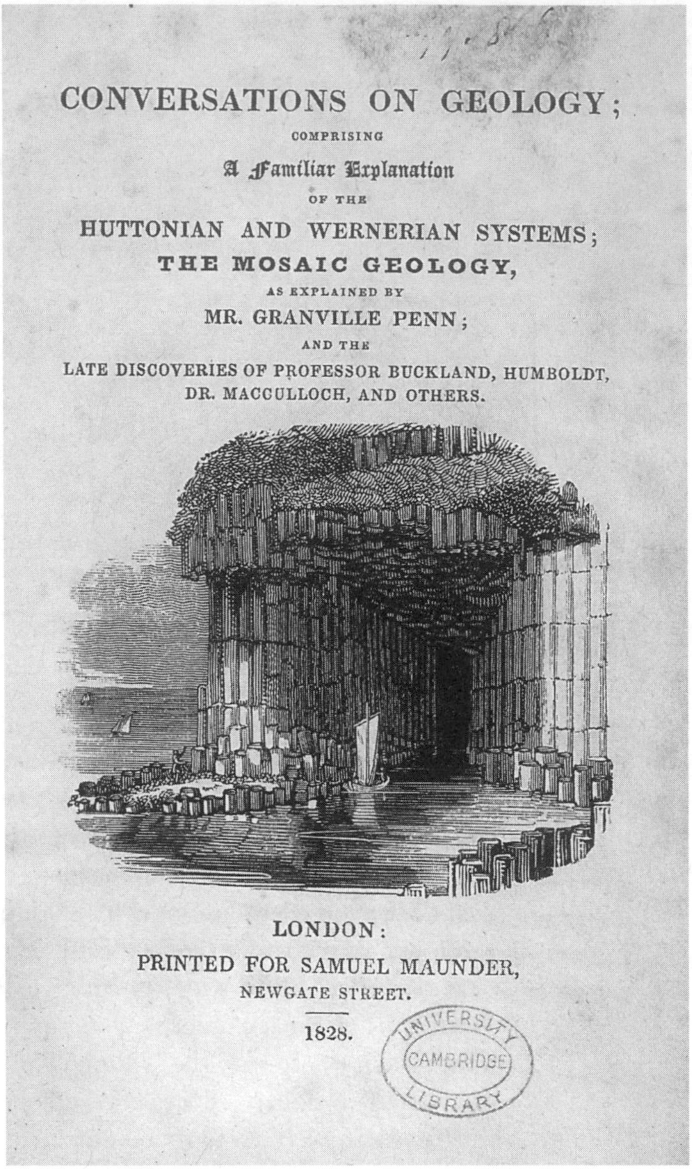

CONVERSATIONS ON GEOLOGY;

COMPRISING

A Familiar Explanation

OF THE

HUTTONIAN AND WERNERIAN SYSTEMS;

THE MOSAIC GEOLOGY,

AS EXPLAINED BY

MR. GRANVILLE PENN;

AND THE

LATE DISCOVERIES OF PROFESSOR BUCKLAND, HUMBOLDT,
DR. MACCULLOCH, AND OTHERS.

LONDON:

PRINTED FOR SAMUEL MAUNDER,
NEWGATE STREET.

1828.

*Figure 3.7.* Title-page of Rennie's *Conversations on Geology* (1828) showing Fingal's Cave.

ture", but a glance at the titles reveals a more traditional message. The Bible is central, physically supporting the books on other sciences (geology, history, geography), just as its content is the basis of all true scientific systems. Note also the prominence given to *Death's Doings* to the left, and, nestling alongside the Bible, Montgomery's devotional epic *The Omnipresence of the Deity* (quoted in Rennie's preface).

Most of the engravings portray famous and scenic geological localities such as the Grotto of Antiparos, Alum Bay in the Isle of Wight, and Fingal's Cave, which also adorns the title-page (Fig. 3.7). Perhaps surprisingly, fossils take up only a small part of the book and are shown in only two engravings. But Rennie was well aware of fossils' spectacular potential, and it is with fossils that the book abruptly ends, incorporating Cuvier's account of the extraordinary mammoth specimen found frozen, flesh and all, in the Siberian permafrost. The three then discuss whether this specimen proves the existence of "an extinct race of arctic elephants", but the mother offers no definite conclusion. On this speculative note she leaves her children to go on a "long visit" somewhere else, leaving behind some geology books which (she hopes) they will now be sufficiently motivated and informed to read without her help.[149]

The children's appetite is whetted most of all by their mother's revelation that the huge bones once thought to be "those of giants, or of *fallen angels*" really belonged to extinct animals. These animals inhabited the "unfruitful" land without Eden, "infertile, barren, and 'cursed'", until they all perished in the Deluge "to give place to a new and blessed land". The divine curse under which they lived appears to be reflected in Rennie's insistence (citing Cuvier but manifestly contradicting him) that most of them were "beasts of prey".[150]

The facing illustration to page 364 (Fig. 3.8) shows one such beast, labelled "Skeleton of a Gigantic Antediluvian Beast of Prey". This is, of course, the peaceable ground-sloth Megatherium, its skeletal form looking especially monstrous since it has been plucked straight out of an engraving showing the specimen in the Madrid museum and deposited (still in the same lumpish pose) in a tropical landscape, with a worried-looking living elephant in the background. Where Ure had set this monster improbably swinging in the trees, Rennie makes it play the predator, setting it upon its mammalian contemporaries.

Among old-earth and young-earth writers alike, fossil monsters were used to bring out the romance of geology. Rennie's target readership made this still more appropriate. Mantell and Lyell had observed in 1826 that a verbal restoration of these creatures might possess "more of the character of a romance" than of scientific fact;[151] but Rennie took this further. These ancient beasts of prey inspire Edward to suggest excitedly that "the ancients were not so fabulous as we think them, in talking of their harpies, griffins, and dragons." His mother seems to agree, pointing out that a huge lizard not unlike "the great dragon of ancient times, or the dragon of St. George" has been discovered in Germany (this was the "Maastricht animal"). Here the analogy between fact and fiction ap-

149. [Rennie] 1828, 367–71.
150. [Rennie] 1828, 362–4.
151. Mantell 1827, 83; [C. Lyell] 1826b, 524.

proaches actual identity.[152] The notion that modern geologists were actually digging up the dragons of ancient fable is given the stamp of Rennie's own authority in the contents page and echoed in the running header: "The Bones of the great Dragon discovered".[153] What more could the young reader wish for?

This strategy, channelling the thrill of dragon-legends into the new science, recalls Mantell's suggestion that the Plesiosaurus "only required wings to be a flying dragon",[154] and may be seen as a reptilian counterpart to Ashe's use of Indian monster-myths and Parkinson's treatment of English folklore. Books like these helped bring about the rebirth of the dragon as a usefully colourful category in mid-nineteenth-century popular science, fulfilling its age-old role as "large extinct reptilian monster" in a newly scientific guise. In the Victorian period St George, the dragon-slaying hero of popular romance and pantomime, would be replaced by "knights of the hammer", the gentlemen of science who set out to subjugate the subterranean world.[155]

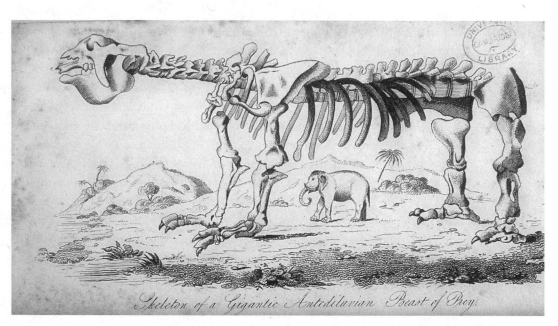

*Figure 3.8.* "Skeleton of a Gigantic Antediluvian Beast of Prey". [Rennie] 1828, facing p. 364.

152. [Rennie] 1828, 365; see Rupke 1983b, 219–20. Compare Mayor 2000, 15–53.

153. [Rennie] 1828, xxii, 365. On the continuing currency of the St George legend see Green 1968.

154. Mantell 1827, 84.

155. Compare Fig. 10.5. On geology as subjugation see J. Secord 1982; Shortland 1994; Sommer 2003.

## Conclusion

By the end of the 1820s, popularizers of both schools were drawing on a new repertoire of fossils to stimulate interest in the science. Hyaenas were old news, and a procession of bizarre extinct reptiles lurched into the limelight. These dragons of the new age, along with the continuing debates over diluvial theory, opened up new possibilities for romance and imagination, which were tested over the course of the decade by writers on both sides of the cosmological divide. The literalist writers we have been examining mined a rich vein of sentiment and spectacle already being exploited by commercial fossil exhibitions. Ure, Young, Bird, and Rennie all offered their readers a gallery of curiosities, combined with *martinien* glimpses of the antediluvian world, and of the Flood that wrecked it. At a time when theatres and art galleries were deluged with scenes of apocalypse, such spectacle was bound to appeal, particularly because it did not challenge widely-held views of Scripture. Provincial museums, which were sprouting vigorously all over Britain, displayed the relics themselves. Hyaena bones from Kirkdale, for instance, could be seen in the Whitby Museum, they were presented in support of Young's theory rather than Buckland's.[156]

In the late 1820s Buckland himself was making less noise, at least in public, than in the years immediately surrounding his Kirkdale analysis. His diluvial theory had come under fire from his geological colleagues as well as from literalists, and in his Oxford lectures he was becoming warier about tying diluvial phenomena down to Noah's Flood, although his public reputation still revolved around his cave researches. Other followers of the new school of geology were also lying low: almost the only recent British books in which a determined effort was made to popularize the science's grand historical aspects were new editions of Jameson's Cuvier (1822 and 1827), a new edition of Bakewell's *Introduction* (1828), and Mantell's two treatises of 1822 and 1826. This last book was the only one apart from Cuvier's latest (and still untranslated) editions of *Ossemens fossiles* (1821–4 and 1825) to incorporate detailed verbal restorations of the antediluvian world and its inhabitants, although speculations about the lifestyles of individual creatures were now becoming more frequent in scientific periodicals. Of the writers listed here, only Bakewell attempted a narrative overview of antediluvian history, and that only in a brief thumbnail sketch. The new story of creation was only beginning to take shape, and remained largely under wraps.

The old story, on the other hand, was enjoying a new lease of life. A heterogeneous body of writing was rapidly developing, helping to cre-

156. Knell 2000, 175.

ate a climate of enthusiasm for the science among a well-heeled readership. Literalist earth-science was widely disseminated, not only in books,
pamphlets, and sermons, but also in museums, public lectures, and literary reviews. It was, in part, this efflorescence which galvanized the
cautious geologists of the new school into writing more popular works
of their own. In the 1830s this situation would escalate into a battle for
middle-class hearts and minds, resulting in a more oppositional stance
among many members of both groups. This particular battle was ultimately won by the new school. As ever, history would be rewritten by
the victors. They would paint a black picture of the literalists as a motley
band of intellectual Luddites, a historical myth that unfortunately survives to this day.

Perhaps more surprisingly, in their own subsequent work the old-
earth geologists would find themselves drawing liberally on the apocalyptic and biblical images associated with literalist earth-science. When,
in 1829, Buckland read a paper to the Geological Society on an unusually
well-preserved pterodactyle found at Lyme Regis, he brought the beast to
life using a technique not normally associated with this formal setting:

> thus, like Milton's fiend, all-qualified for all services and all elements, the
> creature was a fit companion for the kindred reptiles that swarmed in the
> seas or crawled on the shores of a turbulent planet.

> ———————————————————————————"The Fiend,
> O'er bog, or steep, through straight, rough, dense, or rare,
> With head, hands, wings, or feet, pursues his way,
> And swims, or sinks, or wades, or creeps, or flies."
> *Paradise Lost,* Book II. line 947.[157]

Clearly, Buckland still meant to be in the vanguard when it came to
imagining the ancient earth. Earlier in the same paper he mused on the
strangeness of "our infant world", with its "flocks of such-like creatures
flying in the air, and shoals of no less monstrous Ichthyosauri and Plesiosauri swarming in the ocean, and gigantic crocodiles and tortoises crawling on the shores of the primæval lakes".[158] This sentence, like Mantell's
description of the Iguanodon country, points towards more all-encom-
passing "views" than had hitherto been attempted, and may even have
inspired the first fully-fledged "scene from deep time" (De la Beche's
*Duria antiquior* of 1830, Fig. 2.2, depicting these same creatures swarm-

---

157. W. Buckland 1835, 219.

158. W. Buckland 1835, 218. Buckland would later combine these passages into a single
vivid description (1836, I, 224–5).

ing on the Dorset coast).[159] Yet the means by which Buckland revivified his pterodactyle was to dress it in Miltonic garb, dovetailing geohistory with sacred history. This technique recalls Ure's apocalyptic visions and Mrs. R.'s dreams of the antediluvian world, which drew on a rich tapestry of existing associations automatically evoked by scenes and figures from sacred history. By redirecting these associations on a metaphorical rather than literal level, Buckland perhaps hoped to outdo the literalists at their own game. But the opening shot in this campaign had already been fired two years before, by Buckland's star pupil, Charles Lyell.

159. It also inspired Howman's earlier painting of the pterodactyle as dragon (Fig. 2.10): see above, p. 97.

# 4

# Lyell Steps In

During the 1820s Scottish writers were beginning to corner the market in attractive literalist earth-history. The new school of geology, too, is popularly seen as having been "founded" by two Scottish writers, James Hutton and, in particular, Charles Lyell.[1] Lyell's masterpiece, *Principles of Geology* (1830–3), is sometimes credited with having introduced into British science new concepts (such as the great antiquity of the earth) or new methods (such as the use of present-day evidence to make deductions about the past), but these ideas had long been tacit commonplaces among most practising geologists.[2] Lyell's greatest achievements did not lie in matters of theory, but in his ability to synthesize and deploy vast quantities of abstruse scientific data within an elegant, rhetorically compelling work of literature. To assess his significance more accurately, we need to see him not just as a geologist, but also as a man of letters. As I will show in this chapter, Lyell developed several of the narrative techniques

---

1. On Lyell's career see L. Wilson 1972; 1998; *ODNB; DNBS.*
2. Rudwick 1985, 42–6, usefully debunks some persistent myths about the beliefs and concerns of geologists around 1830 (for an example of which see Barber 1980, 222–3).

we have been examining so far, deploying them with new confidence and extending their imaginative scope. His chiselled prose was no mere adornment to his ideas, but an essential part of their appeal. Thanks not only to what he said, but more importantly to the way he said it, Lyell transformed the public profile of geology and its genteel practitioners at a critical stage in the science's development. As a confident statement of geology's disciplinary independence, the *Principles* played a major role both in ongoing boundary-disputes between science and theology, and in wider debates about man's place in nature.[3] In such matters, the pen proved mightier than the geological hammer.

## Apocalypse Regained

Brought up and educated in the south of England, Lyell learnt his geology under Buckland at Oxford University, becoming secretary of the Geological Society of London in 1823. Despite his very English education, however, he became increasingly drawn to an alternative Scottish geological tradition represented by Hutton and John Playfair, and conducted fieldwork around his family seat in northeast Scotland. In 1822 Lyell graduated from Lincoln's Inn as a barrister, but his income from legal work was at best sporadic. From an early age he had harboured literary ambitions, and in 1825 he took the first step towards realizing them: the Scottish publisher John Murray invited him to write for the Tory *Quarterly Review,* the most prestigious of the London periodicals. In October 1827 Lyell used this platform to strike up a battle-cry against literalist earth-science and the clerical domination of English intellectual life.[4] His polemic appeared within his favourable review of the *Memoir of the Geology of Central France* by his friend and scientific ally, the politician George Poulett Scrope. Buckland read a draft of this review and, despite the fact that his own methods came in for vigorous (if veiled) attack, he encouraged Lyell to use the review as a pretext for making an anonymous frontal assault on "the Penn school & the authors of the 'Scriptural Geology'".[5]

For both Lyell and Buckland, excessive adherence to Genesis 1 was the chief obstacle to progress in the science. When Lyell declared his intention to "free the science from Moses", he was thinking as much of the wide currency of biblical literalism as the various schemes champi-

3. On the *Principles* see Rudwick 1990–1; 2005a; 2008 in press; J. Secord 1997.

4. Lyell's review goes against the pattern suggested by Butler (1993, 143) for scientific reviews in the *Quarterly* after 1820, which she claims avoided controversy with "the vociferous clerical lobby".

5. Letter from Lyell to Murray, 7 August 1827, quoted in L. Wilson 1972, 173.

oned by Buckland and others for harmonizing Genesis and geology.[6] The Bible's authority over earth history remained largely unchallenged, at least in public, and was reinforced by literalist writers on geology. Lyell saw no way out for his science other than to combat these writers head-on by demolishing the assumptions on which their approach rested, and denying all links between scripture and geological reasoning—including the diluvialist compromises which he felt Buckland, Conybeare, and others had made in attempting to reconcile the two.[7] True, these men were now shying away from the clear links to Moses which had heralded the science's rise to academic respectability during the early 1820s; but, for Lyell, more was required. Geology and Moses had to be separated forcibly and publicly to set the science on a new philosophical basis. It was for this reason, rather than any personal animosity, that Buckland and his colleagues ended up being tarred with Lyell's broad brush of Mosaic obscurantism; and it was in this spirit, generally speaking, that they took it.[8] With missionary zeal, Lyell set out to educate the public and his colleagues simultaneously.

Lyell's strongest criticisms were levelled against the literalists, whom he perceived as a single "school" and a clear enemy. The last two pages of his *Quarterly* review make the literalist threat quite clear, after observing that Genesis was not intended to teach "minute scientific details":

> We cannot sufficiently deprecate the interference of a certain class of writers on this question who have lately appeared before the public. They are wholly destitute of geological knowledge derived from personal observation [. . . .] While they denounce as heterodox the current opinions of geologists, [. . .] they do not scruple to promulgate theories concerning the creation and the deluge, derived from their own expositions of the sacred text [. . . .][9]

Lyell caricatures his targets, emphasizing their dogmatism and aggression towards the new geology, characteristics which were not shared by most of them. But the most telling comment is Lyell's lofty statement of confidence in his intended audience: "We have no great apprehensions that such writers can ever gain much popularity in a country where philosophical information is widely diffused". This statement should not be taken literally: it is a rhetorical formula, characteristic of this sometime

6. K. Lyell 1881, I, 268.
7. K. Lyell 1881, II, 168–9. See Rudwick 1970; J. Secord 1997, xxiii–xxv.
8. See also L. Wilson 1972, 358.
9. [C. Lyell] 1827, 482.

lawyer. It suggests that Lyell was disquieted at the way in which literalist geology *had* won a large readership—not only among conservative
clerics, but among the intelligentsia (including readers of the *Quarterly*),
whose support the new science urgently needed. His anxiety is reflected
in his strident tone, rivalling that of George Bugg himself, and in his
final insistence that "Too much caution cannot be used against rash or
premature attempts to identify questionable theories in physical science
with particular interpretations of the sacred text".[10]

Lyell aimed to replace these unsatisfactory theories with permanent
principles of geological reasoning. Geology could become a true science
only by adopting the Newtonian concept of unchanging "laws": Lyell's
geologist was to make deductions on the assumption that nature had
always operated with the same laws, the same kinds of causes, and to the
same degree. This last assumption proved hard to swallow and ran counter to Cuvier's widely-followed practice of positing sudden, enormously
violent catastrophes to explain geological phenomena. Those who assumed that nature had operated intermittently on a more violent scale in
the past were later termed "catastrophists" by the Cambridge polymath
William Whewell, as opposed to the steady-state or cyclical geohistory
Lyell's views seemed to point towards, which Whewell dubbed "uniformitarianism". These terms misleadingly imply two distinct schools of
geohistorical theory, but Lyell's quarrel with "catastrophists" was methodological rather than cosmological. However, since his polemics tended
to lump them all together alongside long-exploded theories of the earth
(such as Burnet's), I refer to them here as "catastrophists" when explicat-
ing Lyell's own arguments. I do not use the term "uniformitarian" in this
way because, whatever he may have said in private, Lyell did not present
his views in public as a cosmological theory.[11]

The methodological revolution Lyell hoped to bring about demanded
all his powers of persuasion. The result contributed significantly to the
developing texture of geological spectacle. Lyell saw, for instance, that
his methodology depended on the widespread acceptance of a vast age
for the earth, which his colleagues took as a working assumption but
remained wary of promoting. As Martin Rudwick has emphasized, "it
was their scientific *imagination* that needed transforming".[12] But not only
that of his colleagues: Lyell wanted to dislodge the Bible's imaginative

10. [C. Lyell] 1827, 483.

11. This reticence did not prevent Lyell's opponents from caricaturing his views as preconceived "theories": see Rudwick 1975 on De la Beche's response. Privately, Lyell held a cyclic
view of earth history, as seen in his insistence to Mantell that the Iguanodon would one day
return to Cuckfield (K. Lyell 1881, I, 262).

12. Rudwick 1970, 11.

hold over a large segment of the British public, to show that geology offered grander (and hence, in this age, truer) views of the Creator's work. This ambitious project would be fully realized in his *Principles of Geology* (1830–3); but it was rehearsed towards the end of his *Quarterly* review with the same dual audience in mind.

To show the sufficiency of existing causes, Lyell employed the imagery of scriptural catastrophe, as exemplified by Martin's paintings. In effect, he used catastrophic imagery to undermine "catastrophism". The technique by which he did so, in two pages towards the end of his review, is simple but ingenious, turning on aesthetic as much as philosophical concerns. He first presents the "catastrophists", who assume that existing causes are not impressive enough to explain the great difference between the Tertiary epoch and the present day, and who think that today's geological forces are relatively small-scale compared with former catastrophes. Lyell counters this assumption by observing that present-day forces are really rather impressive. His geological opponents are cast as stubbornly unimpressed spectators at a stupendous display:

> those who can behold the fire of the volcano, and the shock of the earthquake, the waste of the torrent [. . .] who can look around them and be witness to all these signs of change, and still contrast the vicissitudes of former ages with the immutable stability of the present order of things [. . . .][13]

Hutton had employed a similar technique in his 1795 *Theory of the Earth*, appealing to the terrific scale of observable earthquakes and volcanoes in order to suggest that the subterranean forces responsible for these occasional surface effects must be powerful enough to have formed the earth's crust.[14] Lyell took this strategy further, turning it into time travel. In a two-page-long rhetorical question, he asks these same spectators what they would have said "had they been admitted to view the tertiary lakes of Central France in all their original beauty and repose". The scene itself is called up before the mind's eye:

> if they had seen myriads of tender insects frequenting the banks, the crocodile and the tortoise emerging from the water, or the lake-birds swimming on its surface—if they had marked the herds browsing with security amidst forests of palms,—would they have conceived it possible that all this luxuriance of life [. . .] constituted no more than a transient scene?[15]

13. [C. Lyell] 1827, 472.
14. Hutton 1795, I, 139–40. On Hutton's aesthetic strategy here see Heringman 2004, 113–16.
15. [C. Lyell] 1827, 472.

Like Mantell's word-painting of the ancient Weald, this restoration—at once more assured and more comprehensive than Mantell's—is imbued with a sense of picturesque tranquillity. In Lyell's hands, however, this very quality carries rhetorical force: surely the casual observer would never imagine that this peaceful scene was destined to be racked by volcanic activity. The coming destruction is evoked as a literal apocalypse, courting the reader's astonishment in traditional biblical terms in order to heighten the spectacular revelations which only geology can offer:

> Would they have conceived it possible that the various generations of living creatures, then ranging the plain, or swimming the lake, should at length fail—that they were destined to be all "blotted out, and rased from the books of life;" yet that some few, perishing in the waters, should leave their skeletons to be disentombed, in after-ages, from the living rock, and to become "known to men by various names?" How incredible would that prophetic voice have sounded, which should have foretold that these magnificent lakes should one day vanish, their oozy bottom become consolidated, and then, after being buried under repeated streams of liquid lava, again be furrowed out into deep vallies, with intervening hills; and, lastly, adorned and enlivened with a new creation of animals and plants!—that the granite would not only give birth to burning mountains of prodigious magnitude and height, but to a chain of more than three hundred minor cones, and to as many fiery torrents of lava—in a word, that the whole scene, the temperature of the air, the surface of the land, hill and valley, lake and river, with all the countless organic beings who then enjoyed the gift of life, were doomed, in the revolutions of futurity, like the heavens on the opening of the sixth seal, to "depart as a scroll when it is rolled together!" [16]

This last quotation is peculiarly apt to Lyell's purpose, being taken from Revelation 6.14, where it continues: "and every mountain and island were moved out of their places." The exclamation mark is Lyell's, capping this most *martinien* of prophecies.

What Lyell's readers were meant to infer from this performance was that the tranquillity of the lakeside landscape might tempt the casual time-traveller into assuming, wrongly, that nature's energies were now spent. Seen in the geologist's enlarged view, a different interpretation becomes possible: when multiplied indefinitely, familiar agents like volcanism become capable of apocalyptic transformations. In the next section Lyell renovates the old imagery of science as a subterranean quest or vision, recalling the occasional verses circulated among Buckland's

16. [C. Lyell] 1827, 473.

circle. Echoing Cuvier's prediction about bursting the limits of time, Lyell promises more time travel in the near future, when "the geologist" will be able to "restore to our imagination the picture successively presented at remote periods, by the earth's surface and its inhabitants". His students will be able to follow him,

> as Dante in his sublime vision followed the footsteps of his master, and beheld, with mingled admiration and fear, in the subterranean circles environing the deep abyss, the shades of beings who once walked the surface in the light of day, and who still, changed as they were, and unconscious of the present, could draw the veil from the mysteries of the future, and recal from oblivion the secrets of the past.[17]

Dante's Virgil was a particularly appropriate persona for the literary geologist to adopt, being not only an authoritative and informative guide to the underworld, but also a conservative guardian of Classical culture and an epic poet. With such talk of "mysteries" and "secrets", and with the help of a potent mythic and poetic model for his new philosophy and its new caste of leaders, Lyell appeals to his readers' "longing desire to know of things in the remotest ages of the past". This desire, Lyell tells us, springs not from human vanity but from God's wish that we should enlarge our imaginations. Lyell now asserts that the vast reaches of prehuman time revealed by geology are exalting, not humiliating: "if we hold mind to be something distinct from matter, it must be acknowledged that we assert its superiority more clearly by enlarging our dominion over time".[18]

This "dominion" goes beyond the idea of pursuing "our researches [. . .] through the immensity of ages", as Davy had put it in his 1822 speech, now hot off John Murray's press.[19] According to Lyell, the geologist becomes almost superhuman, inhabiting eternity in a virtual sense: "we might [. . .] include within the compass of our rational existence, all the ages [. . .] over which science may enable us to extend our thoughts". The colonial implications of Davy's speech are here brought into the open and used as a justification for the geologist's activity:

> Already is our progress accelerated by the mere knowledge that new worlds are accessible to our research, as when the energies of Europe were awakened by the intelligence that a continent beyond the Atlantic lay open to the enterprise of man.

17. [C. Lyell] 1827, 473.
18. [C. Lyell] 1827, 474.
19. Davy 1827, 51.

taught at Oxford and Cambridge was also gaining popularity, but Lyell felt it made too many concessions to the literalist position. The *Principles* announced the geologists' authority over the telling of earth history, redefining the term "geology" once again. Both conservative literalism and radical transmutationism were excluded as unphilosophical and unworthy of gentlemen. It was in this sense, in managing to elevate geology's public profile, that the *Principles* was widely thought to have raised geology to the status of a science.[28]

Lyell's rhetorical gifts have often been admired, but the rhetoric of spectacular display with which he bolstered his argument has received little attention.[29] Lyell's argument for the impressive effects of existing causes depended on exhaustive demonstration, for which he provided hundreds of recorded events. These comprise a veritable catalogue of disaster stories, as a selection of running titles suggests: "Destruction of Dunwich by the Sea", "Houses Engulphed", "New Islands Thrown Up", "Earthquake in Peru, A.D. 1746", "Earthquakes in Kamtschatka, Martinique, etc.", "Earthquake in Java, A.D. 1699", "Earthquakes in Quito, Sicily, Moluccas, etc.", "Earthquake of Jamaica, A.D. 1692", "Subsidences, Floods, etc."[30] Some of these titles would make eye-catching newspaper headlines or spectacle scenarios (compare Fig. 7.22).

But Lyell did not merely enumerate cases. He made these events dramatically present, building on his earlier evocation of "the fire of the volcano" and other natural spectacles before which he had challenged "the geologist" to remain unmoved. A single paragraph in that review now expanded into several hundred pages, demonstrating the adequacy of existing causes in a myriad of anecdotes, spectacular woodcuts, landscape engravings, and maps (Figs. 4.1 and 4.2). After detailing the effects of recent earthquakes, for instance, Lyell pauses before launching into eighteenth-century examples and asks his old rhetorical question:

> with a knowledge of these terrific catastrophes, [. . .] will the geologist declare with perfect composure that the earth has at length settled into a state of repose? Will he continue to assert that the changes of relative level of land and sea, so common in former ages of the world, have now ceased?[31]

28. J. Secord 1997, xxii–xxiii.

29. On Lyell's rhetorical strategies see Rudwick 1970; 1990–1; Porter 1976; J. Secord 1997; Klaver 1997, 31–47 (including a brief discussion of narrative style).

30. C. Lyell 1830–3, I, 273, 419, 409, 442–7. For a facsimile reprint of this book see C. Lyell 1990–1; C. Lyell 1997 is a one-volume abridged edition.

31. C. Lyell 1830–3, I, 409.

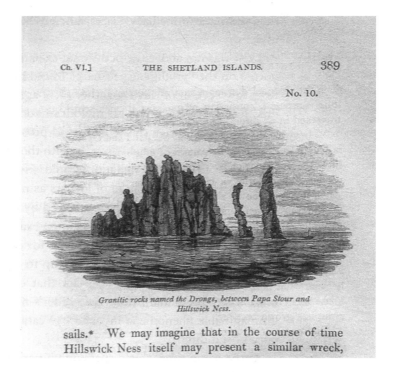

Ch. VI.]  THE SHETLAND ISLANDS.  389

No. 10.

*Granitic rocks named the Drongs, between Papa Stour and Hillswick Ness.*

sails.* We may imagine that in the course of time Hillswick Ness itself may present a similar wreck,

*Figure 4.1.* A typical woodcut in Lyell's *Principles*, illustrating marine erosion with the Drongs, a group of skerries off western Shetland. C. Lyell 1835, I, 389, detail.

He then proceeds to tell yet more earthquake stories. Lyell's narratives are most vivid when they concern recent disasters, such as the Tivoli flood of 1826. He recounts this story with journalistic vigour, telling how the flood

> undermined [. . .] a high cliff, and widened the river's channel about fifteen paces. On this height stood the church of St. Lucia, and about thirty-six houses of the town of Tivoli, which were all carried away, presenting, as they sank into the roaring flood, a terrific scene of destruction to the spectators on the opposite bank*. As the foundations were gradually removed, each building [. . .] was first traversed with numerous rents, which soon widened into large fissures, until at length the roofs fell in with a crash, and then the walls sank into the river, and were hurled down the cataract below.[32]

The asterisk directs us to a footnote indicating that Lyell had got the account from eyewitnesses, the very "spectators" who had watched the "terrific scene". By such authentication Lyell allows us to witness this scene ourselves, albeit from the detached viewpoint of the philosopher. These "graphic and eloquent" descriptions were praised by Sedgwick for

32. C. Lyell 1830–3, I, 197.

Lyell, this contemplation of the sea brings to a close a massive, disparate, and richly detailed collection of images and narratives from all over the world (in Byron's case, Europe) and from different points in its history, hitherto bound together only by the narrator's unifying viewpoint. This final view of the ocean confers a higher unity on all these images by functioning as an "image of Eternity", dwarfing both the tumultuous courses of the successive empires meditated on throughout *Childe Harold* ("Assyria, Greece, Rome, Carthage") and Lyell's catalogue of physical disasters. Through this long view, even earthquakes become agents of reform, "a conservative principle [. . .] essential to the stability of the system", as Lyell explained in the final sentence of volume 1.[41] The political resonances would not have been lost in 1830, when Britain was racked by Reform agitation and fears of revolution.

By this time Byron was still the most famous poet in Western Europe. His untimely death in 1824 from typhoid, caught at Missolonghi when he was engaged in the Greek struggle for independence, had made him even more of a celebrity. Many of Lyell's readers would have known the "Ocean stanzas" by heart. Byron's borrowed voice familiarizes the new leap of imagination Lyell wants his readers to make, from historical particulars out to the flux of time. Thus guided, they leap out and "see" the "boundless, endless, and sublime" reality of deep time, looking through Byron's eyes to see through Lyell's.[42] This strategy inevitably imports something of the pungent philosophic melancholy for which *Childe Harold* was so celebrated, and which becomes particularly intense at this point in the poem: the ocean's eternity is juxtaposed painfully with the narrator's sense of personal transience and of his poem's insubstantiality. It may have been as much Byron as Lyell who led the poet Edward Fitzgerald to write in 1847 that Lyell's "vision of Time must wither the Poet's hope of immortality".[43] Lyell's reader is drawn to view the globe with a Byronic mixture of melancholy and exhilaration. Nevertheless, a delighted expansion of spirit seems to have been uppermost in Lyell's own mind, even if later Victorian poets responded more gloomily. At other rhetorical pressure-points, too, he pulls the reader's gaze back and up from particular details to view the physical world in the light of eternity.[44] These passages exult in the geologist's visionary freedom but also rejoice in the knowledge that the visible, though vast, is but a fraction of the whole truth. Several contemporary critics responded in like

41. C. Lyell 1830–3, I, 479.
42. *Childe Harold* IV.183 (Byron 1986, 200).
43. Terhune and Terhune 1980, I, 566, quoted in J. Secord 1997, xxxix.
44. C. Lyell 1830–3, I, 74, 166; III, 384–5.

manner, reporting "emotions of humility" combined with "elevation of thought".[45]

The sense of intellectual control conferred by elevated viewpoints emerges most explicitly in Lyell's speculations on climate change. To demonstrate the climatic changes brought about by shifts in the relative positions of land and sea, he presents a series of thought-experiments, introducing a sequence of hypothetical changes and accelerating these gradual processes for the sake of the illustration. He begins in the explicitly hypothetical mode of the philosopher:

> Let us suppose those hills of the Italian peninsula [. . .] to subside again into the sea [. . .] and that an extent of land of equal area and height [. . .] should rise up in the Arctic ocean, between Siberia and the north pole [. . . .]

As more of these accelerated changes are hypothesized, Lyell enters into the spirit of his world-making game. His tone becomes commanding, shifting from subjunctive to imperative mood as the changes become more extravagant:

> let the Himalaya mountains, with the whole of Hindostan, sink down, and their place be occupied by the Indian ocean, and then let an equal extent of territory and mountains [. . .] stretch from North Greenland to the Orkney islands. It seems difficult to exaggerate the amount to which the climate of the northern hemisphere would now be cooled down.[46]

We gaze on the scene as if from outer space, a viewpoint reinforced in later editions by a diagram of two alternative alien worlds (Fig. 4.2). Lyell now populates his frozen world with tiny (and cold) geologists whose astonishment and "catastrophist" methodology appears laughable, because we—from our angelic perspective—now know better:

> They who should then inhabit the small isles and coral reefs, which are now seen in the Indian ocean and South Pacific, would wonder that zoophytes of such large dimensions had once been so prolific in those seas; or if, perchance, they found the wood and fruit of the cocoa-nut tree or the palm silicified by the waters of some mineral spring [. . .] they would muse on the revolutions that had annihilated such genera, and replaced them by the oak, the chestnut, and the pine. With equal admiration would

---

45. Anon. 1832a, 39; compare Anon. 1834a. On the reviews see J. Secord 1997.
46. C. Lyell 1830–3, I, 116–17. Klaver (1997, 45) suggests that Lyell's language echoes Genesis 1 ("Let there be light").

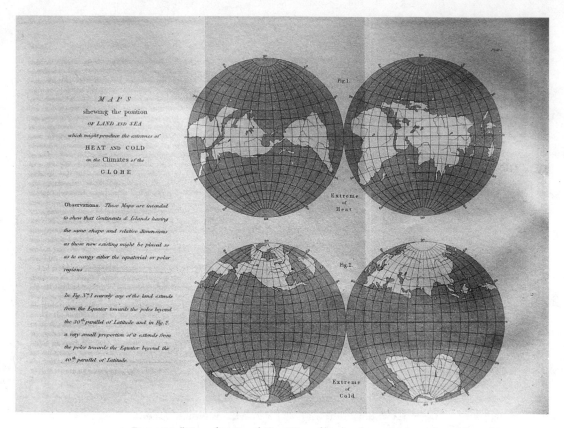

*Figure 4.2.* "Maps shewing the position of land and sea which might produce the extremes of heat and cold in the Climates of the globe". Lyell added these maps to the cheaper edition of his *Principles* to clarify his point about climate for his wider public. C. Lyell 1835, I, plate 1.

they compare the skeletons of their small lizards with the bones of fossil alligators and crocodiles more than twenty feet in length, which, at a former epoch, had multiplied between the tropics; and when they saw a pine included in an iceberg [. . .] they would be astonished at the proof thus afforded, that forests had once grown where nothing could be seen in their own times but a wilderness of snow.[47]

"Catastrophist" methodology is undermined by equating it with the denial of the blindingly obvious (palm trees in the tropics). Moreover, this hypothetical world is brought to life by the presence of reasoning humans and vivid touches like the "wilderness of snow", colouring in the bare diagram shown in Fig. 4.2. It is as if, under Lyell's control, we travel in time to witness it.

He now engineers an alternative future, "supposing" and command-

47. C. Lyell 1830–3, I, 120.

ing further changes to warm the earth (corresponding to the upper pair of globes in Fig. 4.2). This imaginative journey is eased by a picturesque simile:

> If, during the long night of a polar winter, the snows should whiten the summit of some arctic islands, and ice collect in the bays of the remotest Thule, they would be dissolved as rapidly by the returning sun, as are the snows of Etna by the blasts of the sirocco.

Lyell's thought-experiment famously concludes, a few sentences later, by speculating that warmer climes might even result in the reappearance of animals now extinct—although here he is careful to preserve the conditional mood:

> Then might those genera of animals return, of which the memorials are preserved in the ancient rocks of our continents. The huge iguanodon might reappear in the woods, and the ichthyosaur in the sea, while the pterodactyle might flit again through umbrageous groves of tree-ferns. Coral reefs might be prolonged beyond the arctic circle, where the whale and the narwhal now abound. Turtles might deposit their eggs in the sand of the sea beach, where now the walrus sleeps, and where the seal is drifted on the ice-floe.[48]

Here Lyell's prose almost becomes poetry. The fivefold repetition of *might*, partly emphasizing the hypothetical nature of this vision, paradoxically draws us further in by heightening its poetic quality, binding the passage together musically. This is best appreciated by reading the passage aloud. Its musicality is enhanced by the diction, which sustains the lyricism of the "snows of Etna" passage. A hypothetical and alien world is brought vividly into view by Lyell's self-consciously poetic language: *flit* rather than *fly*, *umbrageous groves of tree-ferns* rather than simply *tree-ferns*. The whole passage has a dreamlike quality, not only because of the incongruity of extinct animals returning. The last two images inscribe passivity into the "present-day" scene: as "the walrus sleeps" and the seal "is drifted", so we, as present-day observers, drift into Lyell's dream of a distant summer. This drifting is eased by the sibilant alliteration and persistent long "e" (*sand, sea-beach, sleeps, seal*), and by the last sentence gradually falling into a peaceful iambic rhythm.

This passage on the possible effects of climate change was intended to cast doubt on the idea that the history of Creation inevitably progressed from lower organisms to higher. This "directionalist" model was rapidly

48. C. Lyell 1830–3, I, 122–3.

becoming an article of faith among the new school of geologists. Lyell's opposition to it was unusual, and partly reflects his fear that it left the door open to morally unacceptable speculations about species transmutation. But he was taking a risk in enshrining his most counter-intuitive piece of hypothesizing in the poetic language of the dream-vision, particularly since he had used words like "dreams" and "visions" to disparage alternative hypotheses. Here was a clear opening for a return shot, which the geologist and skilled caricaturist Henry De la Beche took in 1830 with his satirical lithograph sketch *Awful Changes* (Fig. 4.3), circulated among members of the Geological Society.[49]

In this sketch, a professorial ichthyosaur of the future is seen lecturing

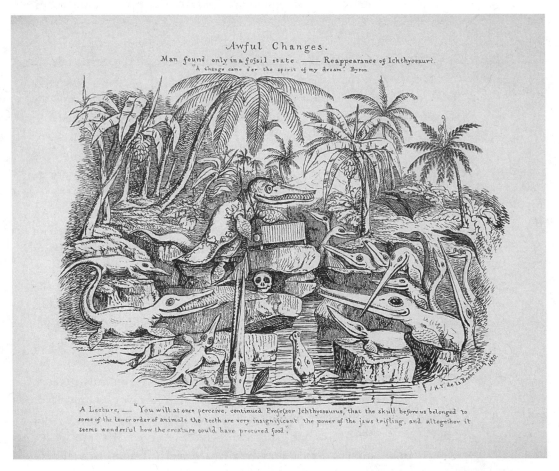

Figure 4.3. *Awful Changes* (1830), De la Beche's satire on Lyell's cyclical geohistory. In a remote future, "Man" has become extinct and the ichthyosaurs have returned to pontificate on his remains.

49. This sketch was known to the anonymous author of a "Popular Course of Geology" who alludes to "the humorous sketch of Mr. Delabeche" and, aware that most readers of the *Magazine of Popular Science* would not have seen the picture, describes it in a footnote (Anon. 1836a, 11). On De la Beche see McCartney 1977; *ODNB; DNBS.*

a saurian audience on the puny anatomy of the now-extinct human.[50] At the top of the picture, beneath the explanatory subtitle ("Man found only in a fossil state.—Reappearance of Ichthyosauri"), is a quotation from Byron's poem *The Dream* (1816), which dramatizes the course of an unhappy love-affair as a sequence of visionary scenes, each prefaced by the line "A change came o'er the spirit of my dream".[51] By excerpting this line De la Beche mocks the "visionary" nature of Lyell's speculations: Byron's refrain is made to allude to the geologist's ability to call up a sequence of scenes. *Awful Changes* can hardly have been intended to mock this possibility in itself: that same year De la Beche himself had designed the first-ever pictorial restoration of extinct saurians in their environment, *Duria antiquior* (see Fig. 2.2). Rather, the caricature casts doubt on the "dreamer" Lyell's ability to pronounce on geohistory. Note that "Professor Ichthyosaurus" wears spectacles: several earlier unpublished satirical sketches by De la Beche caricature Lyell's philosophical myopia by this visual cue.[52]

And yet, as De la Beche surely knew, Byron's poem opens with a prologue on the mysterious potency of dreams, including the following lines:

> What are they?
> Creations of the mind?—The mind can make
> Substance, and people planets of its own
> With beings brighter than have been, and give
> A breath to forms which can outlive all flesh.[53]

For De la Beche, Lyell's speculations were the delusions of an overheated romantic imagination; but these lines illuminate Lyell's ambitions in a kinder light. In an earlier chapter of the *Principles* Lyell had already sung of the geologist's ability to discover other worlds in time just as the astronomer had in space: "to trace the same system through various transformations—to behold it at successive eras adorned with different hills and valleys, lakes and seas, and peopled with new inhabitants". This historical vision amounted to a form of creation. Lyell had concluded his

---

50. On the intended meaning of this sketch see Rudwick 1975; 1992, 48–50. This sketch was ultimately made public in Frank Buckland's *Curiosities of Natural History* (1857) where Buckland claims—presumably on his late father's authority—that it was a good-natured "quiz" upon his father's "geological lectures at Oxford" (F. Buckland 1903, xii), an interpretation which has become standard in popular accounts. It may indeed have been in this spirit that William displayed a large-scale version of *Awful Changes* to his own students, but De la Beche's original intention was to satirize Lyell's "theory".

51. Byron 1980–93, IV, 22–9.

52. Rudwick 1975.

53. *The Dream*, lines 18–22 (Byron 1980–93, IV, 23).

observations by quoting from Barthold Niebuhr's *History of Rome* (translated in 1828), an innovative historical study whose daring reconstructions of ancient history had already been compared by one critic to the "magnificent dreams" (i.e. unwarranted speculations) of Cuvier:[54]

> Meanwhile the charm of first discovery is our own, and as we explore this magnificent field of inquiry, the sentiment of a great historian of our times may continually be present to our minds, that "he who calls what has vanished back again into being, enjoys a bliss like that of creating."[55]

As Childe Harold had been able to

> watch the stars,
> Till he had peopled them with beings bright
> As their own beams; and earth, and earth-born jars,
> And human frailties, were forgotten quite:[56]

so Lyell's geologist blissfully watches and (re)peoples former worlds from his unearthly philosophical height. That refrain from *The Dream* would recur in later popularizations,[57] and the dream-vision would become an established mode of reconstructing the earth's "awful changes".

In this light we may wonder why, considering his enchanting 1827 evocation of Tertiary France, Lyell painted no word-pictures of the ancient earth in the *Principles,* confining his lyrical gifts to present-day processes or alternative worlds and restoring only the Iguanodons of the future. The biblical Apocalypse concluding his 1827 word-painting has been transferred, in the *Principles,* to thought-experiments—to that apocalyptically accelerated sequence of geological changes in which Lyell plays God "and every mountain and island [is] moved out of their places" (Revelation 6.14).[58] An informal comment to Mantell suggests an explanation: Lyell boasted of having a recipe "for growing tree ferns at the pole, or if it suits me, pines at the equator".[59] For Lyell, the epistemological power of geological reasoning was conveyed more forcibly by world-making than by depicting former worlds. The latter had been made by God, whereas

54. Anon. 1829b, 198.

55. C. Lyell 1830–3, I, 74. On Lyell and Niebuhr see Rudwick 1979, 68–9.

56. *Childe Harold* III.14 (Byron 1986, 108). Lyell's own youthful "Lines on Staffa" begin: "Ere yet the glowing bards of Eastern tale / Had peopled fairy worlds with beings bright" (K. Lyell 1881, I, 55).

57. Mantell 1838, 262; Miller 1859, 78; Buckley 1927, 211.

58. These words form part of the same Bible verse with which Lyell's 1827 "apocalypse" concludes ([C. Lyell] 1827, 473).

59. K. Lyell 1881, I, 262.

Lyell's imaginary worlds were the product of the godlike, God-given human imagination in its capacity as "sub-creator".[60]

In fact, this show of power ends up reinforcing the geologist's freedom to re-create scenes from the earth's past. In Lyell's theology of the imagination, all these epochs become incorporated "within the compass of our rational existence".[61] The chapter following Lyell's thought-experiments concludes by echoing Cuvier's peroration on geology's time-bursting potential, as well as publicizing the scientific triumphalism implicit in earlier occasional verses. This is done in literally Virgilian terms:

> Thus, although we are mere sojourners on the surface of the planet, chained to a mere point in space, enduring but for a moment of time, the human mind is not only enabled to number worlds beyond the unassisted ken of mortal eye, but to trace the events of indefinite ages before the creation of our race, and is not even withheld from penetrating into the dark secrets of the ocean, or the interior of the solid globe; free, like the spirit which the poet described as animating the universe,
>
> ———ire per omnes
> Terrasque tractusque maris, cœlumque profundum.[62]

This comparison of the human "mind" with Virgil's divine "spirit" was not intended or received as a form of impiety, since the result of all this unfettered vision and conquest is humility—or, as the concluding sentence in the *Principles* puts it, "a just estimate of the relations which subsist between the finite powers of man and the attributes of an Infinite and Eternal Being."[63]

That this "freedom" is God-given has important implications in Lyell's account. Since his panoramic viewpoint shows "true" geology transcending the traditional imagery of biblical Apocalypse, it "frees" the geologist from Genesis, enabling him to piece together a more "just", if fragmentary, narrative. One of Lyell's subsidiary strategies in the *Principles* was to emphasize the geologist's status as Cuvier's "antiquary of a new order",[64] even as he denied the possibility of ever reconstructing a complete history of the earth. More rigorously than his predecessors, Lyell transformed the familiar metaphors of the "book of nature" and

60. This term is Tolkien's (1983, 139).

61. [C. Lyell] 1827, 474.

62. C. Lyell 1830–3, I, 166, quoting from Virgil's *Georgics*. Many readers would have known Dryden's 1697 translation:
   "Thro' Heav'n, and Earth, and Oceans depth he throws
   His Influence round, and kindles as he goes."

63. C. Lyell 1830–3, III, 385.

64. Cuvier 1813, 1.

"medals of creation" into hermeneutic tools, drawing on his Classical learning and his interest in contemporary historiography and linguistics.[65] For Lyell the strata were like an ancient manuscript, riddled with lacunae and written in obscure characters which gradually yielded to intensive study. As its title suggests, Lyell saw the *Principles* as the necessary groundwork for such study: the enumeration of natural causes making up its first two volumes was intended to form part of the "alphabet and grammar of geology", training the eye to decipher the earth.[66] This rhetoric underlined the geologist's interpretative authority, as hinted in his Dantean analogy of 1827: Lyell was not democratizing scientific knowledge,[67] but helping to define and promote a new kind of scientific expertise. A reviewer in the *Spectator* obligingly observed, "There are sermons in stones [. . .] but they want an interpreter: that interpreter is the enlightened geologist. Such a man is Mr. LYELL."[68] In publicizing the geologist's authoritative role, Lyell was moulding a new public for the new science, and reinforcing new ideas about public education which journals like the *Spectator* were promoting.[69]

## Lyell's Public

The first volume of the *Principles* was written and published during the period leading up to the Reform Bill of 1832, when a (slightly) wider cross-section of society was given the vote. These socially turbulent years saw increasing agitation among the urban working classes, who were not directly represented in Parliament; the radical press played an important part in this struggle. Lyell's status—a Tory gentleman by birth, a liberal Whig by inclination—gave him the opportunity not only to reform geology, but to present it as a force for Reform in its own right from within the ranks of the establishment.[70] Like his writings for the *Quarterly,* the expensive first edition of the *Principles* targeted the conservative leisured classes: Lyell sought to reassure them that geology presented no threat to Christianity and offered no support for atheistic radicalism. But he also believed that correct natural knowledge should be "diffused" to the "vulgar" by an enlightened clerisy of gentlemen.[71]

This attitude had been gaining currency among men of Lyell's class since the late eighteenth century, even more so since the technical break-

---

65. Rudwick 1979.
66. C. Lyell 1830–3, III, 7.
67. Macfarlane 2003, 37.
68. Anon. 1832a, 39, quoted in J. Secord 1997, xv.
69. On these new ideas, both religious and secular, see Topham 1992.
70. J. Secord 1997, x–xxix.
71. On this imperative see Morrell and Thackray 1981, 17–29.

throughs made by the printing and textile industries in the 1820s had allowed publishers to reach out towards less affluent and less educated readerships.[72] Like many liberal educationalists, Lyell had a strongly hierarchical vision of what "popularization" should involve. This sense of the word developed around the same time as the word "scientist" was coined to describe the self-styled experts who sought to set up this audience relation.[73] It reflected an increasing tendency in periodical miscellanies to present science as a body of authoritative, predigested information, communicated via extracts from the leading scientific periodicals, rather than as an ongoing conversation between its readers. In the wake of the Reform crisis, genteel authority became an essential part of editorial strategies for countering the politically subversive forms of science found in the cheap radical press. Equally, it reflected the commercial demands on miscellany editors, who were faced with escalating specialization and a rapidly increasing number of scientific monthlies.[74]

So, in 1834, having safely launched the genteel edition of *Principles*, Murray repackaged it for its third edition as a cheaper, smaller four-volume set (though still not cheap at 24s for the set). The Tory publisher was competing here with the Whig-dominated Society for the Diffusion of Useful Knowledge; Lyell himself had been anxious lest De la Beche's own reflective treatise, published in 1834 by the Useful Knowledge merchant Charles Knight, cut into his own readership by being more affordable.[75] The new edition sold well.[76] Only when it was safely under way did Lyell reveal to Murray that he was working on an elementary work, his projected *Conversations on Geology* (ultimately published in 1838, in textbook form, as *Elements of Geology*).[77] He seems to have been anxious to ensure that his *Principles* and his name were established before he appeared in public as a mere "compiler" (as he disparagingly termed authors of elementary treatises and periodical articles, drawing an implicit contrast with his own authoritative acts of compilation).[78]

Lyell had taken a risk in publicly and simultaneously assaulting Oxford diluvialism, French transmutationism and popular literalism. The

72. On this public see Klancher 1987; J. Secord 2000a, 41–51; 2000b. However, not all men of science agreed that such knowledge should be "diffused" in this way: see Yeo 1984, 14–15.

73. Topham 2000, 560.

74. J. Secord 2000a, 41–69; Dawson *et al.* 2004, 11–14. On the ideological factors in this trend see Barnes and Shapin 1977; on its commercial causes see Topham 2004a.

75. L. Wilson 1972, 343–4. De la Beche had worried Lyell by telling him his treatise was to be "founded upon" the *Principles,* causing Lyell to suspect him of wanting to steal his ideas and sell them off cheaply. In fact De la Beche's *Researches in Theoretical Geology* (1834) countered Lyell's extreme uniformitarianism and proved an influential exposition of the directionalist synthesis.

76. L. Wilson 1972, 343–4, 385, 411.

77. L. Wilson 1972, 408–9.

78. L. Wilson 1972, 343–4.

*Principles* invited vigorous opposition from various angles. In fact the response was overwhelmingly positive, and not only among newcomers to the science. Within the Geological Society, the equation of the biblical Deluge with the cataclysm attested in the rock record soon sank without trace, especially after Sedgwick staged his own "recantation" of what he half-jokingly called a "philosophic heresy" in his retiring Presidential speech in 1831.[79] Sedgwick and others disagreed strongly with some of Lyell's "principles" (notably his denial of progression in earth history); but it was generally admitted that he had changed the status of philosophical debate within the science, publicly reinstating the role of historical interpretation and active imagination in geological thought. Sedgwick commented that geology would be "little better than a moral sepulchre" were it not for its connection with "the imagination, the feelings, [. . .] and the highest capacities of our nature",[80] and in 1833 Bakewell inserted this passage into the latest edition of his *Introduction to Geology* to strengthen his case for speculation.[81] The title of Bakewell's new edition featured a new emphasis on geological "Theory",[82] while De la Beche's anti-Lyellian treatise of 1834 bore the revealing title *Researches in Theoretical Geology*. After a brief, somewhat artificial absence from the rarefied world of the Geological Society and its promoters, speculation and imagination were back in business.

The *Principles* marked a watershed in geologists' attitudes towards popularization. Before 1830, British geology had been in search of a public voice with which to restore former worlds; after 1830, that voice was more or less assured. The change was not due to Lyell alone; his work was as much a symptom as a cause of the new drive towards the "diffusion" of knowledge and the self-definition of scientific elites. But the *Principles* played an important part in bringing this particular science out of its closet of cautious empiricism. Lyell made grand claims for the cultural authority and independence of the new geology;[83] and his whole argument substantiated these claims by engaging the reader's imagination more systematically than had hitherto been attempted by a geologist. His various case-studies, thought-experiments, shifts in perspective, illustrations, and poetic allusions trained his readers to "see" geologically. As Charles Darwin famously told Lyell's father-in-law, "the great merit of the Principles, was that it altered the whole tone of one's mind & therefore that when seeing a thing never seen by Lyell, one yet saw it partially

79. Sedgwick 1826–33b, 313; see J. Secord 1997, xxix.
80. Sedgwick 1826–33b, 314–15.
81. Bakewell 1833, 537–8.
82. On this new emphasis see Bakewell 1833, v–vi.
83. So did some of his reviewers: see Yeo 1993, 99–102.

through his eyes",[84] a statement eloquently borne out by the geological descriptions in Darwin's own *Journal of Researches* (1839).[85]

Many popularizers from now on engaged in similar pursuits. Those from whom Lyell had learnt his own word-painting techniques, notably Buckland and Mantell, now developed their own methods with more boldness and for a wider readership. The spectacular and visionary metaphors into which Lyell had breathed new life began to be employed with a new freedom, reinforcing the geologist's public image as a noble guide who could be trusted to take his followers on spectacular and spiritually uplifting voyages. Indeed, these metaphors soon became clichés: to present them in as fresh a manner as Lyell's required considerable literary panache.

Fortunately, Lyell was not the only writer thus gifted. As popular print and the various mass education movements gained momentum over the next three decades, and as men of science were increasingly encouraged to stage themselves as public celebrities, those geologists who were able to play to the gallery found themselves in great demand. Men such as Buckland and Mantell, as well as newcomers like Hugh Miller, were able to forge lasting reputations as men of letters largely on the strength of their scientific writings. Canny publishers and professional writers (the so-called hacks) took advantage of these possibilities, compiling material already published by others and dressing old ideas in smart clothes. A new flowering of geological popularization was ushered in by a combination of forces: changes in the publishing industry, demographic shifts among the reading public, increasing demand for stratigraphic expertise, Whig-led educational theories, university reforms, the redefinition of gentility around concepts of intellectual leadership, competition from other conservative and radical pressure groups, and the catalysing examples of individuals like Buckland and Lyell, whose *Principles* might be seen as the first fruits of this flowering. The golden age for the writing of earth history had begun.

84. Darwin to Leonard Horner, 29 August 1844 (Burkhardt and Smith 1987, 55).

85. See, for example, C. Darwin 1839, 399–400. For commentary see J. Smith 1994; J. Secord 1997; Herbert 2005.

*Part II*

Staging the Show

# 5

# Marketing Geology

## The Popular Science

On the evening of 23 June 1832 William Buckland stood before a crowded lecture-hall at Oxford University, accompanied by a gigantic fossil skeleton. The beast had recently arrived from Argentina: it was a fine example of a Megatherium, the extinct ground sloth identified by Cuvier. As with the Peale "mammoth" thirty years before, the animal's sheer size was spectacle enough; but Buckland proceeded to bring it to life in his inimitable style. It might appear monstrous, explained Buckland, but it was perfectly adapted to its own environment. Christening the beast "Old Scratch", he painted a vivid picture of its daily routine, digging for roots "beneath a broiling sun":

> He has a spade, and he has a hoe and a shovel in those three claws in his right hand [. . . .] He is the Prince of sappers and miners—I speak in the presence of Mr Brunel the Prince of Diggers. Mr. Brunel eyes him and says "I should like to employ him in my tunnel." "No," say I, [. . .] "he is a canal digger if you please, so I pray give him the first job you have to do." He

[. . .] would drain all Lincolnshire in the ordinary process of digging for his daily food.[1]

The skeletal structure had clearly been designed by a wise and benevolent Creator so that "this Leviathan of the Pampas" could enjoy its portion of earthly existence in ancient Argentina, where now its relics littered the bare plains (to be picked up a few months later by Darwin, who was currently groping towards a very different interpretation of these marvellous adaptations).[2] Some of the Oxford audience shared Darwin's dislike of Buckland's jocular tone and theatrical antics, but Buckland knew how to make palaeontology appeal.[3] At one point, to emphasize the creature's huge size, he invited the geologist William Clift to step through its pelvis and to "come a second time into the world through this cavity in the pelvis of the megatherium",[4] to thunderous applause.

But who was applauding? This was no Anglican academic audience, although some dons were present. Instead of black gowns and mortar-boards, the room was filled with fashionably-dressed men and—shockingly—women. Buckland was up to his old tricks before a much larger and wider public than before, a cross-section of Britain's polite and professional classes spanning most religious denominations and political affiliations. His lecture bore witness to the spectacular "birth" not only of William Clift, but also of a new public forum for scientific display: the British Association for the Advancement of Science. Founded in 1831, the BAAS was a new initiative by the gentlemen of science, organizing annual meetings at different cities and towns in the United Kingdom. Its aim was to make science into a national cultural resource that could transcend denominational barriers, uniting the middle and upper classes under a non-sectarian, apolitical banner.[5] Under Buckland's presidency in 1832, the use of spectacle was deliberately launched at Oxford to attract public attention and display science as a visible, tangible phenomenon. His Megatherium lecture, which lasted until midnight, was the grand finale of the Oxford meeting, bringing techniques he had perfected in the cloistered setting of academic geology out into the public realm. Spectacle remained a central feature of later meetings, and geology continued to take centre stage.

But the BAAS was only part of the picture. As the gentlemen of sci-

1. Manuscript copy of Buckland's lecture-notes, quoted in Rupke 1983b, 242–4.

2. The Leviathan analogy is taken from Buckland's later Bridgewater Treatise: W. Buckland 1836, I, 141–64. On Darwin's fossil-collecting see Herbert 2005.

3. On Buckland's lecture more generally see Rupke 1983b, 242–4; Boylan 1984, 198–200; 1997, 368.

4. Gordon 1894, 126–33, which also contains extracts from Buckland's lecture-notes.

5. The early years of the BAAS are examined in Morrell and Thackray 1981.

ence consolidated their position at the top of the knowledge-producing hierarchy, other groups were becoming increasingly numerous and active. Besides the prestigious learned and scientific societies in the metropolitan centres of London, Edinburgh, and Dublin, and the magnificent fossil museums associated with the top medical colleges (Fig. 5.1), provincial centres across Britain already boasted a thriving network of middle-class philosophical societies and institutions, variously termed "Literary and Philosophical Societies", "Philosophical Institutions", or "Literary and Scientific Institutions". Since the late eighteenth century such societies had enabled early industrialists, engineers, capitalists, and merchants to intellectualize their passion for "progress" by pursuing natural knowledge. In the 1820s and 1830s these organizations formed the solid bedrock of provincial geological practice.[6] Many had their own libraries, museums, and lecture programmes, with geology prominent in all three sites. The BAAS's success both drew on and fed back into the work of these institutions, several of which published high-quality annual volumes of proceedings.

All this was grist to the mill of printers and publishers. Since the slump of 1825–6 they had been wary of relying on the production of expensive new novels, conventionally published in a three-volume format priced at 31s 6d. They now focused increasingly on non-fiction and, in particular, periodicals, encouraging and exploiting the new interest in science and the "useful arts" in a branching proliferation of titles. Between 1815 and 1830 the number of specifically science-oriented periodicals had more than trebled;[7] their number shot up still more after 1836, when taxes on newspapers dropped significantly. Earth history played a prominent role not only in scientific periodicals like the *Philosophical Magazine* (founded in 1798) and prestigious quarterlies like the *Edinburgh Review* (f. 1802), but also in cheap miscellanies like the *Penny Magazine* (f. 1832), in denominational journals such as the Dissenting *Eclectic Review* (f. 1805) and the bi-weekly Free Church newspaper *The Witness* (f. 1840), in illustrated weeklies like the *Illustrated London News* (f. 1842), and even in primarily fictional miscellanies like Charles Dickens's weekly *Household Words* (f. 1850), besides a whole host of lesser-known titles aimed at women or children. The market for non-fiction books was expanding at a similar rate. New books on geology for the general reader appeared in

6. On geology and British philosophical societies see Knell 2000, 49–72; for the wider social context see Morrell and Thackray 1981, 2–16. Case-studies include Orange 1973; A. Thackray 1974; Kitteringham 1982; Morrell 1985; Taylor and Torrens 1986; Torrens and Taylor 1987–94; Taylor 1994; J. Secord 2000a, 195–213; Morrell 2005 (especially 39–71); Torrens 2006a; and the essays in Inkster and Morrell 1983. For a contemporary genteel perspective see [C. Lyell] 1826a.

7. Topham 2004a, 63. On publication trends across the century see S. Eliot 1995; for a contemporary view of the economics of authorship see [Lewes] 1847.

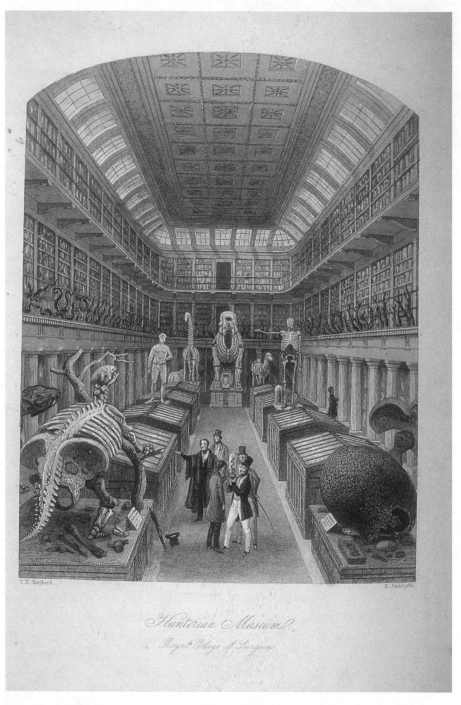

*Figure 5.1.* The Hunterian Museum of the Royal College of Surgeons, London, *c.* 1842, showing spectacular South American fossil mammals in the foreground (Megatherium and Glyptodon) and a giant human skeleton in the background. Anon. 1841–5, I, facing p. 129.

increasing numbers and with varying claims to authority, ranging from Buckland's palaeontological masterpiece *Geology and Mineralogy* (1836) down to smaller-scale productions like the best-selling writings of Miller and children's books by professional writers like Maria Hack and Samuel Clark. Blurring the boundary between book and periodical were serially produced compendia of useful and/or entertaining information, such as Thomas Milner's *Gallery of Nature* (1846), which made much use of writings further up the scale of perceived expertise.

Geology was practised and publicized with new vigour. It became the subject of polite conversation, newspaper articles, fierce public controversy, popular fiction, and even government funding.[8] In the 1830s it seems to have been the most popular science among the chattering classes, at least according to writers in the London literary periodicals.[9] Its popularity long outlasted its first flush of fashionable lustre. In 1844 the *Athenaeum* declared that geology was no longer "idolized" as it had been,[10] perhaps because the latest theories of species transmutation had taken over the London limelight that year with the appearance of the anonymous book *Vestiges of the Natural History of Creation*. But fossils remained central to the ensuing controversy. Besides, London fashions did not always dictate those of other regions of Britain. In 1853 the naturalist Edward Forbes acidly observed that geology remained a favourite topic of provincial dinner-party conversation, fuelled by the "prodigious" quantity of "charming" popular treatises devoured by self-important "town-dandies". Real geology, fumed Forbes, was not about reading and chattering; it was about getting out into the field.[11] The terms of his critique are revealing: it is as if geology were too fashionable for its own good.

Nor did its popularity show any sign of abating in the 1850s. The floodgates had opened with Lyell's *Principles*, but the torrent of geological prose flowed on into the second half of the century, when it was joined by a powerful spate of evolutionary writings, and the drama of earth history—still recognizably the same story—incorporated new theoretical frameworks.[12] The old framework of natural theology and separate creation, however, retained a significant foothold in the mainstream popular-science market long after the new breed of scientific professionals had repudiated it.[13] Publications of this kind continued to excite controversy in some circles. As late as 1873, according to Arnold Bennett's meticu-

8. The British Geological Survey was founded in 1835 after the example of France.
9. [Apjohn] 1833, 409; Anon. 1838; 1844a.
10. Anon. 1844a; 1844b, 1137.
11. [E. Forbes] 1853, 340–1. On Forbes see *ODNB; DNBS*.
12. Bowler 1976; 1988.
13. Lightman 1997b.

lously reconstructed portrait of life in the late-Victorian Midlands, *Clay-hanger* (1910), Miller's seminal treatise *The Old Red Sandstone* (1841, but still in print) was being talked about in the Potteries as a "dangerously original" piece of writing.[14] Bennett may have been exaggerating for nov-elistic effect, but his words do not suggest tired familiarity with a long-established science.

At one level, not much was new about the content of popular geo-logical writing after Lyell's *Principles*. The corpus, although much more voluminous after 1830 than before, occupied the same generic coordi-nates as before: established genres such as travel writing, natural-his-tory compendia, antiquarian works, biblical scholarship, and natural-theology treatises continued to shape and house the growing literature of earth science. A similar mingling took place within the generically capacious covers of periodicals. The poetics of geology had been under construction for some time, and the same old techniques were still used: comparisons with romance, descriptions of huge monsters, folklore allu-sions, devotional rhapsodies, poetry quotations, vivid restorations of the past. Thanks to the scissors-and-paste methods of journalists, reviewers, and other authors, many of the most memorable examples of these tech-niques were now reproduced almost *ad infinitum* in subsequent publica-tions, leading to a considerable degree of conservatism in the imagery of the ancient earth. Only an unusually fortunate pterodactyle, for in-stance, could escape the comparison with Milton's "Fiend" suggested by Buckland in 1829.

But these techniques had occupied only a tiny fraction, if that, of the average geological treatise in 1820. They loomed much larger after 1830. In particular, word-paintings of the deep past became longer, more confi-dent, and more elaborate, mirroring the increasing currency of pictorial restorations in print. The visual iconography of antediluvian caricature, circulated among a handful of gentlemen, was now giving rise to a new genre of more serious, mass-produced fossil restorations.[15] Equally, ver-bal narratives of geohistory and the resurrection of fossil beasts were bursting out of the restricted milieu of handwritten or privately printed humorous verses, and were making their way into sober prose. Not that this was always prose: earth-historians after 1830 quoted from modern poetry much more freely than before, some even using verses they had composed themselves. Nor was it always sober: many authors turned their verbal restorations into dramatic visions of former worlds, marked off stylistically so as to appear islands of semi-fictional discourse within a surrounding matrix of facts. In these purple passages they borrowed

14. A. Bennett 1954, 126, discussed in Harvie 2003, 36.
15. Rudwick 1992.

openly from the legends, romances, novels, and epic poems with whose enchantments they were now competing, and anticipated the techniques of later science-fiction writers. In some cases, episodes of fictionalized science could expand to become the framework for a whole book, spawning new genres from the old melting-pot. When surveying these passages, from cautious restorations to full-scale voyages back in time, it becomes impossible to draw a clear dividing-line between popular science and science fiction.

The contrast with the earlier period is best seen in Mantell, whose work straddles the divide and who, in the 1820s, had already shown himself aesthetically more adventurous than most geologists. His 1822 vision of the Iguanodon country had been confined to a single sentence in the form of a question; in 1826 he had answered that question with just over a page of landscape description, referring readers who wanted more to the separate publications of other geologists.[16] Only a single poetry quotation was to be found in the two books, and only one cautious little vignette of a pictorial restoration.[17] In 1838, however, his popular treatise *The Wonders of Geology* offered detailed poetic meditations and verbal restorations. These included a dramatic nine-page-long sequence of scenes which took his readers through the changes undergone by the landscape of Sussex from the Age of Reptiles up to the Victorian era, first in straightforward third-person narrative, then in the imagined voice of an extraterrestrial visitor—a technique borrowed openly from the *Arabian Nights*.[18] The book was fronted by a melodramatic engraving by Martin showing ancient reptiles at war (see Fig. 7.6) and littered with quotations from modern poetry at climactic moments.[19] Mantell was revelling in the aesthetic possibilities of his science.

Where did his new confidence come from? It was not simply a function of new fossil discoveries, important though these were: most of the creatures mentioned by Mantell had been identified long before. Nor was it purely a matter of his growing confidence in his own scientific authority and literary abilities, though doubtless these helped. Several new contributors to the literature came onto the scene in the 1830s, such as the Somerset fossil-collector Thomas Hawkins in 1834, the Cromarty accountant and ex-stonemason Hugh Miller in 1835, and the Brighton polymath George Richardson in 1838:[20] looking at their earliest geologi-

16. See above, pp. 123–7.
17. Mantell 1822, 305, 175.
18. Mantell 1822, 57; 1827, 82–4; 1838, 369–76.
19. Mantell 1838, 45–6, 87, 168, 491–2, 517, 603, 680.
20. On Hawkins see Taylor 2002; *ODNB*. On Miller see Bayne 1871; Shortland 1996a; Borley 2002; 2003; Taylor 2003; 2007; Knell and Taylor 2006; *ODNB*; *DNBS*. On Richardson see Torrens and Cooper 1985; *ODNB*; *DNBS*.

cal writings we find none of that diffidence with which Mantell had first approached the restoration of the distant past. From the very beginning, these writers displayed the same rhetorical confidence that had taken Mantell so many years to build up. Clearly, a change had taken place in the whole environment of popular geological writing.

The move to a more spectacular and self-consciously poetic style may be seen as part of the changes undergone by the concepts of expertise and popularization since the Napoleonic Wars. If science's social catchment area were to be widened to include the middle and working classes, as many liberal educationalists insisted it should, then it would need to be entertaining enough to attract them and elementary enough to avoid confusion. Scientific practices of all kinds were becoming more specialized, and the new readerships could not be expected to know the basics of any science, let alone one as new as geology; nor could popularizers assume the common context of a Classical education. This perception of a clear gradient of knowledge between expert and public was integral to the new concept of popular science as "rational amusement" provided from on high, delivered from the knowledgeable expert to his ignorant audience. Professional writers and the editors of periodical miscellanies were under ideological and commercial pressures to conform to this model. They reinforced it by seeming to defer to the gentlemen of science and communicate their findings to their public, placing themselves in a mediating position between the two poles of expertise and ignorance.

Of course this model did not work exactly as it was supposed to. For one thing, the literary professionals had agendas of their own which did not necessarily match those of the gentlemen whose names littered their work.[21] Writers who were also practising geologists and collectors, like Miller and Hawkins, deferred to the gentlemen on many (but not all) larger theoretical issues, but foregrounded their own authority when it came to their local areas of expertise. The Edinburgh journalist and folklorist Robert Chambers represents an extreme case, bringing together a genteel line-up of expert witnesses in *Vestiges* to make his own controversial case for species transmutation. Not surprisingly, the gents responded with fury.

But however much individual writers might manipulate the new concepts of expertise and popularization for their own ends, the concepts themselves remained central. All these writers—practising men of science, professional writers, or both—shared a sharpened sense of the need to *educate,* to lead their readers into a new and unfamiliar world of knowledge.[22] The rhetorical stances of the guide, storyteller, and master

21. Topham 2000; Fyfe 2004a.
22. Topham 1992.

of ceremonies, revealing marvels to an awestruck audience, clearly fitted this educational model, inscribing both authority and wonder into scientific discourse. As we have seen, these stances had served a useful subsidiary purpose in some geological writings before 1830. The new premium placed on public education now pushed the rhetoric of spectacular display to the centre of the field. The rest of this book examines how this rhetoric was sustained.

In Part I we were able to survey the development of this rhetoric chronologically, as a narrative. As with any story, some complex issues had to be simplified and some interesting subplots glossed over; but a single line of development was discernible. After 1830, however, the relevant material is not only much more numerous, but presents a branching or free-wheeling set of variations on themes established during the earlier period—increasingly adventurous, perhaps, but not easily reducible to a narrative. In Part II, then, we shall explore this material thematically rather than sequentially. Chapters 5, 6, and 7 examine the various sites and arenas in which geological spectacle took place, paying special attention to urban visual culture. Representations of distant times and places at the theatres and panoramas provided a crucial template for the concept of the "imaginary voyage", and they provide us with valuable insights into how the Victorian mind's eye was expected to work and how texts and images worked together in landscape spectacle. In this light, chapter 8 examines how and why earth-historians wove pictures and poetry into their works. We shall then be in a position to explore their restorations of the ancient earth in chapters 9 and 10, to see how these authors developed increasingly theatrical and panoramic means of staging the past before the mind's eye, culminating in the time-travel narratives and visionary pageants of Hugh Miller.

Miller was perhaps the most successful of geology's early popularizers, a major literary figure in his own right who best embodies the Victorian union of science and literature. His sudden death in 1856 has been taken as the end-point for this book, although many of his geological writings were not published in book form until two or three years afterwards. By this time, many of the other popularizers whose work we are examining had either died or were no longer writing about geology. Thus ended an extraordinary half-century of creative innovation and literary experimentation, resulting in a composite vision of earth history which still dominates our books and screens today. Beneath the state-of-the-art technology of the BBC's docu-drama *Walking with Dinosaurs* (Tim Haines, 1999) lies a sturdy Georgian-Victorian framework.[23] The modern reader unfamiliar with this nineteenth-century background will expe-

---

23. Freeman 2004, 259–63.

rience a sharp shock of recognition when turning from the television screen to Victorian earth-histories.

But this shock of recognition can be misleading. For many people today—certainly for most of those likely to be reading this book—watching a television programme about dinosaurs is not a particularly challenging experience. These ancient monsters, half fact, half fiction, are a familiar part of our world-view. The earth may be billions of years old, but this is too is a cliché: those who relive today Playfair's sense of chronological vertigo in 1788 are often geologists themselves, since they are the ones who actively contemplate the distances involved. Similarly, the idea that a whole class of organisms can vanish from the face of the earth raises few eyebrows today. For many of us these remarkable concepts have become unquestioned items of our mental furniture, rarely dusted down and more likely to be stored in the attic than displayed in the front room. We therefore need to make a monumental effort of imagination when trying to think ourselves into the position of the uninitiated general reader in, say, the mid-1830s. These concepts could not be taken for granted: for most people then (as for many today), they were new and controversial claims, not tired old facts.

Few people encountered geology as a single, monolithic entity or theory, located in a lofty "scientific" sphere. Like other kinds of Victorian natural knowledge, geology provided one of many available forms of rational amusement. It had to compete for attention in a range of arenas—not only in the various forms of print culture, but also in public spaces like museums or lecture-halls, and in private conversations. Its identity was not fixed: in the next section I will show that the uninitiated reader was faced with a variety of conflicting geologies demanding assent. Chapter 6 will then consider which sectors of society the new science was chiefly aimed at, and the range of book formats and literary genres in which it made its appearances.

## Varieties of Earth History

Geology's public face today is often perceived as a solid geohistorical consensus. During the 1830s and 1840s, however, such a consensus was harder to imagine from the standpoint of the literate classes. It is popularly held that when Lyell raised the public profile of the new geology in 1830, his colleagues joined in to promote his version of the science, diffusing it throughout the public realm. Audiences were then compelled either to accept it, since it self-evidently made better sense of "the evidence", or to let their religious scruples blind them into rejecting it. In fact, Lyell's vision was atypical in denying both the directionality of earth history and the possibility that it had been punctuated by at least

one catastrophe dwarfing those observed today. Most of the leading proponents of the new geology in the 1830s favoured a progressionist model
of earth history, in which simpler organisms had been succeeded by more
complex ones, culminating in humans; and many of them continued to
interpret the diluvial gravels and erratic boulders of northern Europe as
evidence of an extremely large flood or (from the 1840s) an episode of
large-scale glaciation.[24] But even within this "progressionist consensus"
there was much room for differences of interpretation, especially when
interpreting diluvial phenomena or defining the boundaries of specific
formations.

The latter activity could escalate into bitter controversy. The retired
soldier Roderick Murchison engaged in several territorial campaigns to
maintain the boundaries of his distended "Silurian System", fiercely re-
sisting encroachments on his fossil empire both from below, by Adam
Sedgwick's "Cambrian" formations, and from above by Henry De la
Beche's "Devonian". These disputes were conducted both within the
Geological Society and at the public forum of the BAAS. In the prestigious quarterlies, review articles by accredited men of science provided
forums for further debate.[25] For the onlookers, gentlemanly geology was
quite obviously a contested area.

Some tenets, of course, were affirmed by all the participants in these
high-level debates. Murchison, Sedgwick, De la Beche, and Lyell all
agreed that the earth was probably millions of years old and that the
fossil saurians they were digging up had become extinct ages before humans had come on the scene. There was also a broad interdenomina-
tional consensus among the gentlemen of science that each species had
been separately created by God. But none of these articles of faith went
unquestioned outside the inner sanctum of the Geological Society. To
gain some sense of just how confusing earth history could be for the
general reader in the 1830s and 1840s, let us tune in to some of the competing voices from the conservative and radical ends of the theoretical
spectrum.

As in the 1820s, literalist earth-historians continued to promote versions of geology according to which humans and giant saurians alike
had been created less than a week after the earth.[26] In the 1830s many of
them became concerned at the growing popularity of old-earth geology,
whose public face had turned away from the self-conscious empiricism
of the Geological Society's early years to an explicitly old-earth science.
Those promoting this alternative Creation-narrative were increasingly

24. On these debates see Boylan 1984; Finnegan 2004.
25. Rudwick 1985; J. Secord 1986.
26. J. Moore 1986.

denying the Bible's scientific authority. To many literalists this was unacceptably elitist as well as misguided, preventing the common reader from accessing natural knowledge through "the poor man's book".[27] A coterie of gentlemen was setting itself up as the only group qualified to interpret the "strange" and "repulsive language" in which they claimed the book of nature was written, just as proponents of the "Higher Criticism" in Germany seemed to be claiming that a deep knowledge of Hebrew, Greek, Syriac, and Ethiopic was necessary to understand the Bible.[28] Barbarous languages and scientific jargon were being wielded to conceal truths that the English of King James had once made clear to all: "Geologists [. . .] would fain keep the uninitiated from equal knowledge with themselves. They throw a mist, therefore, around the science, and frighten away the vulgar with unintelligible endecasyllables and cacophonous names."[29] To many, this was an intellectual equivalent of the Acts of Enclosure, which were now shutting "the vulgar" out of the landscape. Conservative fears about geology's encroaching secularization and elitism were confirmed by the public success of the BAAS, one of whose purposes was to transcend denominational boundaries by keeping a safe distance from particular interpretations of the Bible and promoting a broad-church natural theology instead.[30]

During the 1820s a strongly oppositional stance towards the new geology had been confined to a few literalists only. In the 1830s, however, the new confidence, public visibility and clear anti-literalism of leading geologists goaded many literalists into arguing against the new theories in sermons, lectures, pamphlets, articles, and treatises.[31] They found a ready audience among the Anglican middle classes, where more extreme forms of evangelicalism were increasingly taking root.[32] Literalists shared a common interest in maintaining the Bible's authority, but they still did not form a connected school or movement. Socially and professionally, they were no less diverse than before: contributors in the 1830s included professional writers like Mary Roberts, gentlemen-scholars of independent means like George Fairholme, and antiquaries such as the lawyer

27. "The Bible after this must cease to be, what it has ever been considered to be, the poor man's book" (J. Brown 1838, 50).

28. Sedgwick 1833, 9.

29. Cockburn 1840, 9. Such protests recall those made by Robert Boyle and his colleagues in the late seventeenth century against the elitist jargon of scholasticism; it is no coincidence that many nineteenth-century literalist writers on geology saw themselves as upholding the "true" Baconian values.

30. Brooke 1979; 1991; Morrell and Thackray 1981, 224–45.

31. As before, their tone ranged from the apoplectic (H. Cole 1834) to the urbane (Fairholme 1833).

32. Hilton 1988, 10.

Sharon Turner, the great pioneer of Anglo-Saxon historiography.[33] Clergymen were, however, more numerous than before in the new wave of literalist earth-history. Their number included politically powerful figures such as William Cockburn, Dean of York, and leading naturalists in other fields such as the famous entomologist William Kirby, who was commissioned to write one of the Bridgewater Treatises. The Presbyterian minister George Young, who had been active in promoting geology in the 1810s and 1820s, continued to contribute to the literalist cause: his work, too, took on a more polemical edge in the new climate.[34]

Clerical geologists of the new school came in for special condemnation. Buckland's defiantly non-literalist Bridgewater Treatise of 1836 was branded a "direct and real, though disavowed attack on the Mosaic narrative of the creation, made by a Clergyman of the Church of England".[35] The BAAS, too, consistently came under attack, especially by the pugnacious Cockburn.[36] Here the literalists were joined by the Tractarians or "Oxford Movement", a group of influential Anglo-Catholic clergymen who objected to the secular "religion of science" which the BAAS seemed to be promoting.[37] These figures, including John Henry Newman and Edward Pusey, were more concerned to restore the Church's secular authority than a particular interpretation of Genesis, so the two groups' objections took different forms; but they both won plenty of media attention, and their challenge to the independent authority of the new geology was taken very seriously by its proponents.

At the other end of the ideological spectrum were those who used the progressionist consensus as a basis for theories of species transmutation. If the fossil record exhibited a progression from lower to higher organisms, this narrative could easily be rewritten as an evolutionary story, with new species generated by a natural law rather than by direct divine intervention. With its anti-miraculous stance and its implication that upward mobility was built into the system of nature, transmutationism was strongly associated with radical politics and deistic anticlericalism.[38] It

33. Mary Roberts 1837; Fairholme 1833; S. Turner 1832. For biographical information on Turner see *ODNB*. On Roberts see Lindsay 1996; Gould 1997; Opitz 2004; *ODNB; DNBS*. On Fairholme see Mortenson 2004, 115–30.

34. Cockburn 1840; Kirby 1835, I, 20–43; G. Young 1838; Best 1837; J. Brown 1838; De Johnsone 1838. For biographical information on Cockburn see Orange 1975, 284. On Kirby see Topham 1993, 113–23; *ODNB; DNBS*. Cockburn, Best, Brown, and De Johnsone are not included in the major reference works.

35. J. Brown 1838, 3. On the literalist response to Buckland's treatise see Rupke 1983b, 209–18.

36. See Morrell 2005, 149–52. For a similar attack see [Croly] 1837, 690–1.

37. On this opposition see Orange 1975; Yule 1976, 269–316; Morrell and Thackray 1981, 229–45; Rupke 1983b, 267–73; Yeo 1993, 79, 124–7.

38. The definitive study of these associations is Desmond 1989.

provided radicals with a means of opposing the gentlemen's conservative interpretation of the fossil record without having to explain away or ignore the directionality of the fossil record itself, since the latter was becoming something of an embarrassment to followers of the old eternalist cosmology.

Lyell had unwittingly introduced many British readers to transmutation when he spent many pages refuting Jean-Baptiste Lamarck in his *Principles;* but its first major public appearance in nineteenth-century Britain—in a rather different form—was in *Vestiges of the Natural History of Creation.* This anonymous treatise was written by Chambers and published in London in 1844 to great popular acclaim.[39] Weaving all the physical sciences together into a single argument for "development", *Vestiges* confirmed the literalists' and Tractarians' worst fears about the incipient secularism of the BAAS's religious stance: it claimed to be an exemplary work of natural theology, but it reduced God's creative role to that of a distant First Cause.[40] The more radical writers did away with God altogether, using genteel geology to promote materialism in penny weeklies such as the *Oracle of Reason* (f. 1841).[41]

Most galling for the gentlemen of science was the implicit challenge posed by transmutationists and literalists alike to the new model of scientific authority.[42] The anonymous author of *Vestiges* was clearly not an expert in any of the physical sciences, yet he or she had confidently pieced them together to form a grand theory of his or her own. This apparently irrational behaviour led some gentlemen to suspect female authorship: a prime suspect was the brilliant mathematician and computing pioneer Ada Lovelace, Byron's daughter.[43] In the new order, "compilers" were supposed to show a proper respect for the experts on matters of theory. They were not supposed to engage in high-level speculation themselves, still less to encourage their readers in turn to develop such dangerously independent habits of mind. The literalists' treatment of expert authority was similarly cavalier. Mary Roberts's *Progress of Creation* (1837) begins by eulogizing (Jameson's) Cuvier, but ends by dismissing his analysis of the mastodon in favour of her own judgement: "Cuvier describes this animal as herbivorous, but surely without reason. We can judge of its nature, only by its remains".[44] The absence of self-consciousness or polemic in this and similar passages suggests that Roberts was not openly assaulting the new concept of scientific expertise, but was simply unaware of its

---

39. On the response see Millhauser 1956; J. Secord 2000a. On Chambers see *ODNB; DNBS.*
40. Brooke 1979, 50; Morrell and Thackray 1981, 245; Yeo 1984, 10–11.
41. On these productions see Topham 1998, 259–61; J. Secord 2000a, 308–20.
42. Yeo 1984.
43. J. Secord 2000a, 183–4; *ODNB; DNBS.*
44. Mary Roberts 1837, vi, 243.

implications, maintaining instead the older eighteenth-century model of open debate. Alternatively, she may have been using this "naive" stance as a cloak for determined opposition.[45] In any case, the title-page of her book proclaimed her feminine identity, doubtless affording an excuse for lofty dismissal among the gentlemen of science: I have found no reference to her book in any of their writings. *Vestiges* represented a more damaging affront to scientific etiquette, partly because it was anonymous, but also because it promoted socially corrosive views. The gentlemen were disgusted, closet transmutationists like Darwin watched warily from the sidelines, and the crowds cheered as the fighting continued.

Both versions of earth history, conservative and radical, gained much public attention. Many of Cockburn's pamphlets cost only sixpence, in sharp contrast with the lavish works they attacked, and went into several editions within a year of appearing. More expensive books like Turner's *Sacred History of the World* (1832) went through eight editions in sixteen years.[46] As for the sermons in which literalist earth-history was affirmed, only the most prestigious have been recorded; the total number probably far outnumbered that of the extant printed texts. *Vestiges,* too, was a runaway success for its publisher, going through seven editions in under four years; between 1847 and 1851 eleven thousand copies were sold for 2*s* 6*d* apiece (roughly £22 or $43 today). Much cheaper transmutationist texts were to be found in the radical press, whose publications continued to enjoy a wide circulation. Both varieties of earth history were therefore read and heard by a significant sector of the public.[47]

To maintain this publicity, both literalists and transmutationists made full use of science's imaginative potential. One of the earliest pictorial representations of "fossil man" emanated from radical circles: the talented journalist Charles Southwell incorporated an engraving of the savage negroid ancestor of modern humans into the first issue of the *Oracle of Reason* in 1841, copied from an 1838 article by the radical French naturalist Pierre Boitard.[48] The use of imaginative rhetorical devices has been well documented in the case of *Vestiges:* James Secord has described the opening paragraph, presenting "the earth we inhabit" within the solar system, as "a verbal orrery, calling up the central image of astronomical display". He has also drawn attention to Chambers's construction of a chatty, approachable narrator in imitation of Walter Scott's novels, using homely analogies such as "a saddler's cutting-knife" to describe the head

45. Compare the interpretation of Gould (1997, 37–8).

46. Encouraged by the success of his initial volume, Turner wrote two further volumes in 1834 and 1837.

47. On the sales of *Vestiges* see J. Secord 2000a, 131.

48. [Southwell] 1841, 5. On this image see J. Secord 2000a, 312 (the paperback reprint of 2003 has fuller information). On Southwell see *ODNB*.

*Figure 5.2.* Frontispiece of W. Elfe Tayler's literalist treatise *Geology: Its Facts and Its Fictions.* The background sea-scene is based on *Duria antiquior* (Fig. 2.2). Tayler also sprinkled his book with woodcuts lifted from the works of his opponent Hugh Miller. Tayler 1855, frontispiece.

an outstanding innovator in the poetics of geology, was himself building on the work of literalist popularizers in the 1830s.

These exchanges and continuities confirm that the power-struggle over who was to tell the history of the earth was, by its very nature, a contest for the public's imagination. All the contestants were aware that the winner would need to be a consummate storyteller. Fairholme stated as one of the advantages of a literalist earth-history the fact that "the current of the narrative runs smoothly along, and our minds feel satisfied, and at rest, instead of being constantly suspended in doubt and uncertainty"; modern geological theories, by contrast, were too bewildering to work as narratives.[60] The terms of this critique played along with the BAAS's idea that science should be public, clear, and visible. Indeed, this overwhelming focus on visibility was what the Tractarians found so objectionable about the BAAS's hierophants, who seduced their votaries' gaze away from the invisible truths of Christianity using the "wonders" of the "visible creation" as bait.[61] In the 1850s Newman—by then a Roman Catholic—traced geology's popularity to the imaginative appeal its proponents made, constructing a glorious pageant of Creation

60. Fairholme 1833, 324. A similar sentiment was expressed by Kirby (1835, I, 39–40).
61. [Bowden] 1839, 36.

which made the biblical narrative seem paltry by comparison. Newman knew what he was talking about: back in 1821 Buckland's lectures had "open[ed] an amazing field" to his own "imagination".[62] In an essay of 1859 Newman warned that the "vastness" of the revelations of modern science led its intoxicated devotee to view Christian narratives with "a most distressing revulsion of feeling", in that "his imagination is bewildered, and swims with the ineffable distance of that faith" from science's "exuberant life and reality".[63] Part of Newman's aim in his own theatrical sermons was to "intoxicate" his listeners in another way, pointing them towards the right kind of "exuberant life". This was the age of spectacle, and theatre was the common enabling principle by which all sides drew crowds and won assent.[64]

Proponents of these competing earth-histories often staked out their position by attacking the alternatives. As a result, polemics and dogmatism played an increasingly important role in the literature of Victorian science. Forums for such attacks were carefully selected for maximum impact and audience support. Public arenas like the BAAS and many of the philosophical institutions favoured the gentlemen of science, and old-earth geology was also promoted in review articles by compliant literary periodicals associated with the London intellectual elite, such as the Quarterly Review, the Literary Gazette, and (from the mid-1830s) the Athenaeum. But this was a fraction of the total periodical circulation in a period when publishers and editors were rapidly sniffing out all existing niches and carving out new ones. On issues as charged as the role of science in Christian living, specific newspapers and periodicals soon became associated with particular stances, just as the sermons in specific churches could be expected to take a certain line on attitudes towards Scripture.

Much depended on who the editor or preacher was. These two networks, churches and periodicals, were crucial sites for the formation of knowledge communities in the Victorian era, and they interlocked as the market for denominational journals grew apace. Both networks crackled with controversy, and in the 1830s, 1840s, and 1850s the history of Creation was one subject on which congregations and readerships could be defined.[65] Miller edited the Scottish Evangelical newspaper The Witness from 1840 until his death in 1856, and used it as a platform for promoting old-earth geology and denouncing literalism, "Puseyism", and transmu-

---

62. Gornall et al. 1961–77, I, 109. On Newman see ODNB.

63. Newman 1859, 324–7. For discussion see Klaver 1997, 165–6.

64. On the BAAS's theatricality see Morrell and Thackray 1981, 157–63. On radical theatricality see McCalman 1992.

65. On the importance of oppositional stances to the formation of distinct readerships see Klancher 1987 and Topham 2004b. See also Yeo 1993, 44–6; Topham 1998, 257–8.

tationism. Newman, for his part, used his editorship of the High Church *British Critic* in the late 1830s to attack the tendencies of modern science. Nor were Victorian cities short of actual pulpits: in Oxford, for instance, Newman's sermons as vicar of the University Church, St Mary's, competed with those given by gentlemen of science in college chapels and elsewhere.

Some editors used subtler strategies to steer their readerships through the maelstrom. Under Samuel Wilks's editorship, the evangelical monthly *The Christian Observer* became a prominent forum for debates over Genesis and geology. In 1834 Wilks printed Fairholme's pseudonymous letter "On the infidel tendency of certain scientific speculations", William Conybeare's crisp response, and a counter-response by Fairholme.[66] For giving Fairholme a voice this periodical has been accused of taking an anti-geological stance,[67] but the opposite was the case. Wilks used his position to undermine the literalists in mid-argument, adding lengthy footnotes to defuse their objections to the new school of geology and in one issue adding his own moral poem about a fossil shell.[68] The literalist Henry Cole was allowed to contribute an article on the burial service unscathed by such footnotes, but his polemical literalist treatise earned a withering review from Wilks in the same year. Geologists like Buckland and his American counterpart Edward Hitchcock valued this form of support and recommended these articles to their own readers.[69]

Yet, for all these frantic attempts by editors and clergymen to fix their audiences' opinions, no single variety of earth history was self-evidently persuasive enough to justify later historians' assertions that the debate was clear-cut and settled among "most educated people" in early Victorian times, still less that literalist earth-histories were confined to the margins.[70] Geology was still a very new science: a change of editor, reviewer, or vicar could mean a completely different theoretical stance. Some periodicals were slow to take sides. Between 1830 and 1833 the *Athenaeum* printed a positive review of each volume of Lyell's *Principles,* but also a sympathetic review of Fairholme's *Geology of Scripture;*[71] only in

66. [Fairholme] 1834a; Conybeare 1834; [Fairholme] 1834b. For the attribution to Fairholme see Rupke 1983b, 47.

67. Morrell and Thackray 1981, 236: "the evangelical *Christian Observer* [. . .] accused Daubeny of doubting the inspiration of the Old Testament." In fact this accusation was made by Fairholme, with whose remarks the editor certainly disagreed.

68. Rupke 1983b, 47; for the poem see [Wilks] 1834a.

69. W. Buckland 1836, I, 33 n.; Hitchcock 1851, 230.

70. Lynch 2002b, xi. For a localized indication of the widespread currency of literalist earth-history, see Topham 1998, 257–8.

71. Anon. 1830b; [Apjohn] 1833; [Lindley] 1833.

the mid-1830s did this periodical begin to display clear old-earth sympathies. More seriously, individual writers often seemed uninterested in the theoretical gulf separating literalists from geologists of the new school. One writer for the *Eclectic Review* in 1831 quoted Sedgwick with approval, but on the same page warmly recommended Andrew Ure's literalist *New System of Geology* as "an introduction to geology" of "spirit and right feeling".[72] Footnotes achieved the same effect in popular expositions like Maria Hack's largely Lyellian *Geological Sketches* (1832) and Turner's literalist *Sacred History of the Earth,* both of which recommend the work of Ure and his opponents without registering any dissonance.[73] Roberts's prefatory tribute to Penn and Cuvier in the same breath strikes a similar note. The boundary between literalist and non-literalist geologies was still easily blurred.[74]

In fact, although certain schools of thought were emerging in clear opposition to each other, plenty of writers belonged to none of these schools. One of these was Thomas Thompson of Hull, who contributed to the *Magazine of Natural History* in 1835.[75] This periodical gave a strong voice to the new geology, and tended to review literalist works unfavourably. However, it did print Thompson's article on the biblical names "Than", "Leviathan", and "Behemoth". Thompson proposed that "Behemoth" designated the Iguanodon, which was shown to be contemporary with humans by a passage in the Book of Job ("Behold now behemoth, which *I made with thee*").[76] On the basis of biblical evidence Thompson dated humans right back to the Cretaceous period, the lack of human fossils in the chalk was blamed on the hungry Megalosauri who had been unable to resist crunching up all the available bones of "so tempting a mouthful as a human being".[77] Thompson's interpretation of Genesis 1 is clearly not literalistic in the sense I defined earlier, and he leaned heavily on the work of geologists like Mantell and John Phillips; but his departure from literalism took a very different form to theirs.

Despite the programmatic assertions of the gentlemen of science, theories of the earth were far from dead. For those of a speculative inclination, there were many more ways than Buckland's (or Penn's) to

72. Anon. 1831, 80.

73. Hack 1835, 166–7, 214; S. Turner 1832, 446–58; Mary Roberts 1837, vi. Compare Astore 2001, 93–4.

74. For other examples of boundary-blurring, see Yule 1976, 97–8, 328–30; Michael Roberts 1998, 252–3.

75. Rupke 1983b, 220.

76. Thompson 1835, 318. The literalist J. Mellor Brown (1838, 10) used the same quotation (including italics), but to bring Iguanodon into the period immediately before the Deluge rather than to push back the date of man.

77. Thompson 1835, 318.

fill in the gaps of the tantalizingly laconic narrative of Genesis 1. Miller, for instance, recalled his response to one view reported to him in the 1820s by a friend studying at King's College, Aberdeen: fossils, in this view, were organic remains left over from a pre-Adamite world whose rational inhabitants had long before been resurrected and rapt to higher (or lower) spheres at a Judgement Day of their own, so that no humanoid bones remained.[78] Miller's response shows how powerful these alternative geohistories could be in forming ideas of the deep past:

> How strange the conception! It filled my imagination for a time with visions of the remote past instinct with a wild poetry, borrowed in part from such conceptions of the pre-Adamite kings, and the semi-material intelligences, their contemporaries, as one finds in Beckford's "Vatheck," or Moore's "Loves of the Angels;" and invested my fossil lignites and shells, through the influence of the associative faculty, with an obscure and terrible sublimity, that filled the whole mind.[79]

These Palaeozoic pre-Adamites and man-eating megalosaurs confirm that the solid, secular, public geohistorical consensus from which viewpoint we now smile did not exist in the 1830s, nor for some time afterwards. This consensus, along with the new model of expert authority and scientific "orthodoxy", was under construction by the gentlemen of science, under constant attack by its opponents, and subverted both deliberately and unintentionally by a host of other writers. The British reading public—even those who read specialized science periodicals like the *Magazine of Natural History*—had to make what they could of a welter of interpretations.

The reprinting of older, outdated accounts added to the potential for confusion. In the increasingly lucrative field of popular-science publishing, up-to-the-minute geotheory was not the prime concern of most publishers. As Lyell struggled to complete the second volume of the *Principles*, Jean André Deluc's 1809 *Letters on Physical History* was reprinted; and while readers of the *Witness* learned about Miller's fossil fish, the London publisher J. W. Southgate issued Rennie's *Conversations on Geology* in a third edition, almost unchanged from 1828.[80] As late as 1860, the year after Darwin's *On the Origin of Species* appeared, Nathaniel Hawthorne's *Universal History* (1837) was still being published in London by William

---

78. A similar argument was employed by Isabella Duncan in her book *Pre-Adamite Man* (1860): see Rudwick 1992, 138–40.

79. Miller 1848, [98]–[99]. Compare Miller 1854, 353–4; 1993, 350–1.

80. For one contemporary attempt to situate Deluc's position on the developing Genesis-geology debate see Anon. 1832c, 25–7.

Tegg, introducing the creation of the world to children as an event which "took place about six thousand years ago", without any indication that this had long been a controversial issue.[81] This book sold over a million copies worldwide.

## Conclusion

Popular science in the early-Victorian period was not a coherent or stable entity, but a battlefield or marketplace.[82] Svante Lindqvist has compared eighteenth-century public science to "a mediaeval market fair, with several itinerant troupes erecting their stages and competing for the attention of the public", and his analogy seems, if anything, even more applicable to the early-nineteenth-century scene.[83] In the field of earth history, the public was assailed by a cacophony of competing accounts, many of them lacking labels. Those who wished to see a particular theory fixed as "orthodoxy" were continually dismayed by the persistence of alternative theories, which called for louder popularization on their own part. Proponents of the new model of scientific expertise repeatedly urged this lesson on the experts themselves. An early example was an anonymous writer to the *Magazine of Natural History* who wrote to caution a critic of Ure's *New System of Geology*. "T. E." acknowledged that "most people" (probably including himself) disagreed with Ure's literalist theory, but suggested that Ure's book had its uses in illustrating one of the three current hypotheses for harmonizing the strata with Genesis 1 (the other two hypotheses implied an old earth).[84] Rather than waste time heaping abuse on this book, declared T. E., "men qualified for such speculation" would benefit society more if they deigned to "illustrate the other two".[85]

The gentlemen's response to the literalists bears close comparison with the response of the new breed of "experts" to *Vestiges*. Both varieties of earth history represented "non-expert" challenges, albeit from opposite ends of the theoretical spectrum. In 1845 John Crosse drew a similar lesson from the sudden success of *Vestiges*: "If men competent to the task disdain to popularize science, the task will be attempted by men who are

---

81. [Hawthorne] 1860, 14. This book was first published in 1837: see Roselle 1968, 76–9.

82. Yeo 1993, 45–7; J. Secord 2000a, 437–9, 460; Lightman 2000, 653.

83. Lindqvist 1992, 90–3; compare Orange 1975.

84. These hypotheses had been listed by Conybeare and Phillips 1822, lix–lx. The first hypothesis maintained the literal sense of "day" and a young earth; according to the second (today called the "day-age theory"), the seven "days" represented geological epochs; according to the third (today called the "gap theory"), unknown ages had elapsed between Genesis 1.1 and 1.2.

85. "T. E." 1830, 91, noted in Mortenson 2004, 112. "T. E." was writing in response to "H." 1829.

# Polite Science and Narrative Form

6

If early-Victorian earth history was displayed in an intellectual market-place, two questions press themselves forward. To whom was it sold, and how was it packaged? These questions are closely linked. Identifying the frameworks in which earth history was promoted will help us pin down its target audiences, while a clear understanding of who was participating will allow us to make sense of the forms it took and the sites it occupied. The new geology was not equally available to all: it remained, in large part, a polite science. This is reflected in the range of literary genres and narrative forms in which the science appeared during the early-Victorian period, and the social spaces in which this literature functioned.

## Science for the People?

The second quarter of the nineteenth century saw an unprecedented explosion of new and much cheaper forms of printed science: miscellanies, magazines, newspapers, reprints. The market for scientific spectacle in shows and museums boomed as never before.[1] Access is a highly topical issue today, and historians understandably get excited about the widening availability of public science in the nineteenth century. Scholarship on "cheap print" often has a strongly affirmative tone, popular presentations even more so. The new sciences, at long last, can be seen reaching "the people", and the early Victorians emerge as the architects of our own allegedly egalitarian culture of mass accessibility and democratization of knowledge.

But a splash of cold water helps to restore historical perspective. The new geology remained an overwhelmingly upper- and middle-class pursuit in the early-Victorian period, when these classes represented only 15 to 20 percent of the British population.[2] The science's social catchment area mirrored the new franchise: the 1832 Reform Bill extended the vote beyond home-owners, but only as far as middle-class householders paying a minimum annual rent of £10. Likewise, most of the best-known forums for the new science were restricted to the respectable. Before the 1830s most museums and collections could be visited only by invitation or on application: in both cases, you had to know the right people. Anatomical museums typically excluded women on grounds of delicacy, and public institutions like the British Museum refused tickets to those whose clothes and appearance did not conform to strict standards of personal hygiene.[3]

After 1830 museums became increasingly commercialized, typically charging one or two shillings for entry. Mantell had opened his collection in Lewes in 1829 for free viewing on application, but in 1836 (now in Brighton) its entry fee was a shilling.[4] This represented more than half a day's wages for most tradesmen and servants. Even when museums were free, their opening hours excluded most working people. Weekend openings were rare, although some museums in the 1840s and 1850s began to make serious efforts to attract working men by opening in the evenings.[5] And when the classless ideal seemed at last to be realized, at London's Great Exhibition in 1851, not everyone applauded. Mantell himself, a shoemaker's son and politically no conservative, despised these "dirty women" and "ignorant mobs" and wished they would either go away or become extinct.[6]

1. Altick 1978, 350–89; Morus 1998, 75–98; J. Secord 2000a, 437–70.
2. Even if "middle class" is defined very broadly, including foremen and supervisory workers, the middle classes only amounted to about 23 percent of the population by the 1860s (G. Cole 1955, 55). See also Heyck 1982, 27–8.
3. Anon. 1847, II, 132. On visiting country-house collections, see Carroll 2004.
4. On Mantell's museums see Mantell 1834b; 1836b, 44; Dean 1999.
5. On access and early-Victorian museums see Forgan 1994, 144–7. On the decreasing cost of urban entertainment from mid-century see Burnett 1969, 244–5.
6. Curwen 1940, 273–4. Mantell was, admittedly, seriously ill and depressed at the time.

Many museums, like Mantell's in Brighton, were attached to literary and philosophical societies. These also provided public lectures—but for members or guests only. These institutions were usually founded and controlled by provincial industrialists or wealthy merchants, whose radical stance in the late eighteenth century had by now given way to firm alliances with traditional forms of authority, notably the landed aristocracy.[7] The philosophical societies enabled provincial elites to express their new-found identities as enlightened members of the establishment; their members were concerned to exclude political controversy and maintain respectability. The cost of membership reflects this concern. Joining the Bristol Institution, a crucial site in the development of vertebrate palaeontology, cost £27 2s for the first year and two guineas (£2 2s) for each year thereafter.[8] Even for a literate lower-middle-class wage-earner such as a lawyer's clerk, the initial outlay amounted to a whole year's salary, and the annual subscription represented a month's wages. Subscription fees for the new circulating libraries were also expensive: Mudie's, which stocked science books as well as fiction, charged a guinea per year. The library network widened the range of reading practices among the polite classes, but did little to extend their catchment area downwards.[9]

The BAAS, drawing on existing networks of philosophical societies, was no less exclusive: its lectures were open only to those invited. Like many of the societies, it had originally aimed to exclude women altogether, but hordes of "elegant females" managed to muscle in (quite literally) at an early stage.[10] Some indication of the BAAS's target audience is seen in the kinds of events put on. At the 1832 Oxford meeting, for example, a "regiment" of two hundred horsemen and horsewomen accompanied Buckland on a field lecture like those his students had enjoyed.[11] Buckland was exporting his lecturing methods from academia into the public realm, but his new public belonged to the same social class: most people in 1832 did not own a horse. Nor did the Geological Society suddenly fling open its doors to the less well-off. Membership for Londoners in the 1830s (which included the journal) cost six guineas for admission

7. For an overview see Morrell and Thackray 1981, 2–34.

8. Taylor 1994, 180.

9. St Clair 2004, 247 (on clerks' wages see p. 195).

10. Some were so desperate to attend the evening lectures that, as one of them recalled, the "most extraordinary muscular exertions" were required to ensure a place. Women were forbidden to attend the more specific daytime lectures, so that savants could discourse freely on indelicate subjects. After the York gathering in 1831 Murchison joked to the absent Buckland that a paper on digestion and perspiration had been "read to the *men* of science only [. . . .] the ladies were never treated with a peep into the cloaca, which *you alone* know how to render sweet in the senses of females, and therefore I hold you bound at the Oxford *gala* to enable them inwardly to digest all such matters." For a full analysis see Morrell and Thackray 1981, 148–57, from which these quotations are taken.

11. Morrell and Thackray 1981, 157–8.

plus a three-guinea subscription fee, amounting to four months' wages for a lawyer's clerk.[12]

When we turn to the range of books available, the new variety in pricing and size should not lead us to conclude that access was universal.[13] Thomas Hawkins's *Memoirs of Ichthyosauri and Plesiosauri* (1834) is a clear example of an ostentatiously expensive treatise, an imperial folio measuring 38 × 55 cm (the size of a modern broadsheet newspaper). It sported enormous margins, beautiful engravings, and a price to match: £2 10s. At the other end of the scale, two of the cheapest science books in the 1830s were Samuel Clark's *The Little Geologist* and *Little Mineralogist* (plate 1).[14] These measured a tiny 7 × 9.5 cm and sold for 1s each. For an averagely well-off member of the upper classes, bringing in perhaps £5 a week (such as a retired naval commander on half-pay and with a small independent income),[15] these smart little books might have been tempting as presents for a grandchild; *Memoirs of Ichthyosauri,* costing half his weekly income, would have been well beyond his immediate means. But if our long-suffering lawyer's clerk were to bring home *The Little Geologist* as a gift for his son, his wife would not necessarily be thrilled to see more than half the day's earnings spent on a little pink book about stones. The same clerk could not dream of affording the so-called cheap edition of Lyell's *Principles* (more than two weeks' wages at 24s);[16] while the 2s 6d "people's edition" of *Vestiges,* a bargain for a full-length book in 1847, represented more than a day's work.

Separate taxes on paper, political content, and advertising (the so-called taxes on knowledge) meant that, until the second half of the century, it was hard for publishers to make a profit out of cheap non-fiction. *Chambers's Edinburgh Journal* (founded in 1832 by William and Robert Chambers), a weekly miscellany sold at one and a half pence, was one of the few successes in the cheap-periodical market in the 1830s, partly because fiction was allowed within its pages. Newspapers remained pricey: the daily *Times* cost sixpence, and the bi-weekly *Witness* four and a half pence. For most working people there were always more pressing expenses to be met.[17]

12. Taylor and Torrens 1986, 145.

13. St Clair 2004 gives a superb analysis of early-nineteenth-century literature in relation to questions of access. See also Heyck 1982, 24–8.

14. [Clark] [c. 1838] and [c. 1840]. On Clark see J. Secord 2003b (he is not included in the major reference works).

15. This example is used as an upper- or upper-middle-class reference-point by St Clair 2004, 193–5. For information on everyday costs see Burnett 1969, 189–281; on the pricing of, and access to, books see St Clair 2004, 186–209.

16. On this edition see J. Secord 1997, xiv.

17. On expenditure patterns in the nineteenth century see Burnett 1969, 219–81. On the costs associated with geology, see Taylor and Torrens 1986.

Fig. 6.1 puts "cheap print" into perspective, showing how many weeks' or days' wages these publications represented for our lawyer's clerk, and placing the weekly rent and the price of a four-pound loaf of bread in the descending scale of expenses.[18] In practice, of course, a book representing one week's wages could only be afforded after many weeks' work.

Historians of popular science often (and rightly) point out that price alone does not dictate availability;[19] but it had a considerable effect all the same. Certainly, some individuals did find ways of accessing even the more expensive books. There are famous cases of artisans educating themselves against the odds, heroic models of self-help beloved of later Victorian biographers. The baker and fossil collector Robert Dick, who lived in Thurso on the north coast of Scotland, saved enough to amass a considerable library.[20] But his example is famous because it was not typical, and Dick's neighbours thought he was very eccentric. Alternatively, in areas with enough artisan naturalists, they could club together and save up to buy expensive volumes: some early botanical societies collected an impressive number of books. But this was a slow process, and their libraries could never hope to match those of the wealthy.[21] There was also a national network of Mechanics' Institutes, set up by philanthropic industrialists in provincial urban centres: for a modest annual subscription, working men (but not women) had access to lecture courses and libraries of science books—as long as they were lucky enough to live nearby.[22] Books could be bought secondhand, but not on every street corner. Scientific writings were also excerpted cheaply in miscellanies like the twopenny *Mirror of Literature, Amusement, and Instruction* (1822–49).[23] But an appreciation of Lyell's theories gained by reading a few decontextualized titbits culled from the *Quarterly*'s review of the *Principles* was unlikely to be as rewarding as that gained by reading the *Principles* itself, however fascinating the resulting "transformation of meaning" may be to twenty-first-century historians. In the first half of the nineteenth century only a handful of scientific treatises were reprinted in their entirety for less than 3s, usually after a delay of several years. Buckland's Bridgewater Treatise went on sale for £1 15s in 1836 and was reprinted twice, but even the 1869 "cheap" edition was still very expensive at 15s.[24] Despite

18. For rent and bread prices see Burnett 1969, 209 and 218–19. The book prices in Fig. 6.1 do not take into account the discounted rates sometimes available to subscribers of the most expensive volumes.

19. Fyfe 2004a, 87.

20. Smiles 1878, 51–3.

21. A. Secord 1994b, 278.

22. Inkster 1976; Barnes and Shapin 1977.

23. Dawson *et al.* 2004, 32–3; Topham 2004a.

24. On the pricing of the Bridgewater Treatises see Topham 1993, 140.

| Book | Description | Price | Multiple of wage (lawyer's clerk) |
|---|---|---|---|
| Parkinson, *Organic Remains,* (John Murray, 1804–11) | Royal 4to, 3 vols; many engravings (Fig. 1.5), some coloured | £8 8s 6d | 16 weeks' wages |
| Murchison, *The Silurian System* (John Murray, 1839) | Royal 4to, 2 vols; many engravings, some coloured | £8 8s | 16 weeks' wages |
| Mantell, *Fossils of the South Downs* (Lupton Relfe, 1822) | Royal 4to; many engravings (Fig. 0.1) | £3 3s | 6 weeks' wages |
| Mantell, *Illustrations of the Geology of Sussex* (Lupton Relfe, 1827) | Royal 4to; many engravings (Figs. 3.1, 3.2) | £2 15s | 5 weeks 3 days' wages |
| Hawkins, *Memoirs of Ichthyosauri and Plesiosauri* (Relfe and Fletcher, 1834) | Imperial fol.; many engravings (Fig. E.1) | £2 10s | 4 weeks 5 days' wages |
| Buckland, *Geology and Mineralogy* (William Pickering, 1836) | 8vo, 2 vols; many engravings, one coloured (Figs. 8.2, 9.1, 9.4) | £1 15s | 3 weeks 2 days' wages |
| Most new novels from 1826 on | "Triple-decker" 8vo, 3 vols; some engravings | 31s 6d | 3 weeks' wages |
| Lyell, *Principles of Geology,* 2nd complete edition (John Murray, 1834) | Post 8vo, cheap ed., 4 vols; engravings (Figs. 4.1, 4.2) | 24s | 2 weeks 2 days' wages |
| Mantell, *Medals of Creation* (Henry Bohn, 1844) | Foolscap 8vo, 2 vols; coloured engravings | £1 1s | 2 weeks' wages |
| Lyell, *Elements of Geology,* 2nd ed. (John Murray, 1841) | 12mo, 2 vols | 18s | 1 week 4 days' wages |
| Mantell, *Wonders of Geology* (Relfe and Fletcher, 1838) | Foolscap 8vo, 2 vols; a few engravings, some coloured (Figs. 7.6, 8.11) | 15s | 1 week 3 days' wages |
| Broderip, *Zoological Recreations* (Henry Colburn, 1847) | Post 8vo (Fig. 6.5) | 10s 6d | 1 week's wages |
| [Chambers], *Vestiges of the Natural History of Creation* (John Churchill, 1844) | Post 8vo; no engravings | 7s 6d | 4.3 days' wages |
| Darwin, *Journal of Researches,* 2nd ed. (Murray, 1845) | Post 8vo; engravings | 7s 6d | 4.3 days' wages |
| Miller, *Old Red Sandstone,* 3rd ed. (Johnstone, 1847) | Small 8vo; a few engravings (Fig. 8.3) | 7s 6d | 4.3 days' wages |
| Dawson, *Story of the Earth and Man,* 7th ed. (Hodder and Stoughton, 1882) | Crown 8vo; engravings, gilt-embossed cover | 7s 6d | 4.3 days' wages |

| Book | Description | Price | Multiple of wage (lawyer's clerk) |
|---|---|---|---|
| *Quarterly Review* (f. 1809) | Quarterly review journal, 8vo | 6s | 3.4 days' wages |
| Mantell, *Petrifactions and Their Teachings* (H. G. Bohn, 1851) | Post 8vo (Fig. 6.7) | 6s | 3.4 days' wages |
| [Clark], *Peter Parley's Wonders of the Earth, Sea, and Sky* (Darton and Clark, 1837) | Small format; engravings (Fig. 8.6) | 5s | 3 days' wages |
| Mantell, *Thoughts on a Pebble,* 8th ed. (Reeve, Benham and Reeve, 1849) | Small square giftbook; coloured engravings; gilt-embossed cover (Figs. 6.6, 8.12; Plate 4) | 5s | 3 days' wages |
| Typical weekly rent for small back-to-back cottage, *c.* 1850 | | 4s | 2.3 days' wages |
| Mill, *The Fossil Spirit,* 2nd ed. (Darton and Clark, *c.* 1855) | Small giftbook; gilt-embossed cover (Figs. 8.7–8.10, Plate 5) | 3s 6d | 2 days' wages |
| Typical labourer's weekly rent | | 2s 6d | 1.4 days' wages |
| [Chambers] *Vestiges of the Natural History of Creation,* 7th ed. (John Churchill, 1847) | Small "people's edition" in card covers; no engravings | 2s 6d | 1.4 days' wages |
| Routledge's "Railway Library", f. 1848 | Small paperback reprints | 1s | 0.6 day's wages |
| [Clark], *The Little Geologist* (Darton and Clark, *c.* 1840) | Very small (Plate 1) | 1s | 0.6 day's wages |
| *The Literary Gazette* (f. 1817) | Large-format weekly literary magazine | 1s (8d in 1830) | 0.6 day's wages |
| Quartern loaf of bread, 1820s–30s | 4 lbs 5¼ oz | 10d | 0.5 day's wages |
| *The Times* (f. 1785) | Daily broadsheet newspaper | 7d in 1830 | 0.3 day's wages |
| *The Witness* (f. 1840) | Bi-weekly broadsheet | 4½ d | 0.2 day's wages |
| Milner, *Gallery of Nature* (1846, Orr and Co.) | Serial, sold weekly and monthly (Figs. 6.3, 6.4) | 3d per weekly part | 0.1 day's wages |
| *The Penny Magazine* (f. 1832) | Weekly educational magazine; woodcuts | 1d | 0.05 day's wages |

*Figure 6.1.* The cost of print. On abbreviations for book and paper sizes (e.g. "Royal 4to"), see the appendix. Sources: advertisements bound into the books listed; cover-pages of periodicals; reviews in the *Magazine of Natural History;* J. Secord 1997; 2000a; J. Thackray 1976.

widening possibilities for the persistent or fortunate few, working-class access to works of scientific literature remained limited.

Cost was not the only problem. Working-class literacy rates were low in the early-Victorian period: over 30 percent of the British population were illiterate, and a further 30 to 50 percent only semi-literate. This was, admittedly, partly mitigated by the common practice of reading newspapers and magazines aloud in the pub or at home.[25] Organizations such as the Society for the Diffusion of Useful Knowledge (SDUK) and the Society for the Propagation of Christian Knowledge (SPCK) were striving to stage-manage the entry of inexpensive "safe science" into the working-class home. They hoped to counteract socially disruptive forms of science such as the transmutationism promoted by Richard Carlile and other radical publishers.[26] With the weekly *Penny Magazine* (founded in 1832) the SDUK invited readers to pursue "rational amusement", tempting them in with front-page woodcuts and filling the pages with miscellaneous instruction in prose and verse. One penny, while not an insignificant sum, was certainly affordable for a worker in the 1830s.[27]

But cheapness on its own does not create demand: paperclips and sockholders are cheap, but people do not rush out to buy them unless they feel they need them. Workers in the early nineteenth century were not easily persuaded that they needed the kinds of polite science promoted by bodies like the SDUK. A knowledge of earth history was of limited practical use to most of them, and not many were convinced that it could be a worthwhile form of entertainment. If they did buy printed matter, it was more likely to be ballads or the unashamedly entertaining "penny dreadfuls", periodical miscellanies of sensational fiction whose effects on the working-class mind agonized educationalists. Many workers continued to prefer the more established forms of amusement offered by the pub, a space from which gentlemen were excluded. As Anne Secord has shown, the pub also functioned as a focus for scientific discussion among artisan naturalists, who engaged with genteel science on their own terms and used their own methodologies. Upper-class educationalists hoped that "useful knowledge" would help make the working classes more respectable by stimulating their intellects; but artisans rarely conformed to the social norms implicit in such "respectability". As their genteel correspondents noted with regret, some artisan naturalists could not even keep their rooms tidy.[28]

The promoters of science as "rational amusement" found a more will-

25. On working-class literacy see Heyck 1982, 25–6; Vincent 1989.

26. Topham 1992; Fyfe 2004a, 16–59.

27. S. Bennett 1984.

28. On artisan naturalists see A. Secord 1994a; 1994b, 290–1. For a full account see A. Secord 2008 forthcoming.

ing audience closer to home. By the early-Victorian period, many Mechanics' Institutes had been taken over by the middle classes, who also formed the main readership of the *Penny Magazine*.[29] It is not surprising that forms of science intended to integrate workers into respectable society should have appealed to those who already aspired to these norms of respectability. But philanthropists and educationalists were not the only people serving up polite forms of science for the working classes. There were sound commercial reasons for ensuring that printed science maintained a cultivated tone. Booksellers' shops were frequented by the middle and upper classes, not the working classes. Especially with non-fiction, middle-class readers represented a much firmer consumer base for publishers than the shifting sands of the working-class reading public. Non-fiction could be priced low to make it available to the working classes, but because they could not be relied on to buy it, most publishers kept more than half an eye on their middle-and upper-class heartland.[30] Many in the middle classes now had the vote for the first time, so publishers and writers with ideological axes to grind, radical and conservative alike, vied for their attention and assent. In the early-Victorian period, reconstructions of earth history played a pivotal role in public debates over man's place in nature, God's role in Creation, the authority of Scripture, and science's place in society. These debates had real political consequences, so advocates of the competing geologies outlined in chapter 5 set their sights on the increasingly powerful middle classes.[31] Few, if any, periodicals with significant geological content were aimed at the working classes alone, even radical journals like Charles Southwell's *Oracle of Reason*.[32]

One example will serve to show how writings ostensibly aimed at working men had their better-off readers constantly in view. The first sentence of Miller's *Old Red Sandstone* seems to have a working-class readership in mind: "My advice to young working men desirous to better their circumstances [. . .] is a very simple one." Give up political agitation, recommends Miller, and take up geology instead. The rest of the chapter illustrates this advice with reference to Miller's own career as a stonemason. Fifty-four pages later, however, he asks, "Does the reader possess a copy of Lyell's lately published elementary work, edition 1838? If so, let him first turn up the description of the upper Silurian rocks".[33]

29. Inkster 1980; 1985; S. Bennett 1984.

30. For examples see Topham 2004a, 66; Fyfe 2004c, 78–9.

31. On periodicals and the electorate see Heyck 1982, 36–7.

32. On the importance of middle-class support for various radical causes, see McCalman 1992. On the intrusion of middle-class values into ostensibly artisan radicalism, see Cooter 1984, 201–55.

33. Miller 1841, 1, 54.

Miller is referring to Lyell's *Elements of Geology*, which cost a princely 18*s*, near enough two weeks' wages for the "young working man" addressed at the beginning. In this light the book's first few pages seem designed less to counsel the working man, and more to stage the author before his well-off readers as an impressive model of working-class self-help. At 7*s* 6*d*, after all, *The Old Red Sandstone* was priced well out of the range of most workers.[34]

Thanks to a combination of cultural politics and economic pressures, the spectacle of earth history remained rooted in polite culture, despite the widening potential audience and the new possibilities of cheap print. One of the fundamental norms of this culture was the concept of taste, expressed in terms of aesthetic cultivation and often combined with a commitment to ideologies of progress. This combination helps explain why the upwardly-mobile middle classes found the literary and philosophical societies so congenial: in their hands these institutions became genuinely hybrid, offering lectures in history and literature as well as the sciences. Here geology was marketed as a "poetical" science as well as a useful one, part of a general literary culture.[35] Frederick Francis, for instance, toured the genteel western outliers of London in 1838–9, lecturing at philosophical institutions where he proclaimed geology's power to unfold "trains of sublime thought" which inspired the imagination to "replume its wings, and extend its flight to a loftier elevation".[36] Mantell in Brighton and Miller in Edinburgh sprinkled their public lectures with poetry quotations and extended literary allusions. Miller, in fact, took the opportunity to respond directly to a previous lecture-series on modern poetry, engaging in a debate on literature's relations with science.[37] It has been suggested that Victorian science-lecturers abandoned their predecessors' "rhetorical flourishes" and appeals to "poetic inspiration" in favour of a more utilitarian approach,[38] but this was certainly not the case with geology lectures at the philosophical institutions, the BAAS, or even the Mechanics' Institutes, most of which had become gentrified by the 1830s.[39] Rather than replacing poetry with science in the utilitarian fashion, many lecturers encouraged the two to merge.

34. However, some parts of the book, including the first chapter (Miller 1840), had already been issued more cheaply as articles in *The Witness*.

35. For an example from Liverpool see Kitteringham 1982.

36. Francis 1839, 155–6.

37. For the content of these lectures see Mantell 1838; Miller 1857; 1859 (especially ix, xxxv, 79–87). See below, pp. 384–6, 404–14.

38. Hays 1983, 101–2.

39. For an example of a mechanics' institute lecture see "Lectures on Geology", *Leamington Chronicle*, 23 March 1837 (cutting in Oxford University Museum of Natural History, Buckland Papers, Miscellaneous Geological MSS 1/5). The report contains substantial extracts from a lecture by Thomas Lloyd, drawing heavily on Buckland's Bridgewater Treatise.

From a working-class perspective, the gentrification of artisan institutions—in effect a form of privatization—must have been extremely frustrating. Ian Inkster has put the case bluntly, stating in his seminal study of science lectures that the lecturers themselves became little more than "instruments of dissemination and rehearsal for relatively small numbers amongst the provincial middle class."[40] These audiences were generally motivated less by the practical need to master a new science and more by a desire to broaden their minds and engage in science as "a mode of cultural self-expression", legitimizing their status as provincial elites. In turn, they present a rich variety of material with which today's middle-class social historians may legitimize a professional investment in the theories of Pierre Bourdieu and others.[41] The result has been a number of valuable studies of provincial public science, revealing science's role in middle-class culture as a "foundation of collective belief" achieved through an institutionalized process of "circular reinforcement".[42]

However, the overriding emphasis on issues of power and domination in these studies has meant that the science's aesthetic aspects are often dismissed as a mere side-effect or symptom of power politics.[43] These aspects are central to a full appreciation of how public science worked. Rhetorical tropes and aesthetic forms did not merely decorate the science presented, but helped to constitute it. They served as tools of communication for audiences familiar with *belles-lettres* but not with modern science, and they shaped the significance of science for these audiences.

The importance of pleasure and entertainment in polite science need not be equated with a debasement or dilution of the science itself. Inkster has suggested that the scientific content of lectures after 1840 declined as they became absorbed into "general entertainment", rubbing shoulders with light musical soirées rather than chemistry demonstrations.[44] Suspicion of entertainment as a vehicle for scientific education has a long history, running from Victorian criticisms of Buckland's showmanship and the BAAS's sensationalism right through to Richard Dawkins's fulminations against "whacky" media personalities who "proclaim that science is fun, fun, fun".[45] As Inkster himself notes, however, "the distinction between entertainment and education was more notional than real", and

40. Inkster 1980, 106. See also Shapin 1983, 177 n. 85.

41. A. Thackray 1974, 678.

42. Bourdieu 1977, 167, quoted in Knell 2000, 52.

43. Cautionary voices against "mono-causal" explanations for these social phenomena are raised by Morrell (1985, 23) and registered by Knell (2000, 52 n. 8); for a related critique see Astore 2001, 161–2. On the importance of aesthetics in the bourgeois public sphere since the eighteenth century, see Yeo 1993, 40–3.

44. Inkster 1980, 105–6.

45. Dawkins 1998, 22–3.

recent work has reinforced this view.[46] Scientific entertainment was not a simple matter of middle-class cultural "rehearsal":[47] this term too easily implies a degree of stale elitism and self-congratulation which is often held to characterize "comfortable" middle-class coteries when viewed from the "real world" of gritty working-class science.[48] Such value-judgements help when one is telling a rattling story of artisan resistance, but social privilege and a predilection for poetry and spectacle did not necessarily amount to intellectual stagnation.

Polite science cannot be separated out into distinct components of "scientific content" and "aesthetic form", still less a science like geology, whose historical dimension requires narrative at a fundamental level. In previous chapters we have seen how literary forms helped to constitute geology's public voice; but from the 1830s on the increasing popularity of "rational amusement" among the polite classes, and the widening of those classes, gave the poetics of geology new prominence. With this in mind, we now turn to examine the generic frameworks, publication formats, and narrative paradigms which housed earth history during the early-Victorian period.

## Science as Literature

At the turn of the nineteenth century, geology had been a brand-new science, its practice drawn variously from existing disciplines. Its heterogeneous origins had been reflected in the variety of genres housing it, such as topographical writings, theological treatises, and cosmological narratives. By the Victorian period proper, however, it was typically represented as a discipline in its own right. This situation seems neatly reflected in the rapid rise of the geology textbook and the disciplinary homogeneity of papers presented at the Geological Society.[49] One might expect that the genres which had helped give birth to this science around 1800 would be of no further use in its maturity. In fact, earth history continued to be parasitic on the very same generic repertoire.

As Fig. 6.2 shows, this parasitism took place on both small and large scales. A travel guidebook was now likely to sprout a geological appendix; but the genre itself could also be taken over bodily, resulting in specifically geological travel-guides. Natural-theology treatises often gave room to geology, since it was so useful in demonstrating the Creator's

46. Inkster 1980, 83. See also Morus 1998, 70–98; A. Secord 2002; Alberti 2003.
47. Inkster 1980, 106.
48. Compare Desmond 1989, 7.
49. On textbooks see Morrell 2005, 138–43.

| Existing form | Example in which geology plays a part | Example in which geology dominates or supplants the original form |
|---|---|---|
| Natural-theology treatise | Henry **Christmas**, *Echoes of the Universe: From the World of Matter and the World of Spirit* (1850) | William **Buckland**, *Geology and Mineralogy Considered with Reference to Natural Theology* (1836) |
| Biblical exegesis | Fowler **De Johnsone**, *A Vindication of the Book of Genesis* (1838) | Thomas **Hawkins**, *The Book of the Great Sea-Dragons, Ichthyosauri and Plesiosauri,* גדלים תנינם *Gedolim Taninim, of Moses* (1840) |
| Cosmological narrative | [Robert **Chambers**], *Vestiges of the Natural History of Creation* (1844) | David **Ansted**, *The Ancient World* (1847) |
| Travel guide | George and Peter **Anderson**, *Guide to the Highlands and Islands of Scotland* (1834) | Gideon **Mantell**, *Geological Excursions round the Isle of Wight* (1847) |
| Personal travel-narrative | Charles **Darwin**, *Journal of Researches* (1839) | Hugh **Miller**, *The Cruise of the Betsey; with Rambles of a Geologist* (1858) |
| Dialogue | Humphry **Davy**, *Consolations in Travel* (1830) | Maria **Hack**, *Geological Sketches, and Glimpses of the Ancient Earth* (1832) |
| Prose fiction | Charles **Kingsley**, *Alton Locke* (1850) | John **Mill**, *The Fossil Spirit* (1854) |

*Figure 6.2.* The genres of geology. Most of these books are discussed more fully below.

wisdom; this in turn allowed a clerical geologist like Buckland to use this genre as, in effect, a cover for producing a palaeontological treatise. In some cases this heterogeneity may reflect the contested nature of the discipline: the exegetical tract was the main vehicle for literalist earth-history in the Victorian period, so anti-literalists used the same genre to hit back, or diverted the terms of the debate from exegesis to natural theology by responding from within the covers of a theological treatise. More generally, however, the generic plurality of Victorian geology reflects the varied demands of its target readership. However secure its practitioners felt it was as a discipline, successful popularization in the literary arena meant tailoring it to its audience's tastes and interests, insinuating the science into as wide a range as possible of existing textual forms.

The resulting heterogeneity can be only faintly reflected in a schematic table like Fig. 6.2. One of the problems with such tables is that they present genres as a set of boxes or pigeonholes into which literary works may be neatly slotted. Genres are best seen not as pigeonholes but as sets of broadly-agreed norms (in literary-critical jargon, "horizons of expectations")[50] against which an individual work's distinctive features stand out in relief. In Fig. 6.2, the box into which a work is placed represents only the category to which that work most visibly aligns itself. Whatever may be claimed for its central identity, an individual work of literature floats in a web of intertextuality: at certain points, whether implicitly or explicitly, it draws on conventions associated with other genres. At the most basic level, this may be done by quoting or alluding to texts belonging to different genres, a process which almost goes without saying in the compilatory world of popular-science writing. Promoting a new science entailed displaying its connections to established fields of interest, so it is unusual to find a geological work from the early-Victorian period which does not include some discussion of theology, the Bible, picturesque scenery, geography, or other branches of physical science.

In a book, the generic signals in the main text often exist in creative tension with those in the title-page, epigraph, preface, or chapter-headings. Buckland's *Geology and Mineralogy,* for instance, presents itself on its title-page and epigraph as a natural-theology treatise, but its opening sentence sounds more like the beginning of a travel guide: "If a stranger, landing at the extremity of England, were to traverse the whole of Cornwall and the North of Devonshire; and crossing to St. David's, should make the tour of all North Wales [. . .]".[51] Hawkins's first book, *Memoirs of Ichthyosauri,* is ostensibly an anatomical monograph on fossil marine reptiles, but its main text, too, begins like a topographical work: "In the beautiful and romantic County of Somerset [. . .] are the villages of Street, Walton, Butleigh-Wootton and Kington."[52] Like Thomas Ashe's *Memoirs of Mammoth* it also swerves frequently into heroic-autobiography mode, giving a double meaning to the word "memoirs" in the title; it contains substantial poetry quotations, including one by Hawkins himself; and its epigraph, a long disquisition on the interpretation of Genesis 1, suggests yet another horizon of expectations.[53]

Geology's negotiable position in this supple generic continuum comes across vividly when we look at the trajectories taken by individual writers. Leaving aside his polemical pamphlets, Hawkins's literary career plotted an idiosyncratic course from prose to poetry, and from palaeontology to sacred history. Six years after the publication of *Memoirs of Ichthyosauri* he wrote an equally gigantic sequel, *The Book of the Great Sea-Dragons* (1840), in which fantastic descriptions of ichthyosaurs combine with apocalyp-

50. Jauss 1982, 22.

51. W. Buckland 1836, I, 1. This device is adapted from Conybeare and Phillips 1822, ii–iii, and may be usefully compared to Cuvier 1813, 6–7.

52. Hawkins 1834a, 1.

53. The generic hybridity of *Memoirs* has been examined in detail in Carroll 2006 as evidence of its author's eccentricity.

tic meditations on angels, demons, and the Last Judgement. The fossil record or "Great Book of Dead Times" emerges as, in effect, a newly-discovered book of the Bible in stone. Accordingly, the text was laid out in double columns like a Bible, while the title-page (plate 2) was adorned with red ink, Hebrew lettering, and a Chaldean epigraph. Appropriately, John Martin provided the frontispiece (see Fig. 10.3). Hawkins's prose style, already highly wrought in *Memoirs of Ichthyosauri*, now went beyond that of contemporary epic poets like Robert Montgomery, whose versifications of sacred history were referred to as "hyper-Miltonic" by unimpressed contemporaries. *Sea-Dragons* is mottled with Miltonic vocabulary:

> The long, lank, skinny hands, the deathy paddles of Plesiosaurus, or spotted, or livid yellow and pale, upon them fiend-like he fled: his hide, or black or freckled, or russet, his eyes blood-shot fiery, or green, lizard-like; his teeth, his fangs whetted sharp, gloating upon and crunching the gristles of his dying prey: or fleeting through the Expanse of Ocean, or tempting the Profound, or cresting the Upper Waves, preying, or at watch for prey, or lulling himself upon the wide, the universal deep: coming from the Abysm of Ages, the Gog, or the Magog of Pre-Adamite Earth, Giant of Wrath and Battle, behold! the Great Sea-Dragon, the Emperor of Past Worlds, maleficent, terrible, direct, and sublime.[54]

Prose could hardly bear such syntactic weight or prophetic enthusiasm. After *Sea-Dragons* Hawkins abandoned prose for verse, writing Miltonic epics about angels and demons, beginning with *The Lost Angel and the History of the Old Adamites* (1840): this, like *Sea-Dragons,* was published by William Pickering. Fossil saurians played only cameo roles in his later biblical epics, but these epics contained many more illustrations by Martin.[55]

*Sea-Dragons* can therefore be seen as occupying a generic mid-point between the epic fossil-treatise *Memoirs of Ichthyosauri* and the later epic poems. Its fervid style has been presented by some modern writers as evidence for his mental instability.[56] Certainly, he seems to have had enough behavioural problems to earn him a local reputation as being "near the borderline between eccentricity and criminal insanity": he is recorded as having started local riots, horsewhipped those who crossed him, engaged in obsessive litigation, broken into the British Museum (from which he had been banned), claimed to be "the rightful Earl of Kent", and defended

54. Hawkins 1840a, 24.
55. Hawkins 1840b; 1844; 1853.
56. Purcell and Gould 1992, 107–8; McGowan 2001, 125.

the strawberries in his garden with disproportionate violence.[57] But his early writings on fossils provide no evidence on this front. Viewed in the context of contemporary epic poetry, and alongside developments in English prose under the influence of Thomas Carlyle and cheap sensation-fiction, these treatises are less outlandish than they appear. Hostile critics lambasted his style for its clumsy pomposity, the gentlemen of science sniggered among themselves, but some readers were bowled over: "His language is poetry", gushed the *Metropolitan Magazine*.[58]

For evidence that generic fluidity was a central and accepted feature of popular geological literature, we need look no further than Miller, widely recognized as its most successful exponent. Miller's writing moves easily between geological exposition, philosophical reasoning, local history, topographical description, autobiographical reminiscence, literary criticism, social commentary, and religious meditation. Some of his books are loosely centred on a specific genre—*Scenes and Legends of the North of Scotland* (1835) may be called a folklore collection, *The Old Red Sandstone* (1841) is a form of scientific treatise, *My Schools and Schoolmasters* (1854) resembles an autobiography—but none is remotely typical of its genre, and all the modes listed are found to differing degrees in each of Miller's books.

The periodical press was crucial in fostering the development of such hybrid writing. Most of Miller's books were first published serially, as articles in his bi-weekly newspaper *The Witness*. He wrote well over a thousand other articles besides, most of which were probably never intended for book publication. The need to churn out short essays on diverse subjects meant cultivating a range of styles, leaping nimbly from topic to topic, switching tone where necessary, sprinkling anecdotes, jokes, and quotations where appropriate, and building a strong rapport with his readers by developing a personalized narrative voice. Robert Chambers faced similar demands editing *Chambers's Edinburgh Journal,* which stood him in good stead when writing *Vestiges*.[59] Periodical miscellanies had existed for over a century, but by the 1830s the rapid growth of a non-specialist, middle-class reading public, hungry for knowledge in the face of increasing disciplinary specialization, had led publishers to recast the old format. The new periodical miscellanies moved away from providing forums for debate among readers: scientific information was increasingly commissioned from professional writers or (in the cheaper ones) excerpted from science periodicals.[60] The miscellanies aimed to serve up a

---

57. Taylor 2002; Taylor in *ODNB*.

58. Anon. 1834d, 2. On Hawkins's reviewers see Carroll 2006.

59. J. Secord 2000a, 97–8. On Miller's journalism see Shortland 1996b; J. Secord 2003a.

60. Topham 2004a. On miscellanies see also Butler 1993.

little bit of everything. As the editor of the *Mirror of Literature* announced in 1827, "Fact and fancy; sentiment, poetry, and popular science; anecdote and art; love of nature and knowledge of the world—alternate in its columns".[61] By the 1840s original fiction too had become a staple of many miscellanies. Conventional histories treat Victorian literature as a chaste and dignified succession of single-genre works: novels, narrative poems, lyric sequences, dramas. In fact, more than half of the literature published in the early-Victorian period appeared in the generically promiscuous setting of a periodical, sharing the covers with whatever else happened to be there and providing much entertainment all round.[62] Here science rubbed shoulders with the full range of literary modes, and its promoters helped themselves liberally to these modes in their own writing.

The newsworthy science of geology was no exception. Some of the most attractive books treating this topic were first published in periodicals: Miller's *Witness* articles were not alone here. Mantell's children's book *Thoughts on a Pebble* (1836), chatty and laden with poetry, first saw the light of day as a short essay in *Leigh Hunt's London Journal* in 1834: it was framed as a sequel to Hunt's essay in the previous issue, a set of poetical musings entitled "On a Stone".[63] William Broderip's delightful compendium of animal and fossil anecdotes, *Zoological Recreations* (1847), was first published serially in the *New Monthly Magazine* (1837–43); Charles Kingsley's *Glaucus, or the Wonders of the Shore* (1855), a travel guide addressed to a bored tourist, was an expanded version of an article for the *North British Review*.[64] Other pieces ended up being incorporated into books: Thomas Hood's satirical skit "A Geological Excursion to Tilgate Forest; A.D. 2000", first published in *Hood's Comic Annual,* became the tailpiece of Mantell's *Medals of Creation* (1844).[65] Many more were never published in book form, such as Henry Morley's rollicking piece of time-travel fiction, "Our Phantom Ship on an Antediluvian Cruise" (1851), written for Charles Dickens's fiction-based monthly *Household Words.*[66] All these pieces contain strikingly imaginative representations of geology. Here, as with Dickens's own novelistic techniques, the need to grab (and hold on to) the attention of magazine-readers was a powerful engine of literary innovation.[67]

---

61. Quoted and discussed in Topham 2004a, 65.

62. In the second half of the nineteenth century, periodical publishing rapidly overtook book publishing: see Dawson *et al.* 2004, 10.

63. [Mantell] 1834a; Mantell 1836a; [L. Hunt] 1834. For the attribution to Hunt, see Keene 2007 forthcoming.

64. On Broderip and Kingsley see *ODNB; DNBS.*

65. Mantell 1844, 982–5.

66. On Hood and Morley see *ODNB.*

67. Studies of science in commercial periodicals include Brock 1980; Sheets-Pyenson 1981; 1985; Yeo 1993, 77–87; Cantor *et al.* 2004a; Cantor and Shuttleworth 2004; Henson *et al.* 2004.

The success of periodical miscellanies among the middle classes encouraged publishers to step up the production of miscellanies in book form. Old formats—anthologies, encyclopaedias, natural-history compendia—budded off in new ways to colonize new market niches. Hot on the heels of the *Penny Magazine* came Charles Knight's *Penny Cyclopaedia* (1833–58), available in weekly instalments. Buried within these twenty-seven huge, close-printed volumes were articles on extinct animals, reproducing excerpts from papers by leading geologists like John Phillips.[68] Digests of scientific information flourished: in 1828 the science pages of the twopenny weekly *Mirror of Literature* spawned an annual, *Arcana of Science,* summarizing the year's scientific discoveries.[69] At 4s 6d the *Arcana* was twenty-seven times as expensive as its periodical parent, which points up the importance of middle- and upper-class readerships in the market for popular science.

A cheaper example was Thomas Milner's *Gallery of Nature* (1846), initially available in both weekly threepenny instalments and (for the better-off) monthly shilling instalments.[70] Illustrated by a wealth of woodcuts (see Fig. 6.4), the text covered astronomy, geography, and geology. The end result was a hefty eight-hundred-page tome weighing nearly two kilograms. For over three decades it sold successfully in an upmarket form; later reprints boasted gilt-edged pages and were smartly bound in red or blue gilt-embossed covers. Deluxe compendia like this were typically presented to children as prizes. My own copy was originally given to one John Armstrong by the Blackburn Particular Baptist Sunday School "for Regular Attendance"; no doubt the minister approved of Milner's aim to instil his readers with admiration for the Creator's works. But science books were often used for purposes their authors did not intend.[71] The yellowing organic remains in Fig. 6.3 suggest that the *Gallery of Nature* was also good for exterminating God's less attractive efforts.

The early-Victorian period saw a sharp rise in the number of science publications targeted specifically at women and children. Central to this enterprise was the London educational publisher John Darton, who teamed up with the Southampton bookseller Samuel Clark to produce a range of attractive little science books (as well as the large poster of extinct animals used in Buckland's lectures, seen in Fig. 2.3). In the late 1830s they published *The Little Geologist* and *The Little Mineralogist* (plate 1), written by Clark under the clerical pseudonym "Rev. T. Wilson". One of their greatest successes was *Peter Parley's Wonders of the Earth, Sea, and*

---

68. Anon. 1833–58.
69. Topham 2004a, 65–6.
70. J. Secord 2000a, 463–4; Fyfe 2004a, 237–41. On Milner see *DNBS.*
71. Topham 1998.

306           PHYSICAL GEOGRAPHY.

following streams, which unite in the channel of the Lower Mississippi, and pour down through it into the Gulf of Mexico : —

| | Miles. | | | Miles. |
|---|---|---|---|---|
| St. Peter's | 500 | Kaskaskia | | 300 |
| Penaca, or Turkey | 200 | Maramec | | 200 |
| Iowa | 350 | St. Francis | | 450 |
| Chacaguar | 200 | White | | 600 |
| Des-moines | 600 | Arkansas | | 2500 |
| St. Croix | 300 | Canadian | | 1000 |
| Chippewa | 300 | Neosho | | 800 |
| Wisconsin | 600 | Red River | | 2000 |
| Rock River | 450 | Washita | | 800 |
| Illinois | 500 | Ohio | | 1250 |
| Salt | 250 | Alleghany | | 350 |
| Missouri | 3300 | Monongahela | | 300 |
| Yellowstone | 1600 | Kanawha | | 450 |
| Little Missouri | 300 | Kentucky | | 360 |
| Shienne | 300 | Green | | 300 |
| Quicourt | 500 | Cumberland | | 600 |
| Platte | 2000 | Tennessee | | 1500 |
| Kansas | 1200 | Muskingum | | 200 |
| Osage | 500 | Scioto | | 200 |
| Gasconade | 300 | Waybash | | 550 |
| Jacques | 600 | White River | | 200 |
| Sioux | 500 | Hatchy | | 200 |
| Grand | 500 | Yazoo | | 300 |
| Chariton | 200 | Big Black | | 200 |

The most beautiful tributary of the Mississippi is the Ohio, the *Belle rivière* of the early French settlers, the only large river it receives from the east. No stream rolls for the same distance so uniformly and peacefully ; its banks are adorned with the largest sycamores, its waters clear, and studded with islands covered with the finest trees. All the other great tributaries flow from the west, its confluence with the Missouri, which enters it like a conqueror, and carries its white waves to the opposite shore, presenting one of the most extraordinary views in the world. The country around these vast watercourses is of the most varied description, alternately exhibiting wild rice lakes and swamps, limestone bluffs and craggy hills, deep pine forests and beautiful prairies, the prairies showing an almost perfect level, in summer covered with a luxuriant growth of grass and flowers, without a tree or a bush, only occupied in recent times by elks and buffaloes, bears and deer, and the savages that pursue them. The bluffs of the Mississippi are for the most part perpendicular masses of limestone, often shooting up into towers and pinnacles, presenting at a distance the aspect of the battlements and turrets of an ancient city. In the season of inundation below the mouth of the Ohio, the river presents a very striking spectacle. It sweeps along in curves or sections of circles, from six to twelve miles in extent, measured from point to point, and not far from the medial width of a mile. On a calm spring morning, and under a bright sun, this sheet of water shines like a mass of burnished silver, its edges being distinctly marked by a magnificent outline of cotton wood trees, at this time of the year of the brightest verdure, among which those brilliant birds of the country, the black and red bird, and the blue jay, flit to and fro, or wheel their flight over them, forming a scene which has all of grandeur or beauty that nature can furnish, to soothe or enrapture the beholder. The curvilinear course of the Mississippi is one of its most striking peculiarities. It meanders in uniform bends, which, in many instances, are described with a precision equal to that obtained by the point of a compass. The river sweeps round the half of a circle, and is then precipitated in

*Figure 6.3.* Alternative uses for nature books: a squashed spider in Milner's *Gallery of Nature*. Milner 1880, 306.

*Sky* (1837), also written by Clark, who hijacked the pseudonym of the successful American children's author Samuel Goodrich.[72] Clark added the words "edited by Rev. T. Wilson" to the title-page, doubtless to reassure worried mothers that the contents were theologically sound; this name vanished from the title-pages of later imprints as geology became less controversial. The book was reprinted many times, and in a later deluxe edition its illustrations were coloured by a new automated process (plate

72. J. Secord 2003b. On Goodrich see Roselle 1968.

3). Its style was accessible and engaging: in the preface, Clark deplored the tendency of so much juvenile literature to combine brevity with comprehensiveness, resulting in mere lists of terms. To give his readers room to experience "wonder and admiration", he presented a selection of verbal vignettes illustrating the most impressive parts of natural history.[73] Gutting the book of nature in this way resembled the procedure employed in anthologies and miscellanies, and reflects awareness of the need for effective communication in popular-science writing. Indeed, in the 1830s the very phrase "popular science" was beginning to signify "science effectively communicated" rather than merely "science widely disseminated".[74]

The need to draw readers in gave an important role to the front matter of a book—frontispieces, title-pages, and (later) front covers. Restorations of extinct animals and impressive geological phenomena fronted many geology books from the 1830s onwards.[75] But the mere presence of geology itself counted as a major selling-point for science compendia. Geology takes up almost half of *Peter Parley's Wonders,* and about 30 percent of Milner's *Gallery of Nature,* although both books cover a wide range of science topics. The illustrated edition of Stephen Fullom's overtly natural-theological compendium *Marvels of Science* (1854) defers to the hierarchy of the sublime, opening with religious reflections followed by astronomy; yet its first page, the frontispiece, shows a geological spectacle, a limestone pinnacle at St Andrews.[76] Title-pages, too, show geology being used as bait. Later editions of *Peter Parley's Wonders* contain a coloured example in which the title appears in the mouth of Fingal's Cave (plate 3). The subtitle of Milner's book appears in the midst of Alpine scenery (Fig. 6.4), an association buttressed by the Peruvian mountains in the frontispiece and the basalt columns opposite the contents page, and balancing the geological texts and pictures with which the book closes. Geology was the polite science *par excellence.*

The importance of taste and literary cultivation to polite science is particularly evident in the alignments forged with another kind of miscellany, the literary anthology. Broderip's natural-history essays, written for the *New Monthly Magazine,* were not systematic descriptions of the animals covered, but rather collections of anecdotes and striking facts, sewn together in a high-spirited narrative style and laced with poetry

73. [Clark] 1837, v–viii. For a modern reprint see [Clark] 2003.

74. Fyfe 2004a, 56; Dawson *et al.* 2004, 15.

75. See Figs. 5.2, 6.4, 6.6, 7.6, 7.7, 7.8, 7.25, 8.1, 8.10, 9.3, 9.7, 9.9, 10.3, and E.1.

76. The first edition of 1852 had no pictures. Ebenezer Brewer's *Theology in Science* (1860) also opens with a geological frontispiece, as noted by Rudwick (1992, 136 and 139).

THE GALLERY OF NATURE,
A
Pictorial & Descriptive
TOUR THROUGH CREATION,
by
THE REV? THOMAS MILNER, M.A.

Upper Glacier of the Grindelwald

WILLIAM & ROBERT CHAMBERS,
LONDON & EDINBURGH.

*Figure 6.4.* Alpine title-page of Milner's *Gallery of Nature*. Milner 1880 (apart from the publisher's name, this is identical to the title-page of the 1846 edition).

quotations. *Zoological Recreations* gathered them together into one volume, but the essays remained for the most part only loosely connected. Like Frank Buckland's later volumes of *Curiosities of Natural History*, this was a book for dipping into at leisure.[77] In some parts, especially the chapter on dragons which leads into Broderip's discussion of extinct saurians, the poems come so thick and fast that the page layout looks more like a verse anthology than a natural-history book (Fig. 6.5). Single-author multi-genre miscellanies were also common currency in the early-Victorian period, and George Richardson used this form to promote geology alongside his own fiction, poetry, reportage, and translations of German poetry in two expensive volumes of *Sketches in Prose and Verse* (1838).

Female and (later) juvenile readers were targeted by a new form of lit-

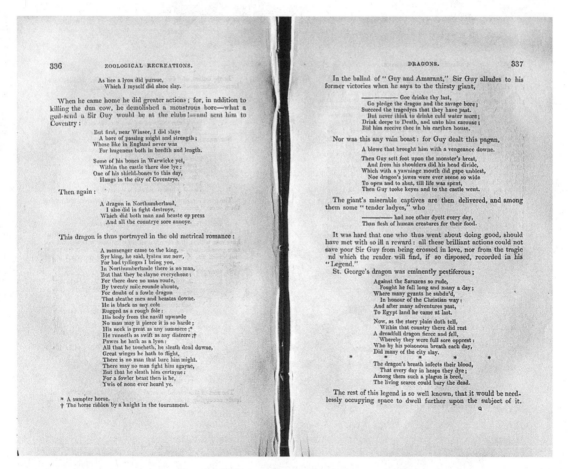

*Figure 6.5.* Popular science as poetry anthology: quotations from dragon romances in Broderip's *Zoological Recreations*. Broderip 1848, 336–7.

77. The first of these was published in 1857 (F. Buckland 1903).

erary anthology which had developed in the 1820s, the giftbook annual.[78] These lavishly decorated little books, of which the *Keepsake* was the best known, were designed to be given as presents. Their contents and physical appearance were designed to crystallize polite ideals of taste and refinement, typically mixing poetry, moral musings in prose, and engravings; some of the work was originally commissioned, some picked from the "beauties" (i.e. best bits) of existing works. The bindings were ornate and colourful, the pages usually gilt-edged. Science was not at first included in these annuals, but publishers producing science for women and children came to employ similar generic conventions.

Mantell's 1834 essay "More Thoughts 'On a Stone'" was ripe for this treatment, with its profusion of sentimental quotations from poets, novelists, and philosophers, pleasingly woven together. Relfe and Fletcher's first edition of *Thoughts on a Pebble* (1836) reproduced the essay more or less as it had appeared in the original miscellany, recasting it as a didactic children's book with a plain paper cover and a dedication to Mantell's young son Reginald;[79] the only major addition was a coloured frontispiece showing the Pebble itself. Subsequent editions pushed the book still closer towards the genre of the giftbook. More and more poetry was included, the Pebble was redrawn (plate 4), and new pictures were appended: colour plates, a decorated title page, and a new frontispiece showing the author and his Iguanodon thighbone (Fig. 6.6). The binding and page layout became more decorative: the eighth edition of 1849 had the square shape of a giftbook, sporting a blue gilt-embossed cloth cover, new colour images, a ribbon bookmark, and gilt-edged pages with ruled borders and wide margins (see Fig. 8.12).

By the mid-1850s the concept of the giftbook anthology had broadened beyond the annuals. Besides brightening up their *Peter Parley* books (plate 3), Darton and Clark now sold a range of "Three Shilling and Sixpenny Gift Books" which claimed to be "especially adapted for School Prizes". Their contents ranged from proverbs, maxims, and edifying tales to historical lore and extracts from choice English poets. Pictures played an important part, as did the covers (plate 5): one advertisement specifically mentioned the "entire new style of binding".[80] The only explicitly scientific title in this list of twenty-six gilt-edged giftbooks was a book designed to introduce youngsters to geology. John Mill's *The Fossil Spirit; A Boy's Dream of Geology* (1854) is framed as a series of "Evenings" in which a wise old man narrates episodes in earth history to an audience

78. Manning 1995; Price 2000.
79. On Mantell's didactic strategy, see Keene 2007 forthcoming.
80. Mill [*c.* 1855], back endpaper.

of orphaned children. Each short chapter begins with a decorative initial letter formed from a fossil, is illustrated by vignettes, and ends with a reflective poem or "song" (see Figs. 8.7–8.10). As in the *Keepsake,* a regular rhythm between prose, pictures, and poems is sustained. At the beginning of the book, immediately after the title-page, two pages of "Select Sentences"—meditative quotations from philosophers and poets—are printed side by side as a kind of miniature anthology;[81] while, as plate 5 shows, the handsome case-binding could not fail to catch the eye.

If properly blended with the exotic, the sublime, and the edifying, geology could be made to fit smoothly into the giftbook tradition, and more generally into the culture of refined sensibility of which the giftbook was a conspicuous sign. The antiquity of fossils not only offered food for reflection, but helped to distance this science from the coarse world of blood, turds, and entrails which made anatomy and zoology so unsuitable for ladies and children. The internal organs of a cat might present clear evidence of God's benevolent design, but, unless you were inured to vivisection, this moral was hard to square with the agonizing and undignified death undergone by the animal during the demonstra-

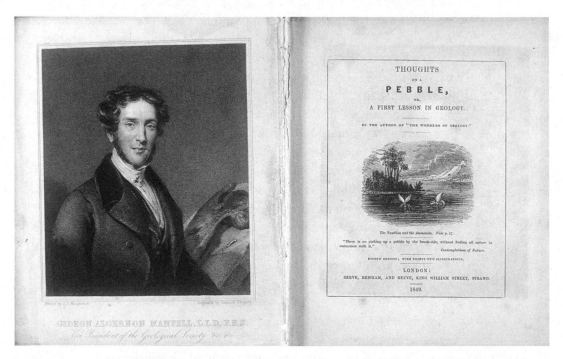

*Figure 6.6.* Frontispiece and title-page of Mantell's *Thoughts on a Pebble.* The frontispiece shows the author with an emblem of his greatest triumph, a thighbone of an Iguanodon beside that of an iguana. The title-page vignette alludes to George Richardson's poem *The Nautilus and the Ammonite,* included later in the book. Mantell 1849.

81. Mill [*c.* 1855]. Mill is not included in the major biographical reference works.

tion. With fossils, as David Brewster pointed out for the wealthy readers of the *Edinburgh Review,* the remains were no longer flesh but stone—"sainted relics, which the most sensitive may handle, and the most delicate may prize".[82] A fossil fish did not smell fishy but was, as Hawkins informed Queen Victoria, "the sacred key to Mysteries sublime".[83] This view was fleshed out in the science columns of newspapers, in magazine articles, and in the contents of compendia and giftbooks, along with the other genres mentioned earlier. Miscellanies on the one hand, and books belonging to multiple genres on the other, played a crucial role in negotiating geology's entrée into polite culture.

All this generic hybridity has important implications for how we view the developing relationship between science and literature. The period between 1780 and 1820 is often seen as a watershed in this relationship, culminating in a divorce. It has been asserted that during this period the meaning of the word "literature" narrowed to include only poetry, drama, and fiction, excluding scientific writing and other non-fiction which had previously formed part of a unitary literary culture.[84] No empirical study has yet been done to test this claim, which is not supported by the *OED*'s citations under "literature".[85] It is, however, apparently upheld by a number of localized studies of Regency literary culture. Marilyn Butler has suggested that the heavy quarterlies were, by the 1820s, beginning to separate reviews of "literature, especially poetry" from reviews of "reformist or scientific thinking"; Topham has shown that the editors of periodical miscellanies like the *Mirror of Literature* spun off their scientific content into separate publications from the late 1820s; and Noah Heringman has argued that writings on rocks and landforms, rooted in the common eighteenth-century literary tradition of topographical writing, split off in the early nineteenth century into separate discourses of descriptive poetry and geological science, each defining itself against the other.[86] A "developing sense of distinct literary and scientific spheres" seems to be confirmed by the comment made by a writer in the *Presbyterian Review* for 1841 who referred nervously to "what may be called the *literature* of the science", as if the two categories were not normally on speaking terms.[87]

However, although this material may appear to map straightforwardly

82. [Brewster] 1837, 39, discussed in Topham 1998, 255.

83. Hawkins 1841, Sonnet LIX (unpaginated).

84. Klancher 1994, 524; Siskin 1998, 6; Heringman 2003b, 6. For a more nuanced account see A. Richardson 1994, 30–3.

85. R. O'Connor 2005a. For an overview of the word's semantic shifts see R. Williams 1983, 183–8.

86. Butler 1993, 143; Topham 2004a, 65–6; Heringman 2004.

87. Dawson *et al.* 2004, 12; Anon. 1841–2, 210–11, cited above, p. 6.

onto our present-day exclusion of scientific writing from the category "literature", we must resist the temptation to draw this equation. Victorian geological writing continued to find a home in many literary genres; furthermore, the quarterlies and other literary journals continued to review science books,[88] and many commentators continued to praise the aesthetic qualities of scientific writing without any self-consciousness. In fact, both these continuities and the sense of "separate spheres" outlined above have their roots in the rapid widening of the British reading public during the early nineteenth century and the breakdown of the ideal of a unitary bourgeois public sphere. With new periodicals and readerships budding off at every turn and scientific specialization running rampant, all-purpose public forums like the *Gentleman's Magazine* became problematic as vehicles for producing knowledge. Instead, as Topham and others have shown, they turned into vehicles of dissemination, serving up science from more specialized sources rather than encouraging readers to make knowledge themselves. The gentlemen of science restricted much of their original research to the learned periodicals. Yet this narrowing-down of the literary sites in which science was produced did not amount to an "abandonment" of literature by science. On the contrary, as the historian Henry Buckle put it in 1857, "literature [. . .] is simply the form in which the knowledge of a country is registered";[89] and he was not just talking about poetry.

Buckle's distinction between "form" and "knowledge" is worth holding on to if we are to avoid muddles about the divergence of science and literature. Since the eighteenth century, the ancient distinction between form and content had taken on a new sharpness, as more forms of knowledge-production withdrew from public literary culture and erected disciplinary boundaries. The knowledge enshrined in a scientific text was increasingly seen as something separate from the literary form which that text assumed. "Facts" were reified as hard nuggets of objective truth distinct from the words communicating them;[90] conversely, aesthetic concerns were increasingly emphasized when judging works produced within or for a general readership (a shift reflected, for example, in the rising status of certain forms of prose fiction). Consequently, the word "literary" was increasingly used in a narrower sense, distinguishing a work's formal, aesthetic qualities from referential qualities such as truth-

---

88. Yeo 1984, 8–9.

89. *OED*, s.v. *literature* 3a.

90. Daston 2001, 75–6; see also Daston 1991–2. In the 1850s this reification played its part in debates over scientific authority, as Huxley and other "professionalizers" contended that practical scientific training rather than book-learning was what qualified a reviewer to judge scientific literature (P. White 2002).

value. In time, these developments would lead to today's problematic separation not only between a text's aesthetic and referential *qualities,* but between literary and non-literary *genres,* with science writing and other non-fiction excluded from the ambit of "literature".

What seems to have gone unnoticed, however, is the glacial slowness of this metamorphosis. By 1820, the date often cited as its endpoint, it had barely begun. Despite the proliferation of new kinds of expertise and authority, the old, inclusive concept of "literary culture" persisted throughout the Victorian era, resulting in a semantic ambivalence reflected in the schizoid titles of some periodicals. The *Literary Emporium; A Compendium of Religious, Literary, and Philosophical Knowledge* (f. 1845) uses the word "literary" in both senses: the main title refers to the variety of literary texts (in the inclusive sense) printed within, while the subtitle distinguishes three different kinds of "knowledge" communicated by this literature, including "literary" knowledge in the narrow sense. To a lesser extent, the same ambivalence affected the word "literature"— hence the self-consciousness of the *Presbyterian Review's* comment on "the *literature* of the science". Of course, generic hierarchies remained, with epic poems at the top and cookery books near the bottom; but both forms still constituted "literature".

The lengthy footnotes which continued to accompany many poems, dramas, and (in particular) novels in the nineteenth century suggest that these genres continued to be judged, at least partly, by the yardstick of correspondence with external truth. Milton's *Paradise Lost* was seen by many as an authority scarcely less reliable than Scripture,[91] while Thomson's *Seasons* continued to be valued in part as a natural-history compendium as late as 1861.[92] Conversely, non-fiction texts continued to be judged partly for their formal or aesthetic qualities. As with words like "geology", "science", and "popularization", programmatic claims over the words "literary" or "literature" made in specific contexts by specific interest-groups should not be taken as reflections of a general pattern or *fait accompli.* Periodical editors came to draw a distinction between reviews of philosophical treatises and reviews of epic poems, not because treatises no longer counted as "literature", but because aesthetic and ref-

91. Pointon 1970, 39; see also Hitchcock 1851, 77–8.

92. This is suggested by the arrangement of Robert Bell's edition of 1861: Bell observes that Thomson "seldom errs in his illustrations of natural history", and wherever Thomson does err, Bell appends the correct information (Thomson 1855–61, II, 9, 89 n., 96 n.). Likewise, almost half of an article on wolves in the *Penny Magazine,* claiming to give their "principal facts", comprises a quotation from *The Seasons* ([Anon.] 1833b). On contemporary debates over the nature and value of poetic description in relation to Thomson, see R. Cohen 1964, 131–87. On footnoted poems and compilatory novels see Leask 1998a; Crawford 2000. On the persistence of didacticism see Heyck 1982, 43–5; A. Richardson 1994; Duff 2001.

erential criteria were used in sharply differing proportions when judging the two kinds of writing, resulting in different reviewing styles.[93]

To take another example, the *Inverness Courier* said of Miller's polemical pamphlet about the state of the Scottish Church that "As a mere literary effort it must rank very high, for the purity and excellence of its language".[94] For this commentator, literary criteria were identical with aesthetic criteria; the word "mere" further suggests that other criteria (e.g. accuracy, timeliness) were felt to be even more important when judging such a work. Nevertheless, this comment was later excerpted in the advertisement pages of John Johnstone's 1847 edition of Miller's *Old Red Sandstone,* along with an extract from a BAAS speech in which the "poetical" qualities of Miller's book were praised by leading geologists.[95] Clearly, Johnstone aimed to sell both pamphlet and book on the basis of their aesthetic appeal. He was not alone: a glance through the quotations used in the advertisement pages of early-Victorian science books confirms that such appeal was the primary bait used to market these books to a polite readership. As Richardson put it, "the publications of Buckland, Lyell, Mantell [. . .] and others, are as much an honour to letters as to science".[96] On the other hand, aesthetic qualities seem to have been less relevant to some of the miscellaneous information printed in bulk by periodicals and encyclopaedias, and (rather later) to papers in some specialist periodicals—although, as Aileen Fyfe has recently shown in her study of cheap evangelical scientific literature, rhetorical subtlety was by no means absent.[97]

Rather than a divorce between science and literature, then, what we see is an increasing feeling among nineteenth-century writers that there were two kinds of scientific literature: that which was intended and/or received as having aesthetic pretensions (usually identified with "popularization"), and that which was not.[98] The question was: where to draw the boundary? If science was an integral part of an increasingly multifaceted and contested literary culture, we need to stop thinking in terms of "given" dichotomies such as "scientific and literary writing".[99]

93. P. White 2002, 84.

94. Miller 1847a, advertisement pages, iv. The pamphlet's title is *The Whiggism of the Old School.*

95. Miller 1847a, advertisement pages, i–iv.

96. G. Richardson 1855, 14.

97. Fyfe 2004a, 107–40.

98. Such formulations run the risk of implying easy distinctions between "aesthetic" and "non-aesthetic" writing. I am not suggesting here that scientific papers or encyclopaedias had no aesthetic content, simply that their authors and reviewers increasingly treated them as if they did not.

99. This dichotomy is used even by scholars who aim to demonstrate underlying continuities: see Cantor *et al.* 2004b, xix.

For a large number of writers, editors, publishers, and readers in the nineteenth century, literature was not restricted to poetry, drama, and fiction, and we need to be alive to the ambivalent meanings which the concept displayed.

## Narratives of Nature?

A particularly clear form of aesthetic organization in science writing is seen when nature is turned into a story of some kind. Scientific storytelling was increasingly used in the Victorian period to negotiate a place for science in public literary culture, and these efforts were informed by the proximity of fictional and factual discourses in the new print culture. This approach was by no means universal: many science books (especially manuals) were designed simply as organized collections of information, with no apparent narrative ambitions.[100] Nevertheless, some of the most vivid examples of scientific storytelling come from the literature of geology.[101] Having sketched the range of book genres and publication formats housing geological writing in general, we now examine the range of large-scale narrative frameworks available for such writing.

I should emphasize that I am using the word "narrative" in its conventional sense, denoting an utterance in which events are linked sequentially to form a story (as distinct from, say, a description). Some scholars use the word in a broader sense as a kind of synonym for "discourse" or "exposition". In a ground-breaking study of twentieth-century biological writing, Greg Myers has analysed two forms of "narrative": the popularizing "narrative of nature", which describes the natural phenomena themselves in an accessible manner, and the specialist "narrative of science" produced by and for scientists, foregrounding the author's expertise and intervention in scientific debates.[102] Barbara Gates and Ann Shteir have adapted Myers's typology to the nineteenth century, adding a further popularizing discourse, the Paleyan "narrative of natural theology", which foregrounds the Creator; and their typology has since been refined in Bernard Lightman's illuminating studies of later Victorian popularization.[103] Such reference-points are helpful, but to use the word "narrative" in this very broad sense does risk obscuring essential distinctions between stories, descriptions, arguments, and deep struc-

100. Vestigial narrative frameworks were sometimes used in such books, but they can rarely be considered "stories".

101. Freeman 2004, 148–52, offers some examples of "prehistoric narrative" from Mantell to Miller. For more focused studies see Merrill 1989; Paradis 1996; R. O'Connor 2003a; 2003b; Carroll 2006.

102. Myers 1990, 141–214.

103. Gates and Shteir 1997b, 11–12; Gates 1998, 38–9; Lightman 1999, 4–7.

tures.[104] There is a world of difference between a text which communicates a certain view of God, nature, or scientific debate, and a text which explicitly turns that view into a story: natural theology, for instance, underpinned much early-nineteenth-century popularization, but only in certain texts did it become the engine of narrative.

The proliferating generic affiliations and publication formats displayed by geology in the early nineteenth century suggest that, if we hope to gain a fuller sense of how science was popularized, we need to delve beyond general discursive formations. We need to unearth the particular narrative frameworks which moulded it into a story. Adapting Myers's insight in another way, we may see the scientific story as a complex speech-act with at least three referents: the author (or other first-person narrator), the audience, and the object of the science's enquiry—which, in geology's case, could be a fossil, the earth, or God Himself. What kind of narrative is generated depends first and foremost on which of the three referents takes centre stage as the story's protagonist(s).

In an autobiographical account or scientific travelogue, the story's primary protagonist is the author. In Darwin's *Journal of Researches* and Miller's *Cruise of the Betsey,* geological information about South America and the Hebrides is framed by the authors' personal adventures, told in the first person and in the past tense. The reader is called on to engage with the author's sense of discovery or novelty as much as with the natural objects themselves. Geology is, in any case, only one ingredient in these narratives: the natives of Argentina and Eigg are observed with as much fascination as are the rocks beneath.[105] However, the heroic image of the geologist developed by the gentlemen of science also offered promising materials for ripping yarns. Hawkins's *Memoirs of Ichthyosauri,* for example, is built on two narrative strands: descriptions and analyses of the fossils alternate with stories about how the author acquired them. The title of the first chapter, "The History of the Ichthyosauri", fuses these two distinct "histories", the descriptive natural history of the animals themselves and the narrative history of Hawkins's discoveries.[106] In the latter episodes, his own heroic sensibilities are contrasted with the ignorant placidity of lesser mortals with West Country accents:

> "Have ye sid my animal sir," said the fossilist Jonas Wishcombe of Charmouth as I called at his house in August to enquire if he had anything

---

104. The problem is exacerbated when the word appears in two senses in the same essay without any shift being signalled (Lightman 1999, 2–8).

105. C. Darwin 1839; Miller 1858. For commentary see J. Smith 1994 and Taylor 2003.

106. Hawkins 1834a, 8.

worth buying;—"I should like vor yer honor sir to see 'un." My heart leaped to my lips—"animal! animal! where!"

"Can't be sid to-day sir—the tide's in."

"What—nonsense! I *must* instantly—come, come along."

"Can't see 'un now yer honor—the tide's rolling atop o' 'un fifty feet high."

"In marl or stone?"

"Why in beautiful ma-arl—and—

"Washed to death"—and I threw myself in despair upon a chair.

How often have I reflected upon the very-Bedlam-impetuosity of my passions at that moment:—the chaffed sea rolling over an Ichthyosaurus and remorselessly tearing it to a thousand atoms—a superb skeleton of untold value triturated to sand by a million pebbles, such was the Promethe-ian idea—torture of my rebel imagination.[107]

The style alone should warn us against treating episodes like this as reliable representations of Hawkins's daily life. With its bursts of exclamation, stylized gestures, and emphasis on the extremities of emotion, this passage comes close to the melodramatic world of the sensation novel, although it was clearly intended to be taken as true. Hawkins's persona, possessed of prophetic powers and heroic ambitions, takes the gentlemen's image of the geologist as romantic knight-errant to a Gothic extreme.

A different kind of story emerged when the earth itself, or the fossils buried in its crust, became the primary protagonist. This stance was implicit in many introductory books on geology, which claimed to offer access to the story of the earth—a conceit informed by the much-cited analogy of the strata as pages of earth history. Fossils became characters or hieroglyphics comprising the "language" in which that history was written. If, as Lightman has suggested, Victorian popularizers of science claimed "to speak for a mute nature", then geology held out the possibility of recovering nature's own words.[108] As one literalist critic put it, "The geologists [. . .] assert that the earth does tell its own tale", although in this critic's view "it is not the earth, but they themselves and their imagination, that speaks".[109] Dickens expressed the geologists' conviction in a characteristically tangible form:

107. Hawkins 1834a, 25. On Hawkins's heroic narratives see R. O'Connor 2003b; Carroll 2006.

108. Lightman 1997b, 207. On the nineteenth-century dream of a purely objective means of recovering nature's own "language", see Daston and Galison 1992.

109. Anon. 1847, ix.

the wind and the rain have written illustrated books for this generation, from which it may learn how showers fell, tides ebbed and flowed, and great animals, long extinct, walked up the craggy sides of cliffs, in remote ages.[110]

Nature's authorial role usually became conflated with that of God, the Author of all things—an apparent paradox which comes across most clearly in an example from the second half of the century, when Henry Hutchinson began a book entitled *The Autobiography of the Earth* by saying that "The story which we are about to read has not been written by man, but [. . .] by the Creator himself".[111] The Bible's own story of Creation, told in Genesis and enlarged by the storylike structures of literalist earth-histories by Sharon Turner and Mary Roberts, is here replaced by a divine text in stone.

Like the heroic image of the geological knight, this merging of divine and natural "authorship" was another rhetorical device for lending unquestioned authority to science itself, as can be seen in an 1837 review of Buckland's *Geology and Mineralogy* in the *Presbyterian Review*. This piece ends in a stentorian tone by addressing the sceptic, referring to the strata as a record of God's creative work:

> Here is the attestation graven in the rock in lines which you cannot rase, and spoken by the voice of creation herself—an authoritative voice which clamour cannot drown, nor persecution silence, and from which ridicule dares not to turn away.[112]

More importantly, geological writers were increasingly taking on the task of transcribing what this "voice of creation" said. Narrative summaries of earth history were becoming increasingly frequent and detailed, extending the pattern laid down in the 1810s and 1820s by Bakewell, Mantell, and others. They often involved vivid verbal restorations or snapshots of the ancient earth at different periods, textual correlatives of the pictorial "scenes from deep time" which were becoming a familiar aspect of popular treatises. Such passages were typically placed at the ends of chapters, moving the text from exposition and analysis to a more immediately engaging vein of storytelling and display. However, although these passages were much expanded from their Regency counterparts, most of them still formed only a small (if rhetorically crucial) part of the

---

110. [Dickens] 1851, 217. Compare Mantell's quotation from Emerson: "All things in nature are engaged in writing their own history" (Mantell 1850, title-page epigraph).

111. Hutchinson 1891, 1, discussed in Lightman 1999, 2.

112. Anon. 1836–7, 246.

text they inhabited.[113] The writer for the *Presbyterian Review* just quoted began his article with a splendid five-page historical narrative of the changes undergone by the earth's surface and inhabitants, which takes up almost one-fifth of the article. This was an unusually high proportion, and like the final peroration reflects its author's confidence in the science's storytelling potential.[114]

This confidence seems to contrast strangely with the book under review.[115] Buckland's *Geology and Mineralogy* contains several vivid verbal restorations of ancient life-forms and reflections on the historical aspect of geology, but any sense of narrative progression in the book as a whole has to be inferred from the order in which topics are discussed. The bulk of the main text comprises two adjacent surveys: first Buckland outlines the successive geological formations, then he describes the proofs of design found in different fossils.[116] This structure resembles that of many geological manuals (with an additional theological gloss), although the usual order is reversed. Most manuals presented fossils in ascending order of complexity, but Buckland began with the "higher" animals, perhaps because of the theological emphasis of his book: Megatheria offered more impressive examples of divine design than worms and plants, and so deserved priority in the text. Conversely, although most manuals presented strata in descending order, reflecting the geologist's unearthing of successive layers, Buckland discussed them in chronological order, perhaps in order to emphasize the gradual preparation of the earth for humans.

However, with the exception of a two-page "history" of a piece of coal, the large-scale historical aspect of Buckland's arrangement was left implicit, ready to be realized by his readers.[117] The dominant scenario projected by the text is that of scientific inquiry and deduction, presenting Buckland scrutinizing the strata in their present form rather than telling the story they contain: "we enter on the examination of strata of the Transition Series".[118] As in the 1810s and 1820s, when his friends had "translated" his lectures into narrative verse in imitation of Erasmus Darwin, readers of his Bridgewater Treatise were now quick to turn its raw materials into an epic of earth history. In George Eliot's *Mill on the Floss* (1860), the fictional Stephen Guest gives the heroine, Maggie Tulliver, an improvised "account of Buckland's treatise", expounding its contents as a "wonderful geological story".[119] That the writer for the *Presby-*

---

113. These passages will be examined more closely in chapters 9 and 10.
114. Anon. 1836–7, 222–6.
115. On Buckland's book see Boylan 1984, 214–18; Topham 1992; 1993; 1998.
116. W. Buckland 1836, I, 34–102, 135–523.
117. W. Buckland 1836, I, 481–3.
118. W. Buckland 1836, I, 60.
119. G. Eliot 1980b, 334 (VI.ii). For commentary see Topham 1998, 256.

*terian Review* did the same for that journal's readers serves to underline the importance of the periodical press in fostering the development of scientific narrative.

Explicitly didactic literature presented a quite different distribution of narrative emphasis. Scientific "Conversations", for example, often told the story of a fictional child's education by its mother; but, as we saw when examining James Rennie's *Conversations on Geology,* the untutored child clearly stood in for the reader, a device which proved useful for teaching correct behaviour as well as correct science.[120] The fictional nature of the dialogue form had never been in question, but this aspect was increasingly underlined as the "Conversations" form developed. Jane Marcet's *Conversations on Chemistry* (1805) had imitated the layout of the Classical dialogue, presenting a series of direct speeches separated in blocks and marked by speech-prefixes indicating the speaker's name.[121] The same arrangement had been used by both Rennie and the mineralogist Delvalle Lowry in the 1820s. The literalist Fowler De Johnsone gestured towards dramatic form in his use of this layout, adding introductory scene-setting passages to his dialogues between Truth and Infidelity, although Truth's speeches eventually swamped the conversation and turned it into a sermon.[122]

A more novelistic format was favoured by other writers, resembling more the arrangement of Fontenelle's still-widely-read *Conversations on the Plurality of Worlds*: the stylized blocks of speech were worked into a continuous narrative, usually in the first person. Humphry Davy's *Consolations in Travel* (1830) was an influential example, although this was unusually exploratory and reflective for a didactic work, and was certainly not aimed at children. Maria Hack's natural-theological conversations, beginning with *Harry Beaufoy; or, The Pupil of Nature* (1821), were still more novelistic: here the moral and scientific education of the almost sickeningly well-behaved Harry was recounted in a smooth third-person narrative.[123] Its sequel, *Geological Sketches, and Glimpses of the Ancient Earth* (1832), shows Harry's conversations with his mother and father taking place in specific localities, such as a walk among the hills. As in romance fiction, each chapter begins with a verse epigraph, and even the illustrations are smoothly incorporated within the story:

> Mrs. Beaufoy then, taking a small map out of the portfolio which had, from the days of his early childhood, often contributed to Harry's pleasure

---

120. On this genre see Myers 1989 and 1992; Gates 1998, 37–44.
121. Fyfe 2004b.
122. De Johnsone 1838.
123. On Hack see *ODNB*.

or improvement, said: "Here is just the thing for you; a little sketch, drawn purposely to show the boundaries and situation of the different volcanic regions of the earth. [. . . .]"

"Oh, thank you! this *is* just the thing, indeed."[124]

As the title and subtitle of *Harry Beaufoy* suggest, the primary protagonist of books like this was the person being educated, regardless of what subsidiary "stories" he or she was shown discovering in the natural world. The fact that such characters stood in for the reader is suggested by Kingsley's strategy in *Glaucus* (1855), a chatty introduction to the geology and natural history of the seashore.[125] Kingsley's implied reader is alarmingly specific: *Glaucus* is addressed to an adult male reader, imagined as a bored tourist from the city and directly incorporated into the fictional frame. The book begins like a talking tourist-guidebook:

> You are going down, perhaps, by railway, to pass your usual six weeks at some watering-place along the coast [. . . .] You are half tired, half-ashamed, of making one more in the "ignoble army of idlers," who saunter about the cliffs [. . . .][126]

Kingsley here adapts two conventions of the "Conversations" genre. In the hands of Rennie and Hack, this form enlivens the imparting of information from knowledgeable author to ignorant reader by discreetly projecting that relationship onto fictional characters. Kingsley retains the liveliness of direct speech, but simultaneously maps these fictionalized characters back again onto author and reader. The novice's expressions of shame—another convention of the genre, figuring the reader's ignorance and willingness to be guided[127]—are thus handed over, with a disconcerting directness, to the reader. This is the rhetorical equivalent of a hefty shove out of the front door.

The importance placed by conversation-narratives on the direct speech of a fictive "expert" aligns this framework with other, more fantastic kinds of didactic fiction. In the first chapter of Davy's *Consolations in Travel,* the narrator's companions go off to a party and leave him sitting alone in the Colosseum in Rome. He falls asleep and dreams that an invisible "Genius" catches him up into space and shows him visions of the progress of civilization, then transports him to Saturn, where he

---

124. Hack 1835, 158–9.

125. On this book see Merrill 1989, 215–35; Myers 1989.

126. Kingsley 1855, 1. A similar strategy occurs in William Higgins's "Geology of the Watering-Places" (1842, 318–19).

127. Myers 1989, 190 (though Myers, here discussing Ruskin, sees the element of shame as an "extreme version" of the more usual ignorance topos).

beholds pulsating, tentacled beings engaged in otherworldly cogitations amid glacial masses of cloud. Finally, the Genius shows him the blazing spiritual inhabitants of a comet. Like the blest souls Dante meets in the *Divine Comedy,* these beings are reincarnated philosophers, having progressed upwards in the scale of creation since they lived on earth as men.[128]

Instruction by a supernatural being who reveals truths about human destiny in visionary form to the amazed protagonist is a didactic literary framework going all the way back to Classical times. In early-Victorian Britain, its definitive Christian expression was found in Milton's *Paradise Lost,* where the archangel Raphael shows Adam visions of the future history of the world. Its best-known Enlightenment manifestation was the Comte de Volney's *Ruins of Empires* (1791), whose radical message of secular progress had been eagerly dramatized by the poet Percy Shelley in his dream-vision *Queen Mab* and given a geological twist in Byron's *Cain,* with its voyage among the extinct worlds in Hades.[129] Such works sired an entire European family of hypersensitive and hyper-Miltonic young men seeking enlightenment, from those of the so-called Spasmodic School (e.g. P. J. Bailey's 1839 verse-drama *Festus*) to the protagonist of Robert Hunt's mystical novel *Panthea* (1849).[130] The visions their instructors offered them often took the form of grand sequences of historical scenes. For instance, the Hungarian Imre Madách's cosmic drama *Az ember tragédiaja* (*The Tragedy of Man,* 1859–60) consists of a sequence of visions shown by God to Adam in which he sees himself, Eve, and Lucifer reincarnated in different roles at different epochs of human history, ending with a technocratic dystopia followed by an icy scene of human extinction.[131]

The unfolding narrative of deep time had strong roots in this literary tradition. The common didactic framework, going back beyond Milton and Volney, suggests an unexpected kinship between the tortured romantic hero, questing through space and time for inaccessible truths, and the protagonist of a treatise like *Harry Beaufoy.* Is the difference more one of emotional tone than intellectual kind? Will young Harry grow up to strike a bargain with the Devil? Part of the point of Hack's book was to discourage excessive enthusiasm in children: Harry is presented as having a "curious and romantic eye" and a "poetical tempera-

128. Davy 1851, 18–65 (identical to the 1830 text). For commentary see Knight 1967.
129. Volney 1979; Shelley 1989–2000, I, 265–423; Byron 1986, 901–21.
130. On *Panthea* and similar fictions see Millhauser 1956; J. Secord 2000a, 467–70. On *Festus* see McKillop 1925 and Birley 1962. On Bailey and Hunt see *ODNB.*
131. On Mádach and other cosmic dramatists see Esslin 1994.

ment", but "it was not the wish of his parents that their son should be a poet; and [. . .] they endeavoured to regulate it".[132] Yet the comparison helps us to make sense of two phenomena which seem odd to a modern eye: the enormous demand for didactic literature on the one hand, and the sheer pedantry of the instructor-figures in heaven-storming epics like *Festus* and *Panthea* on the other. These are only surprising if we continue to assume, along with a vocal minority of Georgian poets, that didacticism and imagination are mutually exclusive.[133] These categories have always been contested, and we must not forget the imaginative power of novelty. For many early-nineteenth-century readers, didactic literature about a science like geology was a first step into a new world of intellectual stimulation—perhaps even leading to addiction, "like the opium-eater, and his drug", as Hawkins put it in the final sentence of *Memoirs of Ichthyosauri*.[134]

For a modern reader, perhaps the most appealing example of geological didactic fiction is Mill's *Fossil Spirit*. Mill claimed an educational purpose in his preface, and the book's initial scenario—a Hindu priest lecturing to a group of orphans—is explicitly didactic. But, within this outer shell, Mill implanted a subsidiary framework enabling him to make full use of the fantastic possibilities of first-person narrative while simultaneously narrating the entire history of life on earth. As in Madách's cosmic drama, reincarnation powers the movement through different eras: the priest tells a story (in the first person) of an Indian Fakir whose soul had passed through various animal forms since life on earth began, and who had been granted by God the ability to remember these transmigrations in order "to show him how the Creator had brought good out of destruction, and life from death".[135] The resulting autobiography is punctuated by the priest's own songs, and by a need to explain the basic principles of geology; but these interruptions are no more cumbersome than the songs and explanations of locality or character background which punctuate the historical novels of Walter Scott and the Gothic romances of Ann Radcliffe.[136] The story offers an ingenious means of giving a voice to nature, and realizes Lyell's paean to the visionary power of geology: "we might [. . .] include within the compass of our rational existence, all the ages [. . .] over which science may enable us to extend our thoughts".[137]

132. Hack 1835, xvi–xvii.
133. A. Richardson 1994 and Duff 2001 offer convincing routes out of this dichotomy.
134. Hawkins 1834a, 51.
135. Mill [c. 1855], 3.
136. On embedded verses in novels see Crawford 2000, 123–5; Price 2000, 65–104.
137. [C. Lyell] 1827, 474.

Mill's subtitle, *A Boy's Dream of Geology*, points up his story's kinship
with the dream-vision, a form which maintained its popularity in a range
of genres from occasional verses to novels. Kingsley had used a similar device to Mill's in the "Dreamland" chapter of his novel *Alton Locke*
(1850), whose protagonist dreams that he recapitulates the history of life
in his own person.[138] More fundamentally, Mill's narrative extended the
idea that the inspired geologist (like the antiquary) could "see" former
worlds as if in a visionary trance. From "seeing" these worlds it was only
a short step to travelling into them. The early-Victorian period saw a
new drive to construct literary forms of geological time-travel, often
using the dream-vision as a template—but only in brief, self-contained
episodes lasting a few pages at most.[139] Not until the era of H. G. Wells
would British writers use time travel as a large-scale narrative framework for a whole book.[140]

The narrative stance which did most to foster the development of
time-travel episodes was that of the author as guide, leading the reader
through the natural world[141]—or its surrogate, a museum or gallery. The
more serious visitor to a fossil exhibition might take along a geological
treatise to help them interpret the displays;[142] many illustrated geological
works functioned as virtual cabinets of fossils, with their lavish plates depicting fossils in minute detail and the text providing a tour among these
specimens. This stance is implicit in the title of Milner's *Gallery of Nature*
and explicit in its subtitle, *A Pictorial and Descriptive Tour through Creation;*
the text itself only sporadically maintained this stance, but the many engravings were designed to give the reader glimpses of the natural world.
The title-page reinforced this with its large, panoramic Alpine scene (see
Fig. 6.4). As James Secord has observed, the tour framework could serve
as an attractive alternative to genuine storytelling in the post-*Vestiges* era,
suppressing the spectre of evolutionary narrative.[143]

This framework was equally useful for literalists like Samuel Best
and James Mellor Brown, who strove to maintain spectacular appeal
while avoiding the grand narrative of progressionist earth-history. Best
and Brown praised Buckland's *Geology and Mineralogy* as a metaphorical

138. Kingsley 1967, 337–57 [chapter 36]; see J. Secord 2000a, 469 n. 100. Dream-visions of
transmigration continued to be didactically useful in the late-Victorian period: in *My Neigh-
bour's Shoes* (A.L.O.E. 1881), also targeted at boys, an objectionable lad dreams himself successively into the bodies of those he has harassed during the day, including his sister, her cat, and
a sparrow, and is tormented in these forms by a fairy who has taken over his own body.

139. See chapters 9 and 10 below.

140. One of the earliest Continental examples was Pierre Boitard's time-travelling dream-
vision *Paris avant les hommes*, published posthumously in 1861 (Boitard 1861; see Rudwick 1992,
166–70).

141. Morrell 2005, 231–3; Knell and Taylor 2006; Keene 2007 forthcoming.

142. Topham 1998, 235; J. Secord 2000a, 444.

143. J. Secord 2000a, 462–3.

museum, "a cabinet of facts [. . .] skilfully arranged and beautifully polished",[144] through which, to our "pleasure and astonishment", Buckland "leads us [. . .] and points out in them as we pass the wisdom and goodness of the Creator".[145] Brown developed this metaphor in detail:

> A museum of fossils is a field of rich and pleasing reflection to a thoughtful mind—and who could wish for a more agreeable and intelligent companion in his survey than the author of the Bridgewater Treatise on Geology? There is something affecting in walking among those ancient relics—something which irresistibly makes a solemn impression on our feelings. It is like wandering through the catacombs of a departed world.[146] I look round the chamber, and see myself surrounded with animal generations now extinct. There lies the Ichthyosaurus on his lias bed! I shudder at the giant reptile, and rejoice that he is now still forever. There is the Mammoth:—"his bones are as strong pieces of brass; they are like bars of iron. His teeth are like the upper and the nether millstones." The Dinotherium, the Iguanodon, the Plesiosaurus, and the Pterodactylus, are there also; and I think of Behemoth, and the Unicorn, and of Leviathan, that "king of all the children of pride." On earth are not now their like.[147]

The passage continues by paraphrasing parts of the descriptions of Behemoth and Leviathan in the Book of Job (chapters 40–1), using the language of King James to transform Buckland's treatise into a virtual museum of biblical monsters. The Bible, in fact, replaces Buckland as the true "guide" to the natural objects displayed. This was, I think, a deliberate reaction to the authoritative self-image favoured by the gentlemen of science themselves, and which Lyell had presented by analogy with that most authoritative of all cosmic dream-visions, Dante's *Divine Comedy*. In 1827 Lyell had predicted that in future years his readers would be able to follow their geologist-guides into the prehuman past (a territory on which the Bible was silent), just "as Dante [. . .] followed the footsteps of his master" Virgil into the underworld.[148] Brown was working to restore the Bible's centrality as a guide to earth history, so he demoted Buckland's status from guide or interpreter to "intelligent companion".

144. J. Brown 1838, 3.

145. Best 1837, 42.

146. Perhaps an allusion to Edward Young's famous lines in *Night Thoughts* (Night IX, lines 127–8) on the antediluvian world: "Of *One* departed World / I see the mighty Shadow" (E. Young 1989, 260). This poem, first published in 1745, was still enormously popular in the first half of the nineteenth century. James Montgomery, Thomas Porch, and Thomas Hawkins had already used the same lines as epigraphs to their respective depictions of the antediluvian world (J. Montgomery 1813, title-page; [Porch] 1833, ix; Hawkins 1834a, 51).

147. J. Brown 1838, 39–40. The structure and much of the diction closely resembles the equivalent passage in Ure's treatise (1829, liii).

148. [C. Lyell] 1827, 473, discussed above, p. 169.

Many geology books, of course, were not merely virtual guidebooks, but actual guidebooks. Geologically informed tourist-guides like the Anderson brothers' *Guide to the Highlands and Islands of Scotland* (1834) and Mantell's *Geological Excursions round the Isle of Wight* (1847) were designed not solely for the armchair tourist, but also to function in tandem with sites visited by an active reader.[149] The same was true of museum guidebooks, which became a popular form for geological textbooks to assume. Richard Owen's *Geology and Inhabitants of the Ancient World* (1854) was the official guide to the Crystal Palace monsters, while Mantell's last book, *Petrifactions and Their Teachings* (1851), was a five-hundred-page geological treatise structured as a guide through the British Museum's fossil collections. Woodcuts and engravings here not only served as surrogate objects, but also helped the visitor or tourist to see the objects and displays in the correct way.

As with the early mammoth-guidebooks of Ashe and Rembrandt Peale, many of these guidebooks used various narrative devices to present the objects on display either as a connected story or as a series of objects with stories attached. Mantell enlivened the British Museum's ichthyosaur displays by quoting three long autobiographical passages from Hawkins's *Memoirs of Ichthyosauri,* dramatizing the fossils as the objects of rabid enthusiasm by an eccentric collector. These extracts are relegated to footnotes, as if they are mere digressions; but, as the page reproduced in Fig. 6.7 shows, they are so long as to squeeze out most of the main text on those pages. Mantell makes his motives clear: "The following account [. . .] is too graphic [. . .] to be omitted", "I cannot resist the insertion of the following racy account", "we will now leave Mr. Hawkins to tell the story".[150] Mantell may have learnt this technique from Richardson, who had been the curator of the Mantellian Museum in the 1830s, and whose miscellany *Sketches in Prose and Verse* was, in part, a literary companion to the museum. Its two volumes included virtual tours of the exhibits (including a poem about the Iguanodon), a Roman romance purporting to explain the provenance of a bronze Cupid in the museum, and a story in pseudo-mediaeval narrative verse about one of Mantell's field-trips.[151]

Such books did not exist in isolation, but worked symbiotically with museums, exhibitions, and geological localities in the British countryside. Once again we find ourselves at some distance from the old view of popularization as the diffusion of information from active expert to

149. Compare Higgins 1842, 317–56.

150. Mantell 1851, 347–8 n. 2, 378–9 n. 4, 381–2 n. 1. Mantell was decent enough to append a notice advertising Hawkins's original treatise ("this splendid and scarce work"), of which some copies were still unsold (Mantell 1851, 347 n.).

151. G. Richardson 1838, 1–27, 189–240, 289–95.

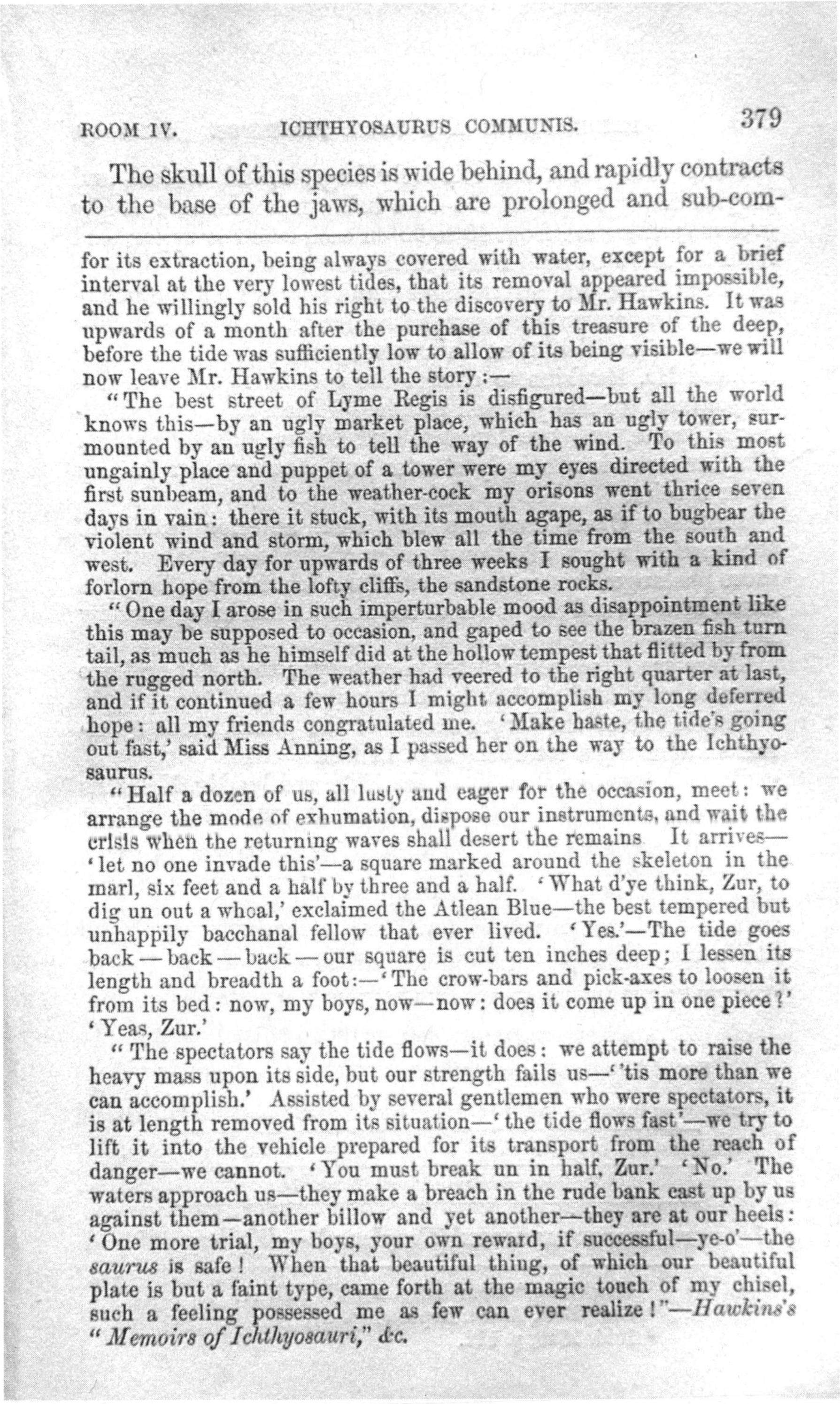

The skull of this species is wide behind, and rapidly contracts to the base of the jaws, which are prolonged and sub-com-

---

for its extraction, being always covered with water, except for a brief interval at the very lowest tides, that its removal appeared impossible, and he willingly sold his right to the discovery to Mr. Hawkins. It was upwards of a month after the purchase of this treasure of the deep, before the tide was sufficiently low to allow of its being visible—we will now leave Mr. Hawkins to tell the story :—

"The best street of Lyme Regis is disfigured—but all the world knows this—by an ugly market place, which has an ugly tower, surmounted by an ugly fish to tell the way of the wind. To this most ungainly place and puppet of a tower were my eyes directed with the first sunbeam, and to the weather-cock my orisons went thrice seven days in vain: there it stuck, with its mouth agape, as if to bugbear the violent wind and storm, which blew all the time from the south and west. Every day for upwards of three weeks I sought with a kind of forlorn hope from the lofty cliffs, the sandstone rocks.

"One day I arose in such imperturbable mood as disappointment like this may be supposed to occasion, and gaped to see the brazen fish turn tail, as much as he himself did at the hollow tempest that flitted by from the rugged north. The weather had veered to the right quarter at last, and if it continued a few hours I might accomplish my long deferred hope: all my friends congratulated me. 'Make haste, the tide's going out fast,' said Miss Anning, as I passed her on the way to the Ichthyosaurus.

"Half a dozen of us, all lusty and eager for the occasion, meet: we arrange the mode of exhumation, dispose our instruments, and wait the crisis when the returning waves shall desert the remains  It arrives—'let no one invade this'—a square marked around the skeleton in the marl, six feet and a half by three and a half. 'What d'ye think, Zur, to dig un out a whoal,' exclaimed the Atlean Blue—the best tempered but unhappily bacchanal fellow that ever lived. 'Yes.'—The tide goes back — back — back — our square is cut ten inches deep; I lessen its length and breadth a foot:—'The crow-bars and pick-axes to loosen it from its bed: now, my boys, now—now: does it come up in one piece?' 'Yeas, Zur.'

"The spectators say the tide flows—it does: we attempt to raise the heavy mass upon its side, but our strength fails us—''tis more than we can accomplish.' Assisted by several gentlemen who were spectators, it is at length removed from its situation—'the tide flows fast'—we try to lift it into the vehicle prepared for its transport from the reach of danger—we cannot. 'You must break un in half, Zur.' 'No.' The waters approach us—they make a breach in the rude bank cast up by us against them—another billow and yet another—they are at our heels: 'One more trial, my boys, your own reward, if successful—ye-o'—the *saurus* is safe ! When that beautiful thing, of which our beautiful plate is but a faint type, came forth at the magic touch of my chisel, such a feeling possessed me as few can ever realize !"—*Hawkins's* "*Memoirs of Ichthyosauri*," &c.

---

*Figure 6.7.* "[W]e will now leave Mr. Hawkins to tell the story": a huge footnote in Mantell's guidebook to the British Museum's fossil collections. Mantell 1851, 379.

passive reader: many of these books required physical stamina and active mental engagement with the outside world.

## Conclusion: Books and the Culture of Display

So-called mass approval for geology between 1830 and 1856 was courted along predominantly middle-class lines, and this sheds light on the spectacular and heavily aestheticized form it often took. If the BAAS and philosophical institutions exemplified the polite takeover of public science in the early nineteenth century, we should not be surprised to find literary culture—the hallmark of taste and refinement—becoming inextricably entangled with museums, displays, and science lectures. As Inkster has shown, one of the main reasons polite Victorians attended lectures was to stimulate and refine their literary appetites, and lecturers (both at universities and in public) were expected to recommend particular books to their audiences.[152] Literary appetite and digestion alike were stimulated in conversations at reading clubs, *conversazioni,* and soirées, and scientific authors like Mantell and Buckland became literary lions.[153] Buckland plugged his forthcoming book not only in a lecture at the BAAS's Bristol meeting in 1836, but also when visiting a nearby friend's house, where he improvised an informal lecture in the drawing-room.[154] The most important lectures were reported in detail in periodicals like the *Literary Gazette,* where they took on new narrative guises.[155]

Celebrities in the more established branches of literary culture responded with varying degrees of enthusiasm. Henry Brougham was sceptical of the BAAS's tendency to make "science a matter of popular excitement and show",[156] while the Irish poet Tom Moore mocked the Bristol gathering in a futuristic satire entitled "Anticipated Meeting of the British Association in the Year 2836". Here the assembled "*savants* and dandies" pontificate on the fossilized remains of bishops ("Episcopus Vorax") and peers ("*Aristocratodon*"): the progressive nature of earth history has resulted in the extinction of the outdated, predatory forces of conservatism. But Moore also mocked the way in which this ideology of progress was used theatrically to generate wonder:

> "Admire," exclaim'd Tomkins, "the kind dispensation
> By Providence shed on this much-favour'd nation,

---

152. Inkster 1980, 80 and 89; F. Buckland 1858, xxxix.
153. J. Secord 2000a, 155–90; Alberti 2003.
154. Gordon 1894, 28; Rudwick 1985, 170; Topham 1998, 253–4.
155. See, for example, Anon. 1841b.
156. Brougham to Sedgwick, September 1838, quoted in Morrell and Thackray 1981, 163. See also Orange 1975, 279–83.

> In sweeping so ravenous a race from the earth,
> That might else have occasion'd a general dearth [. . . .][157]

Other literary celebrities contributed celebratory verse to keep the show going. The grand old man of English poetry, William Sotheby—whose age and style made him seem the relic of a former literary world—attended the Cambridge meeting a few months before his death in 1833 and wrote a commemorative panegyric in Augustan heroic couplets:

> Can I pass o'er, as one to science known
> Thy willing toils enlighten'd MURCHISON:
> Whether thou pierce beneath the billowy robe,
> That by the Deluge cast, o'erspread the globe, [. . . .]
> Or where time graved on his sepulchral page
> The indelible impress of that early age,
> When the scaled Saurians stretch'd their hideous length
> And Mammoths labour'd with unwieldy strength.
>   The muse sagacious LYELL! would record
> The course by thy advent'rous steps explored;
> Whether thou climb the Alps, or view below
> Through their ice-arch perpetual rivers flow;
> Or trace the gradual progress of decay,
> As time's soft footstep wears the rock away [. . . .][158]

Sotheby's diluvialism may have misrepresented Murchison, but by presenting these men as heroic travellers in time and space, this poem endorses the BAAS's ostentatious theatricality. For Sotheby and Moore alike, genteel geology and its practitioners were fit subjects for "the muse", underlining the centrality of both literature and science in the pageantry of high culture.[159]

Science lectures were not only used to promote books; they also became books themselves, recycling visual aids in printed form. The result could be spectacular: Thomas Webster's enormous coloured strata-diagram tucked into Buckland's *Geology and Mineralogy* was over a metre long.[160] The proximity of reading to listening helped maintain a rhetoric

---

157. T. Moore 1915, 665–6, discussed in Klaver 1997, 161–2. Mantell satirized the Tories in a similar manner in his 1831 pro-Reform poem "The Age of Reptiles" (Dean 1999, 107–8). On more recent uses of the word "dinosaur" to imply moribund conservatism, see S. Montgomery 1991, Haste 1993; W. Mitchell 1998.

158. Sotheby 1834, 15–16. On Moore and Sotheby see *ODNB*.

159. Compare William Whewell's presentation of the history of science (in 1837) as an epic pageant (Yeo 1993, 156–7). The persistence of theatrical metaphors in present-day history of science is critically examined in Lindqvist 1992. On science and literature as complementary aspects of Victorian high culture see P. White 2002; 2005.

160. W. Buckland 1836, II, plate 1.

of orality and immediacy in popular-science writing, employing exclamations, direct addresses to the reader, chatty first-person-plural narrative, and the fiction of a virtual guided tour. The lecture-course format was used as a structuring device even for texts that were never read aloud.[161] In the age of the Literary and Philosophical Institution, this permeability was only to be expected. Exhibits, lectures, *conversazioni,* and the books themselves often shared the same institutional framework, the same intended audience, and sometimes even the same roof.

These intricate connections are best illustrated by the case of Mantell, whose first lecture-series in Brighton in 1834 allegedly sprang from his inability to tell individual visitors to his museum all they wanted to know about the displays.[162] The popularity of museum and lectures alike encouraged local patrons to invest in founding the Sussex Institution, under whose auspices Mantell hosted *conversazioni* and gave further lectures, including field-trips "à la Buckland".[163] The institution also provided a library and reading room, and the museum's new curator, Richardson, acted as a guide, ensuring that visitors got as much as possible out of the exhibits.[164] In turn, the anecdotes and guided tours with which Richardson brought the museum to life took printed form in *Sketches in Prose and Verse.* The poet Horace Smith also contributed fantastical dream-visions and panegyrics to the growing cluster of stories surrounding Mantell's collection.[165] Richardson, who played a central role in transforming Mantell's 1837 lecture-series into the book *The Wonders of Geology,* was now in a good position to write a geology book of his own. His *Geology for Beginners* was published in 1842, which ruffled Mantell's feathers.[166] By then, unfortunately, the institution's chief patron had died suddenly, and the collection had had to be sold to the British Museum, where Richardson continued to curate it until his death in 1848. Mantell's decision to write a detailed guidebook to the British Museum's fossil collection stemmed in part from his personal investment in many of the specimens.

The literature of geology was inseparable from the wider context of scientific spectacle. It fed on and into these networks of performance, display, and conversation, where amusement and education mingled. Thus far, however, we have examined only the sites and structures in which earth history was actively promoted. Outlining these larger frameworks

---

161. For example, Anon. 1836a.

162. Dean 1999, 130.

163. Letter from Mantell to Benjamin Silliman, 18 June 1834, quoted in Dean 1999, 131–2.

164. On curators as museum guides see also Taylor 1994, 187–8. On Richardson's curatorial career see Torrens and Cooper 1985.

165. Smith's Mantellian poems include Magdalen College, Oxford, MS 377, I, 157–9; Alexander Turnbull Library, Wellington, New Zealand, MS-1956; Daubeny 1869, 123–6.

166. G. Richardson 1842; for the ruffled feathers see Mantell 1844, vii–viii.

of active popularization has been a necessary first step towards understanding how former worlds were brought to life for the public. But the lack of any clear boundary between entertainment and edification in these productions is mirrored by a corresponding lack of boundary between "scientific" and other forms of spectacle. Geological phenomena were displayed within, and in competition with, an enormous range of theatrical reconstructions of the landscape and assorted natural disasters, from pantomime stage-sets to outdoor simulations of volcanic eruptions. Most of these sites would not be considered "scientific" today, and few were intended to promote geology; yet they all contributed to the science's public appeal.

These sites now demand our attention for two reasons. First, they provided geology's early-Victorian popularizers with a rich source of imagery for reconstructing exotic landscapes, the distant past, and historical change. Second, and more fundamentally, these displays helped to constitute the Victorian concept of the mind's eye, and they claimed the ability to transport spectators in imagination to the distant sites they represented. As such they came to define what Gillen D'Arcy Wood has called "virtual tourism", imaginary travel for an untravelled public.[167] To understand how the ancient earth was staged in the Victorian imagination, we must first explore how virtual tourism worked in general. This means taking a broader perspective in the next chapter, looking at the phenomenon's late-eighteenth-century origins as well as its Victorian efflorescence. Texts, especially poetry, came to play a much larger role in the process than is usually assumed, reflecting wider changes in the concept of poetry itself and shedding an unexpected light on geology's claim to be the most "poetical" of the sciences.

167. G. Wood (2001, 184) uses this term of illustrated editions of Scott's poems; I here extend it to cover a range of literary, visual and architectural sites.

# Time Travel and Virtual Tourism
# in the Age of John Martin

Spectacular entertainments were a prominent feature of Victorian city life. Rooted in the human need for amusement, they also fed a demand for information about phenomena beyond immediate reach: distant lands, ancient times, stars, the afterlife. Since the eighteenth century, as the word "Enlightenment" suggests, this need had manifested itself more and more as an appetite for visual stimuli which made the invisible visible. Natural philosophy had long been implicated in this demand, and the boundary between savant and showman was not always clear. Menageries like London's Exeter Change exhibited exotic animals to the untravelled city-dweller, while public experiments in electricity and chemistry revealed mighty, invisible forces at work in nature. Telescopes, like William Herschel's at Slough, offered socially acceptable visitors direct visual access to worlds scarcely visible to the naked eye. Microscopes revealed further invisible worlds, inhabited by fearsome monsters.[1] Edifica-

1. On these shows see Altick 1978; Schaffer 1983; Ritvo 1987, 205–42; Golinski 1992; Schaffer 1996; Morus 1998. See also the essays in *BJHS* 28 (1995) on eighteenth-century lecturing.

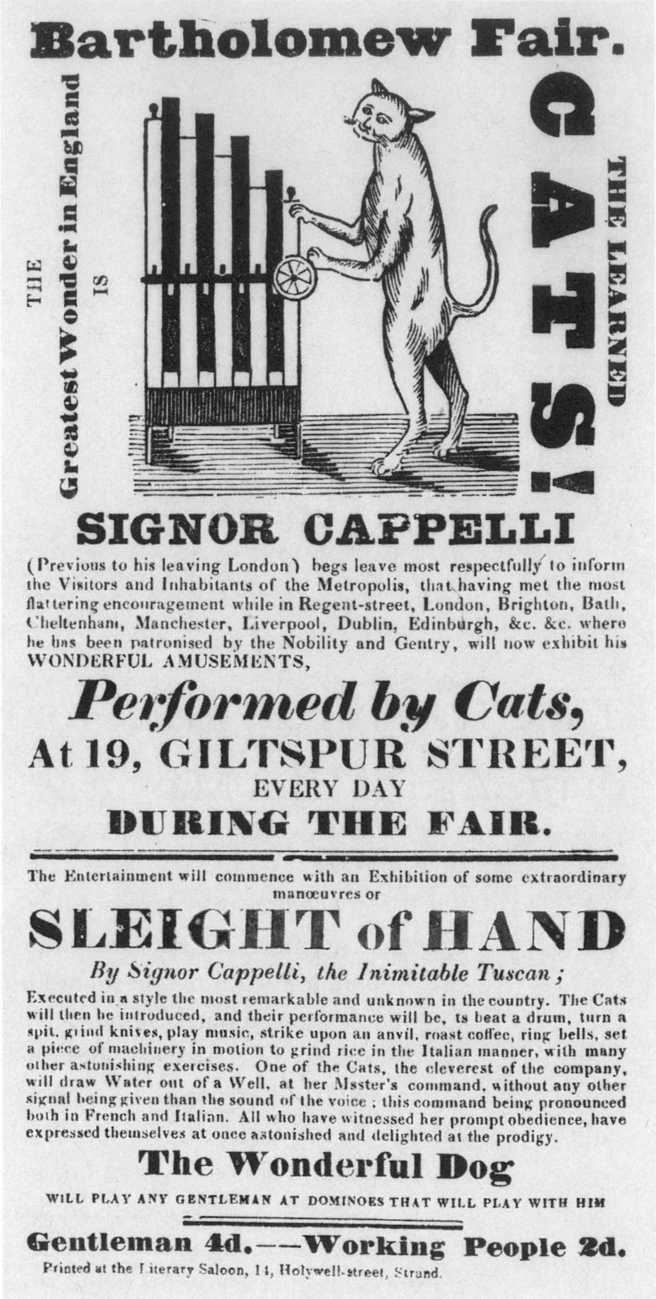

Figure 7.1. Advertisement for a troupe of performing cats, also featuring the domino-playing "Wonderful Dog". The entry fee for working people was just twopence, making it much more widely accessible than most of the theatrical displays discussed in this chapter. Printed handbill, 1832.

tion could be gleaned from the most frivolous-seeming demonstrations. Even Signor Cappelli's troupe of "Learned Cats", who roasted coffee and played the organ (Fig. 7.1), was billed by one journalist as proof of "the progress of education and the march of mind".[2]

2. Anon. 1829c; Altick 1978, 307.

The landscape played a central role in satisfying the demand for contact with unfamiliar objects. The Georgian polite classes encountered geology as one facet of a more general culture of landscape spectacle.[3] For those with money and time, tourism catered for this demand. The weird geological formations of the Derbyshire Peak were much visited during this period; the Grand Tour, taking in the Alps and culminating in Rome, offered still more.[4] But travel was time-consuming and risky, and with Britain and France at war for much of the 1790s, 1800s, and 1810s, the Continent was effectively out of bounds. Middle-class wage-earners could rarely afford such luxuries in any case, although in the Victorian age the railway network would make travel within Britain increasingly affordable and convenient.[5]

Public demand for contact with distant places was more reliably met by travel writing and topographical poetry. This literature cradled the seeds of Victorian virtual tourism. Poems like Thomson's *Seasons* invoked "the Muse" to display scenes for the mind's eye to drink in. Such imaginary views often assumed theatrical trappings, reflecting a long-standing conceit that the visible world was a theatre, a backdrop for the drama of history. To understand how the mind's eye was expected to view these imaginary scenes, we first turn to the theatre itself, and to its direct descendants, the panorama and diorama. Here, among the ancestors of the cinema, is where virtual tourism began in earnest. This chapter begins by introducing these sites and their links with contemporary geological display, for which the art of John Martin became a shared reference-point. We then examine how poetic texts worked alongside the displays to produce the illusion of verisimilitude: their use of Byron, the most popular poet of the period, will serve to focus this discussion. Finally, with the specific case of apocalyptic spectacle and its implications of time travel, we return to the geological imagination and its complex relation between fact and fancy.

## The Panoramic Imagination

The late eighteenth century saw momentous changes in the construction of theatrical space, exemplified by London's Drury Lane Theatre under David Garrick. More attention was devoted to the scenery, the audience was removed from the edges of the stage, and the overhead candelabra gave way to lamps hidden behind wings. The stage was now

3. Hamblyn 1996; Heringman 2004.
4. On travel and tourism see Buzard 1993; Chard and Langdon 1996; J. Brewer 1997, 630–41; Chard 1999.
5. On the costs of railway transport see Burnett 1969, 215–16.

isolated from the audience as an illusionistic representation of reality, its proscenium arch framing a window on the world. This effect paralleled the staging of natural scenery in landscape gardening, domestic architecture, and tourism itself, where the search for "picturesque" views resembling the art of Claude Lorrain or Salvator Rosa demanded some stage-management on the spectator's part.[6]

In the 1770s, Drury Lane's scenery and costume design were run by the Alsatian landscape artist Philip James de Loutherbourg. This pioneer of what we now call "special effects" set new standards of illusionistic verisimilitude. One of his extravaganzas (a genre blending songs, scenery, and slapstick) was The Wonders of Derbyshire (1779), a commedia dell'arte adventure set in the haunted scenery of the Peak. It comprised a series of grand views of geological features in front of which Harlequin and his friends capered in the usual manner. Later examples included Omai, or A Trip Round the World (1785), based on the voyages of James Cook. That such spectacles could substitute for tourism itself is suggested by the independent publication of a "programme" to The Wonders of Derbyshire in the form of a little guidebook to the sites depicted: its Derbyshire author treated the show as a virtual journey through his county.[7]

De Loutherbourg's best-known experiment was his mechanical theatre, the "Eidophusikon, or Representation of Nature", which he built in his home in 1781 (see Fig. 7.17).[8] There were no actors, but plenty of painted and movable scenery, sound effects, music, and an ingenious use of lights. In a stark departure from theatrical convention, the auditorium was darkened. The stage was only ten feet wide, but, with the lights down, the audience had no clear external point of comparison to break the illusion. Instead of plays, the Eidophusikon showed a sequence of topographical scenes, culminating in a scene from Milton's Paradise Lost featuring Pandemonium, the palace erected by the demons in hell. At 5s initially (about £44 or $87 today), the Eidophusikon's entrance fee was high: it offered a respectable audience the thrill of theatrical display without the need to enter a theatre. In the late eighteenth century, theatres were rowdy and uncomfortable; their reputation among the middle classes as dens of vice was exacerbated by a growing suspicion of any fiction, let alone theatrical fiction.[9] To satisfy middle-class appetites for topographical spectacle, new forms of theatrical non-fiction were developing, somewhat along the lines of The Wonders of Derbyshire. Panoramas,

6. Barrell 1972, 5; Baugh 1987, 110–17; G. Wood 2001, 19–66. See also Lindqvist 1992, 84–93.
7. R. Allen 1961; 1962; Altick 1978, 120–1; B. Smith 1985, 114–19; Heringman 2004, 249–51. On de Loutherbourg see ODNB.
8. On the Eidophusikon see [Pyne] 1821; R. Allen 1966; Altick 1978, 121–7; Hyde 1988, 115–16.
9. Altick 1978, 184–5. On anti-fiction attitudes see Gallaway 1940; for an example of the connection between fiction and theatre in the minds of their detractors, see "Priscus" 1838.

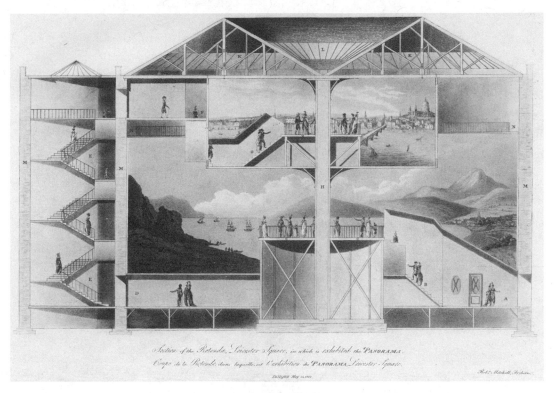

*Figure 7.2.* The interior of Robert Barker's panorama rotunda in Leicester Square, with assorted ladies and gentlemen viewing the two scenes on display. This aquatint was painted by Robert Mitchell, the building's architect. R. Mitchell 1801, plate 14.

dioramas, and other such "rational amusements" took scene-painters, technicians, and spectators out of the theatres and into new buildings which dominated middle-class metropolitan entertainment for almost a century.[10]

In 1787 Robert Barker, an Irish portrait-painter living in Edinburgh, patented a 360-degree frameless landscape-painting which he called "La Nature à coup d'Œil" (bastard French for "nature at a glance"), but soon renamed with the catchier coinage "panorama" (neo-Greek for "all-embracing view"). It was housed in a special rotunda, designed by the Aberdonian architect Robert Mitchell, which let in the daylight through an upper aperture (Fig. 7.2). Viewers stood at some distance from the canvas; the absence of a frame and the painting's extreme verisimilitude produced the illusion of being present at the site depicted. Barker's Panorama opened in 1794 in London's Leicester Square, which remained the hub of Britain's panorama industry for over half a century. Similar rotun-

10. The most detailed study of these sites is still Altick 1978, though he only covers London. Mannoni 2000 covers a wide range, as do Stafford and Terpak 2001 in less detail. The best histories of panoramas and dioramas are Hyde 1988; R. Wood 1993; Oettermann 1997. For more literary approaches see Galperin 1993, 34–71; G. Wood 2001, 99–120.

*The geometrical Ascent to the Galleries,*
*in the Coloseum, Regent Park*

*Figure 7.3.* "The geometrical Ascent to the Galleries, in the Colos[s]eum, Regent's Park". This aquatint from 1829 shows the building and its panorama of London still under construction, with a gentleman and lady standing precariously on the high walkway. The central structure would feature London's first passenger lift. Aquatint, 1829.

Figure 7.4. "A View Near the Colosseum in the Regent's Park"; one of a series published by the tailor Benjamin Read to advertise "The Present Fashions". Aquatint, 1829. City of Westminster Archives, Ashbridge 123.2.

das were built in Paris, Berlin, and Petersburg, not to mention Edinburgh itself (see Fig. 10.2), showing new views in each season. During wartime they functioned somewhat like the newsreels of the 1940s, depicting crucial episodes in recent campaigns, while at other times they transported visitors to exotic locations around the world: Venice, the Alps, Corfu, Jerusalem, Spitzbergen, New Zealand. In 1829 the most famous of all panoramas opened at the Regent's Park Colosseum, where it formed the centrepiece of a huge neoclassical entertainment complex.[11] This was a permanent view of London as seen from the top of St Paul's Cathedral (Fig. 7.3), designed by Thomas Hornor. It remained on display until 1864 and was continually updated to incorporate new buildings. The Colosseum was so popular in high society that the fashionable tailor Benjamin Read used it as a backdrop in a series of advertisements for new clothes (Fig. 7.4). A similar principle was followed by the cosmorama, where tiny

11. On the Colosseum see Hyde 1982.

*Figure 7.5.* A typical piece of Victorian virtual tourism: "Hamilton's Delightful Excursion to the Continent and Back, within Two Hours!" The show probably involved a moving panorama or diorama. Printed poster, 1860s.

life fire had been started by John Martin's crazed brother, Jonathan, who was acting under divine instruction in a dream.[25] The "British Diorama" seized on the event's spectacular potential, drawing liberally on Martin's own apocalyptic style of painting. In a final irony, one of the lamps spilt

25. On Jonathan Martin's career see Balston 1945; *ODNB*. There is no evidence that John Martin was mad.

burning turpentine on the transparency, and within hours the "British Diorama" was ruined far more thoroughly than the Minster itself.

Disaster scenes were further enhanced in the 1830s, when Daguerre inaugurated the "double-effect" diorama, painting the transparency on both sides and illuminating each in turn. Aided by sound effects, this "before-and-after" technique was used to stage an avalanche overwhelming an Alpine village, a Roman basilica ruined by fire, and the eruption of Mount Etna. In direct competition with these shows was the Cyclorama, added in 1848 to the Colosseum complex. This realized the Lisbon earthquake of 1755 with the help of terrifying sound effects, including excerpts from the grimmer bits of Mozart, Handel, and Beethoven, performed by a mechanical "orchestra". The thrill of terror also dominated the supernatural end of the market. Magic lanterns—transparencies projected onto a white wall in a darkened room—had long been valued for their ghostly effects, and by the 1790s the phantasmagoria had made its appearance in revolutionary Paris: using sliding glass transparencies, ghosts could be made to appear, approach, then vanish or retreat.[26] In 1802 the device was shown in London, with the theatrical apparatus of a drop-curtain; it was widely imitated, especially by theatre companies looking for realistic phantoms. In the 1830s this form of transparency saw more respectable service, providing lecturers with visual aids and presenting picturesque "dissolving views" or "phantom views", where one scene seemed to melt into another.

The early nineteenth century saw a host of "oramas", but the kineorama, hemirama, pleorama, astrorama, udorama, myriorama, padorama, physiorama, stereorama, and diaphanorama were mostly new words for old tricks. The underlying principle, rooted in the new dramaturgy, was to provide what has been called "the shock of the real": visual immediacy, usually combined with a strong claim to truth.[27] These aims also informed contemporary landscape-painting, including the relatively new genre of "historical landscape", of which Martin was the best-known exponent: human figures were dwarfed by sublime landscapes, colossal buildings, and weather effects, just as scene-painting and special effects were dwarfing actors on the London stage. Theatre, rather than the Royal Academy, was the mainspring of Martin's panoramic visions of apocalypse. His detractors, idealist "anti-theatrical" critics like Wordsworth, Charles Lamb, and John Ruskin, deplored the physicality of his images and mocked them using the vocabulary of melodrama, panorama, and phantasmagoria.[28] His admirers used the same vocabulary

26. Mannoni 2000, 136–75.

27. G. Wood 2001.

28. Anon. 1834b; see Meisel 1983, 167.

in his praise, which reflected Martin's aim of rendering visible—almost palpable—the most melodramatic of past events. Martin reconstructed most of the epoch-making catastrophes in sacred and secular history: the Deluge, the fall of Nineveh, Belshazzar's Feast, the plagues of Egypt, the Crucifixion, the Last Man, the Last Judgement, and many more. His new method of mezzotint engraving enabled these visions to be reproduced and disseminated throughout the world.[29]

Panoramic forms of display were taken up in some museums, notably William Bullock's Egyptian Hall, which in 1822 housed an Arctic panorama complete with live Laplanders to enhance the illusion.[30] His innovative animal displays, too, took on panoramic form, opening in 1812 under the neo-Greek title "Pantherion" (i.e. "all animals"). As his guidebook explains, the stuffed animals were surrounded by modelled trees and foliage, "assisted by an appropriate panoramic effect of distance, which makes the illusion produced so strong, that the surprised visitor finds himself suddenly transported from a crowded metropolis to the depth of an Indian forest".[31] Bullock made the colonial implications clear:

> [Thanks to] the extension of our Colonies throughout the habitable world [. . .] the writer feels confident, that if his exertions are seconded by the Public [. . .] he will very shortly be enabled to make a Collection of Natural History far surpassing any thing of the kind at present in existence.[32]

The conventions of the panorama also penetrated specifically geological exhibitions in the second quarter of the nineteenth century. The Scarborough Museum's rotunda, which opened in 1829, featured a section of the Yorkshire coastal strata painted around the inside of the dome,[33] while an unlocalized handbill from 1838 advertised the "Typorama", a twenty-five-foot-long "Modelled View of the Undercliff, *ISLE OF WIGHT*", which accompanied a fossil exhibition.[34]

Mantell, meanwhile, was busy applying Buckland's visually rich lecturing style to his own museum, illustrating fragmentary specimens not only with maps and sections, but also with skeletal reconstructions and lifelike restorations of the animals themselves, hoping to render the fos-

29. Feaver 1975.

30. Altick 1978, 273–5. On ethnographic displays see Qureshi 2004.

31. Bullock 1812, 2. On this exhibition see also Wonders 1993, 30–1; Yanni 1999, 27–8. On this form of response to natural-history displays, see Carroll 2004.

32. Bullock 1812, iv–v.

33. Knell 2000, 69–70, 168–9. This section (which Knell contends was not the work of John Phillips) was repainted in reverse in 1906. For images see Osborne 1998, 315–17; Osborne and Bowden 2001, 41. On the increasingly creative use of space in Victorian science museums, see Forgan 1994; 1999; Yanni 1999.

34. Bodleian Library, Oxford, John Johnson Collection, Dioramas Box 3.

*Figure 7.6.* The classic image of Victorian saurians eating each other: Martin's mezzotint *The Country of the Iguanodon,* printed in Mantell's *Wonders of Geology.* The legend beneath reads, "The Country of the Iguanodon Restored by John Martin [. . .] from the Geological Discoveries of Gideon Mantell". Mantell 1839, frontispiece.

sils "intelligible to the uninstructed observer".[35] Versions of two of the pictures he commissioned still survive. The most famous is John Martin's *Country of the Iguanodon,* executed sometime between 1834 and 1837 and surviving as the mezzotint frontispiece of Mantell's *Wonders of Geology* (Fig. 7.6).[36] Here Mantell introduces Martin as "the eminent painter of 'BELSHAZZAR'S FEAST'", signalling the continuity between Martin's meticulously reconstructed historical apocalypses and this nightmarish vision of ancient warfare.[37] Martin's scene was itself based on an earlier picture, *Reptiles Restored,* made for Mantell by the panorama-painter and natural-history illustrator George Scharf.[38] The surviving watercolour (plate 6) was painted in 1833; according to the lower marginal note, it served "as a Sketch for a Picture 3 Yards long" which may have hung in Mantell's museum.

*Reptiles Restored* does for the Sussex Weald what De la Beche's *Duria*

35. Mantell 1851, 8; G. Richardson 1838, 8–10. On Mantell's collection see Norman 1993.

36. It is often stated that Martin's original painting hung in Mantell's museum (e.g. J. Secord 2004, 145), but I have so far been unable to track down any evidence for this.

37. Mantell 1839, v. *Belshazzar's Feast* was painted in 1821 and is reproduced in Plate 8.

38. On Scharf (George Scharf Sr.) see *ODNB.*

*antiquior* had done for Dorset three years earlier (see Fig. 2.2), and its composition effectively realizes the scene described at the end of Mantell's *Illustrations of the Geology of Sussex*. Barring a reptile lurking to the left of De la Beche's picture, the terrestrial fauna of the Age of Reptiles had never been seen in colour before. The tropical greens and yellows of Scharf's picture give this landscape a distinctly exotic air. As Richardson describes it in his virtual tour of the museum, "the colossal Iguanodon [. . .] appears to reign undisputed monarch of the wild and wondrous scene", presiding over "some mighty Nile, or still mightier Mississippi".[39] In contrast to Martin's miserable monsters, this monarch seems contented. Richardson spells out the lesson:

> Yet these giant forms tremendous,
> Creatures wondrous, wild, stupendous,—
> Each and all were framed for bliss;
> Form'd to share, without alloy,
> Each its element of joy,
> By the Power that rules to bless,
> All were made for happiness![40]

Lions might well lie down with lambs: in Scharf's pastoral scene, a crocodile chats happily with a plesiosaur down by the riverside. This served as a model not only for Martin's scene, but also for a drawing by George Nibbs which Scharf lithographed in 1838 for Richardson's own "virtual tour" of the museum (Fig. 7.7). Here the Iguanodon has become a lizardly gossip, swapping saurian stories with two ichthyosaurs; only in the distance (to the left) can any warfare be discerned.[41] On the other hand, Richardson's later *Introduction to Geology* (1842) featured an engraving by Martin which extended the universal predation of *The Country of the Iguanodon* to show no fewer than thirteen reptiles writhing in mortal combat on a small patch of ground (Fig. 7.8). The contrast between Scharf's idylls and Martin's nightmares shows how not only the deep past, but the very same scene, could be rendered in opposite ways depending on which brand of romance was required.[42]

Early-Victorian showmen rarely attempted to compete with these authoritative restorations. One of the few examples documented was exhibited at the Colosseum's "Gallery of Natural Magic" in 1839. The main part of the gallery housed demonstrations of electrical, telescopic, and

39. G. Richardson 1838, 9.
40. G. Richardson 1838, 7.
41. On Nibbs's picture see Rudwick 1992, 76–8.
42. See Rudwick 1992, 80–1.

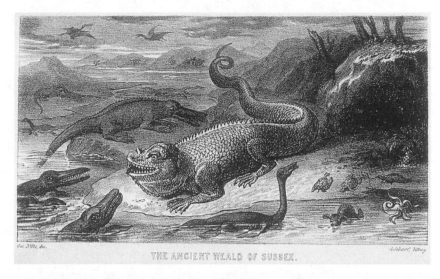

*Figure 7.7.* "The Ancient Weald of Sussex", George Scharf's engraving for George Richardson's miscellany *Sketches in Prose and Verse* (2nd series). Note the smile on the Iguanodon's face. G. Richardson 1838, frontispiece.

*Figure 7.8.* "The Age of Reptiles", John Martin's steel engraving for Richardson's textbook *Geology for Beginners*. If you look carefully you can see thirteen unhappy reptiles. G. Richardson 1842, frontispiece.

microscopic apparatus; then, according to the advertisement in Fig. 7.9, visitors were conducted to "the Caverns" to see a phantasmagoria. Between these extremes of truth and fiction, the caverns also displayed "the GEOLOGICAL REVOLUTIONS OF THE EARTH" in a "series of scenes" running sacred, human, and earth history together: Creation, "The Period of the *Iguanadon, Megalosaurus, Pterodactyles,* &c.", Paradise, the Deluge, and finally the Nile delta, "indicating the progress of civilization in the

THE
# GALLERY of NATURAL MAGIC,
AN ENTIRELY NEW EXHIBITION,
AT THE
## Colosseum, Regent's Park,
COMBINING
*The most varied Amusement with the most refined Instruction.*

THE ILLUSIONS and MAGICAL EFFECTS of OPTICAL SCIENCE now presented are of the highest order, and such as have never before been exhibited to the Public.

When the weather will permit, the astonishing effects of the SUN'S RAYS when CONDENSED, will be shown by the

### Most Powerful Apparatus in Europe.

The greatest heat known is obtained, and the hardest and most refractory substances are seen to drop like wax before a fire.

PHOTOGENIC DRAWING, or SHADOW-FIXING, will be occasionally shown, by an elegant apparatus, with all the improvements suggested by the discoverers of this singularly beautiful art.

The remarkable SPOTS IN THE SUN, or other extraordinary Phenomena, may be examined by instruments of superior powers and capabilities.

### The Largest Electrical Apparatus in the World

Is now for the first time presented to the Public. The terrific power of this most magnificent instrument must be seen to be duly appreciated; the most interesting experiments where electrical energy is required, will be exhibited on a scale of splendour never before witnessed. The Charge from a Battery will be sent through FIVE MILES OF COPPER WIRE. The plate of this stupendous machine measures seven feet in diameter, which gives upwards of eighty square feet of electric surface for excitation, and is mounted in a perfectly novel and most scientific manner, for obtaining the full effects of this wonderful agent.

IN THE MICROSCOPE ROOM an unequalled ACHROMATIC SOLAR MICROSCOPE, *on a disc of 256 square feet,* will be seen in operation at One, Three, and Five o'clock—possessed of powers, and with scenes of living objects, surpassing all others. This splendid instrument will be shown in *all weathers,* as a light of great intensity is employed in the absence of the Sun, by which this instrument is rendered sufficiently powerful, without the presence of any apparatus that can render an explosion possible.

THE CIRCULATION IN AQUATIC PLANTS, together with the most difficult opaque TEST, and transparent objects, may be examined and studied by means of costly *Achromatic* Instruments of exquisite workmanship. *The whole of the Diamond Beetle* may be here seen *at once,* 28 inches in length, by an instrument that has no equal in public exhibition.

IN THE MAGIC MIRROR will be shown the surprising effects of Optical Combinations, by which means a succession of beautiful scenery, of a perfectly novel character, is presented to the view.

The Visitor will then be introduced to the Caverns of the Colosseum, where the GEOLOGICAL REVOLUTIONS OF THE EARTH will be exhibited in a series of scenes, viz.

| | |
|---|---|
| The CREATION. | PARADISE. |
| The Period of the *Iguanadon, Megalosaurus, Pterodactyles,* &c. | The DELUGE. |
| | The NILE, indicating the progress of civilization in the renewed world. |

THE WORLD OF SPIRITS will also give up its PHANTOMS, at a WITCHES' SABBATH in the HAUNTED CAVERN.

This splendid collection of Instruments of Science forms altogether an unrivalled Exhibition, and is well worthy the attention of the scientific, the serious, and all classes of the community, who, in their pursuit of knowledge, seek to combine instruction with the most intellectual amusement.

*Open from Twelve till half past One—Two till Half-past Three—and Four till Half-past Five. Admission, One Shilling.*

G. H. Davidson, Printer, Tudor Street, Blackfriars.

*June 1839.*

*Figure 7.9.* Advertisement for the Colosseum's "Gallery of Natural Magic", an exhibition of optical technology. Printed handbill (1839), Bodleian Library, John Johnson Collection, Scientific Instruments Box 1.

renewed world". Like many other early-Victorian representations of human history and earth history (see Figs. 9.2–9.3), this sequence embodied the optimistic ideology of Progress.[43] Its medium is not specified in the advertisement: these scenes may have comprised a static set of murals,[44] but, in the context of the gallery's emphasis on technological wizardry, it seems more likely that they formed a set of dissolving views, in five acts like a play. The magic lantern which produced the subsequent phantasmagoria could easily have provided such views. It is striking, in any case, that this display was felt to belong in an exhibition of "natural magic"— in other words, among optical effects and illusions. Perhaps the mere act of viewing scenes from the distant past was felt to amount to a form of scientific necromancy, bringing the dead to life like Frankenstein.

"Frankensteinic" was how one contemporary summed up the best-known example of Victorian palaeontological spectacle, the life-sized sculptures of extinct animals built in 1853 for the Crystal Palace Gardens at Sydenham (Fig. 7.10, plate 7) by the artist Benjamin Waterhouse Hawkins, in collaboration with the palaeontologist Richard Owen.[45] Together they redesigned what Owen now called "dinosaurs", rejecting the lizard-like creations of Mantell and Buckland in favour of a more chunky, mammalian look. James Secord has called the whole Crystal Palace project the apotheosis of a short-lived "utopian union between commerce, education, and reason";[46] but it was also the apotheosis of virtual tourism. Like its main competitor, the Regent's Park Colosseum, it invited visitors to wander through reconstructed mediaeval cloisters, Assyrian palaces, lead mines, and limestone caverns; but it outdid the Colosseum with its life-sized models of people from across the globe, and the monsters of a former world lurking in the woods (plate 7). All these displays combined to embody the progress of civilization and the commanding power of British science. The monsters were (and still are) staged in a temporal panorama, allowing millions of visitors to stroll from one epoch to the next.[47]

These theatrical developments seem to lead naturally from the literature we have analysed in Part I. Geology had already shown the earth to be, in Lyell's phrase, a "theatre of changes", bringing its practitioners into "as immediate contact with events of immeasurably distant periods, as with the affairs of yesterday."[48] This formulation replicates the response

43. Bowler 1976; 1989.

44. As suggested by J. Secord 2000a, 440.

45. On Hawkins and Owen see *ODNB*.

46. J. Secord 2004, 166. On these monsters see also Doyle and Robinson 1993.

47. Rudwick 1992, 140–50; J. Secord 2004. On the time-travelling dimension of the Crystal Palace as a whole, see Whewell 1852, 14–15.

48. W. Buckland 1836, I, 202.

to the panoramas: "we are every day not only informed of, but actually brought into contact with remote objects."[49] Nevertheless, most nineteenth-century scenes from deep time were staged not in gardens, theatres, or museums, or even on paper, but in people's minds. Panoramic displays offered new models for the imagination. As Claude Lorrain's paintings had shaped eighteenth-century tourists' views of nature, so William Makepeace Thackeray now found Constantinople to be a real-life "Stanfield diorama".[50] The terms associated with these shows passed into common speech at the level of metaphor and simile. The phantasmagoria, for instance, provided a useful mental model for the perplexing visual sensations introduced by the high-speed railway, or the chaos of modern urban life: Walter Scott described London in 1828 as a "phan-

49. Review from 1826, quoted in Wilcox 1988, 37–8.
50. [Thackeray] 1846, 98–9; see also Wilcox 1988, 41–2. On the "picturesque-panoramic" mode of touring, see Buzard 1993, 172–92.

*Figure 7.10.* "The Crystal Palace and Gardens", a coloured panoramic view drawn by Benjamin Waterhouse Hawkins and printed by George Baxter. From left to right, the animal sculptures shown are Megalosaurus, Hylaeosaurus, two Iguanodons, and a Labyrinthodon. The palace itself can be seen in the background. British Museum Prints & Drawings, 1901-11-5-53.

tasmagorial place" where "the objects of the day come and depart like shadows".[51]

These metaphors reflected a renewed interest in sciences of the mind. The rise of new theatrical moving-picture technologies in the nineteenth century gave fresh substance to the eighteenth-century pictorialist concept of the imagination, according to which the business of language (especially poetry) was to stimulate pictures before the mind's eye.[52] A handful of now-famous idealist writers, such as Coleridge and Wordsworth, were beginning to contest this pictorial model;[53] but pictorialists like Addison remained firm favourites with the reading public, so the old model maintained its grip well into the nineteenth century. When Thackeray attempted to evoke Constantinople for his readers, for instance, he warned that "if you were never affected by a theatre, no words can work upon your fancy".[54] Now that the magic lantern was selling dreams for all to see, dreams themselves could be characterized as taking place on some mental stage, the "magic-lantern pictures of a doze".[55] Here is Dickens reflecting on his memories of Italian city life:

> The rapid and unbroken succession of novelties that had passed before me, came back like half-formed dreams [. . . .] At intervals, some one among them would stop, as it were, in its restless flitting to and fro, and enable me to look at it [. . .] in full distinctness. After a few moments, it would dissolve, like a view in a magic-lantern [. . . .][56]

Daydreams became panoramas: in *Adam Bede* (1859), George Eliot asks her readers to imagine "what sort of picture Adam and Hetty made in the panorama of Arthur's thoughts on his journey homeward".[57]

By contrast, Hugh Miller used a similar analogy to express the difficulty of grasping the whole view described by a favourite writer: "the imagery seemed broken up into detached slips, like the imagery of a magic lantern [. . .] and my mind was filled with gorgeous pictures".[58] In his autobiography Miller recalled how, when lying delirious with smallpox, he had tried to control the hallucinations that arose before him "as scene succeeds scene in the box of an itinerant showman". One scene had

---

51. Altick 1978, 219; Scott 1890, II, 160.

52. On the eighteenth-century model see R. Cohen 1964, 212; W. Mitchell 1986, 23–4; McNeil 1986, 183–93. On sciences of the mind see Flint 2000; A. Richardson 2001. Castle 1988 offers a particularly insightful analysis of the phantasmagoria's role in structuring reverie.

53. W. Mitchell 1986, 23–5.

54. [Thackeray] 1846, 99.

55. G. Eliot 1994, 194.

56. Dickens 1998, 77, cited in Flint 2000, 147.

57. G. Eliot 1980a, 441.

58. Miller 1847b, 135.

reproduced "every line and tint" of the incantation scene in Carl Maria von Weber's operatic melodrama *Der Freischütz,* which Miller had seen in Edinburgh fourteen years previously. Musing on the *terrae incognitae* of the human mind, he drew an analogy with the new art of photography:

> Of that accessible storehouse in which the memories of past events lie arranged and taped up, they [the metaphysicians] appear to know a good deal; but of a mysterious cabinet of daguerreotype pictures, of which, though fast locked up on ordinary occasions, disease sometimes flings the door ajar, they seem to know nothing.[59]

These analogies fed back into the business of writing. Dickens's narrative technique is often anachronistically called "cinematic", but closer models may be found in the moving pictures of his own day, as signalled by chapter-titles like "A Rapid Diorama".[60] Literature was increasingly judged by the theatrical impressions it left on "the reader's mental retina", as Harrison Ainsworth put it, "with descriptions to supply the place of scenery".[61] Such developments did not apply to fiction alone. Theatre was bowing to the growing demand for factual information about distant times and places, while the literature of fact was becoming increasingly theatrical. But this in turn raises a pressing question: in a culture dominated by visual entertainment, could texts hope to play more than a merely decorative role?

## Verse and the Virtual-Tourist Guidebook

In the nineteenth century this question was not just academic. The new drive for simulation and virtual tourism met stiff opposition from many literary professionals who saw this as compromising the privileged position of the poet. The attitude often termed "Romantic anti-theatricalism" lasted well into the Edwardian period and was held by writers of varying political persuasions, from Wordsworth and Lamb to Ruskin and Miller. Although they differed in their estimations of specific forms—Ruskin and Miller, for instance, were great admirers of the panorama—these figures shared an antagonism towards what they saw as the market-driven tendency of theatre and art to cater for the senses alone, abandoning the suggestive potential of allusion and leaving nothing to the imagination. Poetry's sovereign power to call up images in the mind was being debased by inadequate realizations of those images. As Miller put it:

59. Miller 1854, 332; 1993, 328–9.
60. Meisel 1983, 64; Flint 2000, 146–7.
61. Ainsworth 1836, xii–xiii, discussed in Meisel 1983, 65–6.

> In perusing our fine old dramas, it was the truth of nature that the vividly-drawn scenes and figures [. . .] suggested; whereas the painted canvass, and the respectable but yet too palpable acting, served but to unrealize what I saw, and to remind me that I was merely in a theatre.[62]

Such protests chimed in with educationalists' worries about the cognitive dangers of excessive visual sensation, and the links between sensationalism and sensuality.[63] This disapproval has since been echoed by several theatre historians who, interested mainly in drama as text, have seen the rise of the Regency pantomime and Victorian melodrama as indicating a sharp decline. Then as now, this text-centred view fails to take special effects and scenography as seriously as they deserve.[64]

Nevertheless, from the viewpoint of high literature, texts were certainly treated in a cavalier manner, even the verse-dramas of leading poets. Where the text got in the way of the spectacle, scissors were applied mercilessly and other texts pasted in where appropriate. Take *Manfred*, Byron's hugely popular cosmic drama of guilt and rebellion set in the Alps.[65] Byron had not intended it for performance, but in 1834 it was staged at Covent Garden by Alfred Bunn as a scenic extravaganza. Songs sung by various spirits were realized by dioramic painted views of Mont Blanc, the underwater world, a storm at sea, and an orrery showing a comet blazing through space. The choreography and décor of the Alpine scenes replicated Martin's recent *Manfred* paintings, with a magic lantern producing a realistic Witch of the Alps. Manfred's visit to the underworld was choreographed to reproduce Martin's engraving of Pandemonium in *Paradise Lost*. Whereas Byron's play ended with the dying Manfred's struggle for self-mastery, Bunn's version concluded with an avalanche diorama accompanied by lines snipped from Byron's Oriental romance *The Giaour*.[66] *Manfred* had become a successful disaster-movie.

Alternatively, verse-drama could become a vehicle for antiquarian reconstruction. Charles Kean's 1853 production of Byron's Assyrian tragedy *Sardanapalus* was conceived as a restoration of ancient Nineveh. The design was based on the recent excavations by Austen Henry Layard but drew heavily on Martin's Mesopotamian apocalypses, especially in the final "panoramic view of the Burning and Destruction of Nineveh".

62. Miller 1854, 331; 1993, 328. For further discussion see Meisel 1983, 30; Paley 1986, 134–5; Carlson 1994, 134–75; Peters 2000, 294–302.

63. Gallaway 1940; Altick 1957, 108–15; A. Secord 2002, 30–2.

64. On such critiques see R. Allen 1961, 58; Booth 1975, 8. Present-day parallels include James Cameron's magnificent restoration of the *Titanic* in his 1997 film of that title: see G. Wood 2001, 13–14.

65. Byron 1986, 274–314.

66. E. Brown 1966, 144–5; Carr 1973; Howell 1982, 98–105; Meisel 1983, 169–81.

For Kean, a member of the Society of Antiquaries, lifelike restoration *was* dramaturgy. He claimed that it had not previously been possible to stage *Sardanapalus* "with proper dramatic effect, because, until now, we have known nothing of Assyrian architecture and costume." Kean's chief aim was "to render visible to the eye, in connection with Lord Byron's drama, the costume, architecture, and customs of the ancient Assyrian people",[67] and he stuffed his programme-notes with information about ancient Nineveh, correcting many of Byron's "factual errors". To George Henry Lewes's rhetorical question, "Was Byron only a pretext for a panorama?",[68] Kean and most of his audience would have answered, "Of course." The acting received short shrift, but as time travel the show was a roaring success. According to the *Times,* viewers left the theatre "looking forward to the invasion of our island by Julius Cæsar as something in the remote future".[69] Adam Sedgwick was among those impressed by this "gorgeous and striking scenery, that has been so unexpectedly dug from the very bowels of the earth": these words were Kean's, but they point to the show's appeal for a geologist.[70]

If poetry was becoming a mere picture-hook, it is easy to see why the literati were uneasy. Their dissatisfaction echoes on today in highbrow critiques of screen adaptations which take liberties with the sacred texts of Jane Austen or Arthur Conan Doyle. At the heart of such protests is the assumption that pictorial matter should illustrate a text, and not the other way round. As an analytical tool for today's historian, this rhetoric of "faithful illustration" is not helpful: images did not necessarily serve texts.[71] The word "illustration" was undergoing a semantic transformation in the early nineteenth century. In 1800 it signified a text or utterance which rendered a thing visible to the mind's eye, and was applied to pictures only in a secondary sense (as in "illustrated with engravings"). By 1842, however, when the first issue of the *Illustrated London News* appeared, pictures had hijacked the term's primary meaning in many circles, reflecting the same shift in attention from mind's eye to body's eye which the literary critics so deplored. By this time, technological advances had enabled a massive rise in the publication of printed images, whose vulgar intrusion into newspapers was equated by Wordsworth

67. Undated programme-notes quoted in Bohrer 2003, 179. On Kean see Schoch 1998; *ODNB.*

68. Altick 1978, 186; compare G. Wood 2001, 41.

69. Anon. 1853, 7, noted in Bohrer 1994, 214. For other contemporary responses see E. Brown 1966, 197; Howell 1982, 78–9; Bohrer 2003, 178–82 .

70. Undated programme-note quoted in Meisel 1983, 182. For Sedgwick's response see Clark and Hughes 1890, II, 257.

71. For critiques of this rhetoric see Treadwell 1993; Sillars 1995, 11–13.

with a return to the primitive pictograms of the caveman—the ultimate irony of "progress".[72] Commercial pressures aside, the primacy of the word was also being challenged by new theories of visual education such as that of Johann Heinrich Pestalozzi, whose ideas informed the Crystal Palace project: Waterhouse Hawkins told the Society of Arts in 1854 that his monster-sculptures were intended to convey knowledge directly through the eye.[73] The panoramists were moving in a similar direction, claiming that "a Panorama alone" could adequately convey a view which surpassed all description.[74]

These trends have been seized on by historians tracing the origins of that nebulous construct, "modernity", which has sometimes been simplified as a series of cultural victories: the visual over the verbal, fragmentation over continuity, speed over leisure, mass consumption and instant thrills over elite reflection.[75] Ever since Walter Benjamin's seminal essays on the subject, photography and cinema have remained the key reference-points for such studies. Since the panoramas and their kin were the direct ancestors of this technology, they have typically been seen as "an early step in the modern supersession of poetry by spectacle", "suggesting the desire for cinema long before its realization".[76]

But this perspective has its problems. The idea of "mass visuality" tempts us to take the panoramists' word for it when they claimed that their shows captivated "all classes of Spectators".[77] However, in sharp contrast to the "Learned Cats" (see Fig. 7.1), the entrance fee to a panorama was normally a shilling, a tidy sum in 1830: these shows were dominated by polite society and the leisured classes. A "modernist" perspective also risks obscuring the importance of leisure itself in the panorama experience. Barker's original (and short-lived) *franglais* coinage, "La Nature à coup d'Œil" ("nature at a glance"), has been taken to mean that panoramas were supposed to discharge all its information in a single "instant thrill", like a modern television advertisement;[78] but, although the thrill of the "reality effect" seems to have been immediate, it is physically impossible to take in a panorama at a single glance. The Colosseum's panorama of London even provided little telescopes in

72. Anderson 1994; Sillars 1995. On Wordsworth's reaction see G. Wood 2001, 172.

73. J. Secord 2004, 140–2.

74. Anon. 1833a, 3–4; see G. Wood 2001, 117–18.

75. On modernity, speed and fragmentation see Kern 1983. Crary 1990 gives a subtle analysis of visual culture and modernity.

76. G. Wood 2001, 117; Aguirre 2002, 28.

77. This claim by the New York panorama proprietor John Vanderlyn is quoted in Aguirre 2002, 33.

78. G. Wood 2001, 119 (see also 5–6). On modern advertising see Kitis 1997, 304.

the viewing galleries.[79] The diorama, with its seated audience and moving pictures, comes slightly closer to the modernist ideal; but, as with the lingering camera-shots of early cinema, the pace was still extremely slow compared with most present-day equivalents.

If anything, prolonged scrutiny seems to have heightened the illusion. Viewers' imaginations often supplied missing details such as the murmuring of crowds or the sounds of battle far below.[80] Thomas Dibdin described his astonishment on first viewing Robert Ker Porter's panorama of *The Taking of Seringapatam* (1800):

> It was as a thing dropt down from the clouds—all fire, energy, intelligence, and animation. You looked a second time, the figures moved, and were commingled in hot and bloody fight. You saw the flash of the cannon, the glitter of the bayonet, the gleam of the falchion. You longed to be leaping from crag to crag with Sir David Baird [. . . .][81]

This is a long way from the "modernist" idea of the panorama as a "single frozen moment in time".[82] Leisurely viewing allowed time for the mind's eye to engage in dialogue with, rather than being swamped by, the sense-impressions admitted by the body's eye. In this respect the new media displayed a certain continuity with earlier conventions of artistic viewing, inviting active engagement on the spectator's part to realize the picture's narrative content.

Look, for instance, at plate 8. Martin's Babylonian apocalypse *Belshazzar's Feast* packed an initial punch in the welter of lurid reds and oranges which burst upon the viewer, but this wash of sensation was only the first stage in a "reading" process. This picture represents three acts in a tragic drama.[83] Perspective and vortices propel the eye sequentially around the scene, beginning with the fiery Hebrew letters spelling the king's doom on the left-hand wall. The rays of light from the letters lead the eye downstage right, then further upstage into the distance to witness the "wonder and dread" of the royal *ensemble* and the massed crowds, and on towards the Tower of Babel and wrathful storm-clouds in the background. Finally, the eye returns downstage and lights on the central figure of the prophet Daniel, who is interpreting the divine inscription for Belshazzar, telling him that he is to be slain and his kingdom taken over by the Medes and Persians. Daniel's gesture reunites the left-right opposition of

79. Hyde 1988, 89.
80. Altick 1978, 149–50.
81. T. Dibdin 1836, I, 146.
82. G. Wood 2001, 111; Stafford and Terpak 2001, 99.
83. This narrative was spelt out in [J. Martin] 1821, and is discussed in Meisel 1983, 21-2.

flaming letters and doomed king. This temporal perspective enhanced the painting's illusion of life and motion. Spectators who knew the Biblical stories were perfectly prepared to "read" canvases in this way: one commentator wrote of Martin's *Fall of Nineveh* (see Fig. 7.20), "A story on canvas was never better told."[84]

On the face of it, this seems to remove any need for text, apparently confirming the claims made in the nineteenth century by anti-theatrical critics and proponents of visual education, as well as by today's historians of "modernity". Nevertheless, texts were integral to early-nineteenth-century visual culture, and were not just pretexts for it. The new middle-class audiences could not be expected to know all the conventions of reading a painting, so the narrative structure of *Belshazzar's Feast* was specified by Martin himself in an explanatory booklet accompanying the exhibition, guiding the untutored eye around the picture.[85] Guidebooks and exhibition catalogues played a crucial role in the consumption of spectacle and the management of virtual tourism. They have traditionally been treated by historians of art and visual culture as little more than sources for reconstructing an artist's intentions; their functions, and still more the text-to-image relations they inscribe, have been almost entirely ignored.[86]

Textual accompaniment was particularly necessary for geological displays such as the palaeontological "panorama" at the Crystal Palace. Many visitors were unfamiliar with the narrative principle of successive geological epochs, so—in defiance of Pestalozzi's textless ideal—various official and unofficial guidebooks were published to show how to read this story and identify its protagonists.[87] For the many illiterate visitors, they might as well have been monsters drowned in Noah's Flood.[88] Not everyone was convinced by the restorations, with their anachronistic background foliage (Fig. 7.10, plate 7): the botanist John Lindley felt that the gardens as a whole had realized "Martin's vision", all except the absurdity of "Saurian monsters in Penge Wood".[89] Texts offered a more reliable means of bringing the creatures to life. Miller, for example, was credited with producing "dark, dreamy visions like a Martin" in his ver-

84. Feaver 1975, 102.

85. [J. Martin] 1821.

86. This is true even of Andrew Wilton's groundbreaking 1990 study of J. M. W. Turner's literary activities. An exception is Aguirre 2002, 35–6.

87. Official guidebooks include Owen 1854 (for the monsters) and Phillips 1854 (for the whole show). Compare Carroll 2004, 52–3.

88. This is allegedly how many visitors did "read" the exhibition: see Taylor 1997, xxxvii–xxxix; J. Secord 2004, 158–9. Even the official guidebook to the gardens was uncertain, informing visitors that the monsters on show had lived "Long ages ago, and *probably* before the birth of man" (Phillips 1854, 157, my emphasis). Terror was another unintended response to these visual stimuli: see Rudwick 1992, 148–51.

89. [Lindley] 1854; compare Boime 2004, 563–4.

bal restorations of the ancient earth, and his description of the Iguanodon in the *Sketch-Book of Modern Geology* (1859) was presented by one reviewer as a necromantic incantation for use at the Crystal Palace. The reviewer introduces an excerpt from Miller's description as follows:

> Most of our readers must have seen the restorations of the chief extinct animals in the gardens of the Sydenham Palace, and amongst these they must have particularly noticed the huge and horrid Iguanodon [. . . .] Now Hugh Miller makes the plaster restoration of Mr. Hawkins to walk, and with his vivid words he reclothes and refits the enormous bones of the same animal ranged in the gallery of the British Museum. Step forth, them, thou revivified monster! [90]

Behind Miller's "vivid words" lies a long-running exchange between visual and literary culture. This is seen most clearly in the ways in which spectacle guidebooks used quotations from poetry, the definitive medium for "vivid words". This exchange has been almost entirely ignored by historians of the "sister arts" debate, who have tended to view the relation between language and painting most often in terms of a struggle for priority or a drawing of conceptual boundaries, rather than the pragmatic question of how texts and images worked together.[91] Using verse to accompany a landscape painting was increasingly common in early-nineteenth-century exhibition catalogues: these often played on the familiarity of the verses quoted, so that word and image ended up "illustrating" each other.[92] For example, J. M. W. Turner's *Battle of Waterloo* (1818) illustrated the Waterloo stanzas in Byron's *Childe Harold's Pilgrimage,* some lines of which were quoted in the catalogue entry to illustrate the painting.[93] Although the painting depicts a single scene, its associations are defined and enlarged by the title's reference to an entire battle; this process is paralleled by the quoted fragment which recalls Byron's entire Waterloo sequence. Mind's eye and body's eye engage in dialogue to bring Turner's image to life.

Poetry quotations of this kind became more frequent in the 1820s, especially in the guidebooks to commercial displays. Martin's art bridged the gap between academic exhibitions (such as Turner's) and commercial spectacle, and both his paintings and his published engravings were typi-

---

90. Anon. 1859, 446. The ensuing description is quoted and discussed below, pp. 406–7. Compare the unofficial Routledge guide to the Crystal Palace monsters, quoted at length in Freeman 2004, 1–2.

91. The classic study is Hagstrum 1958; the wider debates about primacy and definition are examined in W. Mitchell 1986.

92. Meisel 1983, 32.

93. Wilton 1990, 146.

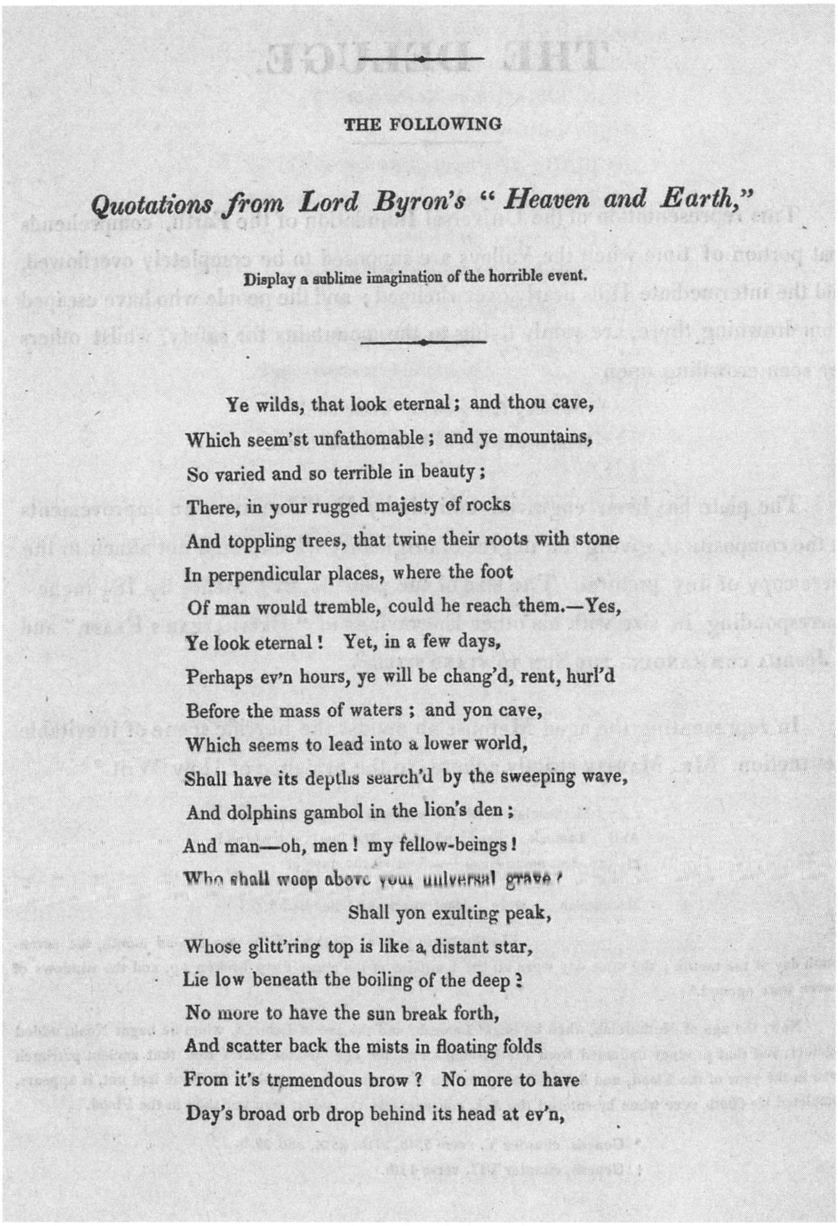

THE FOLLOWING

*Quotations from Lord Byron's " Heaven and Earth,"*

Display a sublime imagination of the horrible event.

Ye wilds, that look eternal; and thou cave,

Which seem'st unfathomable; and ye mountains,

So varied and so terrible in beauty;

There, in your rugged majesty of rocks

And toppling trees, that twine their roots with stone

In perpendicular places, where the foot

Of man would tremble, could he reach them.—Yes,

Ye look eternal! Yet, in a few days,

Perhaps ev'n hours, ye will be chang'd, rent, hurl'd

Before the mass of waters; and yon cave,

Which seems to lead into a lower world,

Shall have its depths search'd by the sweeping wave,

And dolphins gambol in the lion's den;

And man—oh, men! my fellow-beings!

Who shall weep above your universal grave?

Shall yon exulting peak,

Whose glitt'ring top is like a distant star,

Lie low beneath the boiling of the deep?

No more to have the sun break forth,

And scatter back the mists in floating folds

From it's tremendous brow? No more to have

Day's broad orb drop behind its head at ev'n,

*Figure 7.11.* A page from the booklet accompanying Martin's *Deluge* engraving (see Fig. 3.4). [Martin] 1828a, 4, British Museum Prints & Drawings, X.7.5.

cally accompanied by separate shilling guidebooks. These were littered with poetry quotations, the sources of which ranged from the Bible to Martin's own literary friends and contemporaries.[94] The proportion of verse to prose was so high as to turn these booklets almost into anthologies (Fig. 7.11). Two-thirds of Martin's 1825 *Deluge* booklet consisted of

94. [J. Martin] 1821; 1822; 1825; 1828a; 1828b; 1832.

extracts from Byron; the swollen two-page epigraph to his *Fall of Nineveh* guidebook gathers verses from the Book of Nahum; and his *Fall of Babylon* guidebook compiles extracts from Isaiah and Jeremiah into a sonorous litany of doom.[95] These guidebooks coincided with the emergence of giftbook anthologies, to which Martin himself contributed engravings. The giftbooks' material and aesthetic links with nineteenth-century poetry and fiction have recently attracted critical attention;[96] but they present still closer links with exhibition guidebooks. Here the poets danced attendance to the artist.

The poets were not necessarily hostile to this treatment. The guidebook accompanying Martin's 1828 *Deluge* mezzotint (see Fig. 3.4) ends with forty lines in verse by Bernard Barton, "Recollections of Mr. Martin's Print of the Deluge". Barton had addressed this poem to Martin as an "illustration of thy magnificent print"; it is an unusually rapid example of the "circuit of communication" cherished by book historians.[97] In an age when the concept of "illustration" was tending towards the purely pictorial, it was important for Martin to stress the subservient role of his quoted texts. The guidebook to his *Fall of Nineveh* engraving of the same year is bursting with excerpts from his friend Edwin Atherstone's recent poem *The Fall of Nineveh,* but Martin specifically denied that his picture was "a mere illustration of the Poem".[98] The same was true of his treatment of Byron. Martin's *Deluge* was undoubtedly inspired in part by Byron's diluvial verse-drama *Heaven and Earth* (1822), and Martin quoted large chunks of this drama in his 1828 guidebook;[99] but, as the subtitle to page 4 indicates (Fig. 7.11), Martin's aim was to reconstruct "the horrible event", not to illustrate a text. Even Bible quotations were subservient to this purpose. Accuracy and verisimilitude were his concerns: the Bible's veracity and Byron's imaginative vividness made them powerful auxiliaries to Martin's restoration of a vanished scene.

Martin's anxiety to maintain interpretative primacy partly reflects his desire to be seen as a great artist, not a mere tradesman. It was harder for panorama-painters to aspire to such ambitions, celebrated though some were. Many panoramas were team efforts, their guidebooks usually written by their proprietors. Here, too, poetry was set to work at illustrating scenery, but this power relation was sometimes complicated by gestures towards the poet's imaginative ownership of the scene described.

95. [J. Martin] 1825, 1–2; 1828b, 3–4; 1832, 2.

96. Manning 1995; Price 2000. On Martin's contribution to the giftbooks see Feaver 1975, 103, 110–11.

97. [J. Martin] 1828a, 3, 6–7. On this mezzotint see Feaver 1975, 92; Rudwick 1992, 21–4.

98. [J. Martin] 1828b, 16.

99. [J. Martin] 1828a, 4–5, 8; see also [J. Martin] 1825, 1–2.

The Leicester Square Panorama sold sixpenny guidebooks at its own exhibitions. These shared a common tripartite structure (which Martin imitated): a diagram of the scene, a "Description", and finally an "Explanation". The diagram is numbered, with a key; the "Description" sets the scene with topographical information and reflection and ends with a historical outline; and the "Explanation" focuses on the details numbered in the diagram, again mixing information with reflection.

From the 1820s on, poetry quotations proliferated in these guidebooks. Byron, again, was a favourite: sentimental Byronism was more easily abstracted from his verse once the man himself was dead.[100] In the guidebook to the Geneva panorama, designed by the new proprietor, Robert Burford, in 1826–7, Byron appears four times to describe the view. He was subsequently wheeled in to illustrate Niagara and Mont Blanc;[101] other favourites included William Sotheby, Samuel Rogers, and James Montgomery.[102] This concentration of verse has implications for the literary status of these displays. At the panorama, as Gillen D'Arcy Wood has observed, "the Georgian elite's love affair with the visual poetry of landscape reached the urban middle-class": prospect poems and Grand Tours were condensed into a painted scene.[103] But Wood's suggestion that the panorama displayed a "natural antagonism to poetry" is not sustained by the evidence of the guidebooks. Alpine scenes, far from being "scenes devoid of narrative or lyric suggestion", were layered with poetic and historical associations which tourist and virtual-tourist guidebooks made explicit.[104] The panorama was staking a claim to the poet's status not by repudiating poetry, but by incorporating it.

What kind of text-to-image relation do these quoted verses present? According to a schema developed by Roland Barthes, "anchorage" occurs when words denote or locate an image, and "relay" occurs when both contribute related units to a larger structure.[105] The former category is exemplified by the captions to press photographs, and one might be tempted to apply it to the verses quoted in the ephemeral panorama-guidebooks. But the relation between a spectacle and its guidebook is physically looser than the relation between an image and a text which both appear on a single printed page. Panoramas' "anchorages" are represented by the labels to the numbered key in the guidebooks' diagrams,

---

100. Byron (d. 1824) is conspicuously absent from earlier panoramas of Venice and the Alps, where one would expect him (Anon. 1819; 1821a).

101. Anon. 1827, 3, 8; 1833a, 3; 1837a, 3.

102. Anon. 1835, 3–4, 6, 9; 1837a, 4–5.

103. G. Wood 2001, 103.

104. G. Wood 2001, 116–17.

105. Barthes 1977, 38–41. On text-to-image relations in illustrated fiction see Sillars 1995 (see especially 17–27 for a critique of Barthes's taxonomy).

not by the guidebooks as a whole—and even these labels do not refer directly to the display. The poetry is harder still to pin down. That it elicited an instant emotional response in a sentimental formula may be true, but this does not tell us much. Panoramas did serve up landscapes for the eye to consume, but these were complex commodities whose text-to-image relation bears close attention. Their guidebooks were more than mere "background information" or "annotations" to the displays, or a purely "objective" counterbalance to the spectator's vertiginous absorption in the scene.[106] Rather, they facilitated and modified this absorption.

Take the guidebook to the Geneva panorama. Fig. 7.12 reproduces the beginning of the "Explanation" section: for the first item, Lake Leman, a sober narration of geographical facts is preceded by a quotation from *Childe Harold*. This might be seen simply as a versified label for the panorama's image of the lake. There is, however, an almost Magrittean discrepancy. The verse describes the lake at night, whereas both the panorama and the guidebook's prose depict a sunlit scene. Bearing in mind the panoramists' concern for accuracy, we cannot dismiss this discrepancy as clumsiness—particularly in view of the third poetry quotation on this page. Here the narrator mentions that storms sometimes occur over the placid lake depicted in the panorama, and Byron, the poet of tempest, bursts in upon the prose: "The sky is changed!"[107] Both Byron quotations call up vivid images before the mind's eye which represent the scene under quite different atmospheric conditions to those depicted. As such they add an imagined temporal dimension to the static display. Panorama visitors already experienced illusions of small-scale motion (perhaps the waves in this case), but the Byron quotations invoke larger shifts, working as a mental diorama.

A more literally dioramic passage appears in the same guidebook's "Description", where the narrator's lyrical prose blends with Rogers's verse to evoke the shifting lights of sunset:

> almost each moment some change takes place; they first assume a golden tint, which, as the sun declines, gradually warms into pink, and finally into a glow of crimson, of indescribable beauty. Mont Blanc [. . .] shines forth its brilliant light when the lower mountains are enveloped in total darkness.
>
> "I love to watch in silence till the sun
> Sets; and Mont Blanc, array'd in crimson and gold,
> Flings his broad shadow half across the lake; [. . .]

106. Galperin 1993, 44; Leask 1998a, 173–4; G. Wood 2001, 101–2. Aguirre (2002, 35–6) acknowledges the guidebooks' strategic role.
107. *Childe Harold* III.92 (Byron 1986, 131).

*No. 1.—Lake Leman.*

"Lake Leman woos me with its crystal face,
The mirror where the stars and mountains view
The stillness of their aspect in each trace,
Its clear depth yields of their far height and hue."—BYRON.

The Lake of Geneva, or "Lacus Lemanus" of the ancients, occupies the lowest part of the valley, which separates the Alps from the Jura mountains, and extends, in the form of a crescent, between the canton de Vaud and Savoy. Its length, on the Swiss shore, is eighteen leagues; on the Savoy, fourteen leagues. Its greatest breadth, which is between Rolle and Thonon, is three and a quarter leagues; in the neighbourhood of Geneva it is not very deep; but in some places, particularly opposite Mellerie, it is 190 fathoms, which is seventy-five deeper than the Baltic.* According to M. de Luc, it is 1126 feet above the level of the sea.† Besides the Rhone, there are forty-one small streams empty themselves into the lake, and when they are swoln by the thawing of the Alpine snow during the summer months, occasion it to rise five or six feet above its ordinary level. Lake Leman has long passed for one of the finest lakes in the South of Europe, and is perhaps only equalled by that of Constance. Voltaire gives the preference to it above all others.

"Que le chantre flateur du tyran des Romains,
L'Auteur harmonieux des douces Georgiques,
Ne vante plus ces Lacs et leurs bords magnifiques,
Ces Lacs que la Nature a creusés de ses mains
Dans les campagnes Italiques
Mon Lac est le premier. EPITRE A GENEVE.

The water, excepting at the influx of the Rhone, is as pure as crystal, and—"Deeply darkly beautifully blue;" it seldom freezes more than a few paces from the shore, and between Geneva and the great sand bank. During the summer months, the Bise, or regular north-east wind, occasions a gentle and agreeable agitation in the water; but, when the wind is east or west, the most sudden and destructive tempests frequently occur.

"The sky is changed! and such a change! oh night,
And storm, and darkness, ye are wondrous strong,
Yet lovely in your strength, as is the light
Of a dark eye in woman! Far along,
From peak to peak, the rattling crags among,
Leaps the live thunder! not from one lone cloud,
But every mountain now hath found a tongue,
And Jura answers, through her misty shroud,
Back to the joyous Alps, who call to her aloud."—BYRON.

Excursions of pleasure on the lake, and to the numerous villages and towns upon its borders, form the principal amusement of the Genevese; and, towards evening, the water is studded with small pleasure and fishing boats from various parts. The author of an agreeable work, ("Alpine Sketches,") describes an excursion of the kind in the following glowing terms:—"In the evening we hired a gondola, and rowed upon the lake; the water was so perfectly transparent, that the smallest object was discernible at the bottom, and the air possessed a balmy sweetness peculiar to these climates. The terraces of numerous chateaus, intermixed with luxuriant vineyards, reach along each edge; while beyond, the trees, rising in amphitheatre one above another, darkened with the deep foliage of the pine and cedar, break at once upon the glittering snow-clad mountains and glaciers, which, through sixty miles distant, seem impending in dazzling splendour over the smooth surface of the waters, in which they are beautifully reflected. Of all the scenes in nature, none can exceed this charming spot, where, though an eternal winter reigns within your view, you are enjoying, under a cloudless sky, the warmest beams of a southern sun, tempered by a refreshing breeze, which almost always plays upon the lake; and which, coming from the distant mountains, brings with it the fragrance of ten thousand aromatic shrubs, over which it passes, and meets you with a balmy sweetness gratifying to every sense."—There are twenty-nine species of fish found in the

* Dr. Goldsmith states the depth of the Baltic at 115 fathoms.
† Mr. Shuckburgh, 1152. M. Pictet, 1134.

6

*Figure 7.12.* A page from the booklet accompanying Robert Burford's panorama of Geneva. Anon. 1827, 8.

But, while we gaze, 'tis gone! and now he shines
Like burnished silver; all below, the nights.—[. . . .]
When, like a ghost, shadowless, colourless,
He melts away into the heaven of heavens;
Himself alone revealed, all lesser things
As tho' they were not!                    ROGERS.[108]

The panoramas have been seen as "unlimiting the bounds of painting"; [109] these textual illustrations widened its bounds still further. As catalysts for the touristic imagination, they enabled one panorama to incorporate a series of virtual dioramas and dissolving views to vary the scene in the mind's eye. *Childe Harold* was so well known that many viewers had ready-made mental images of these scenes, which the quotations triggered off. These passages belie the "modernist" claim that the panorama hijacked the conventions of the prospect poem "for the purposes of visual entertainment rather than lyric effusion".[110] The image may have been predominant, but poetry and painting worked together to produce it.

This symbiosis was not restricted to the "mental diorama" model. A poetry quotation could also function as a lens or frame through which the reader "sees with the poet's eyes".[111] This commonplace analogy implies more than merely borrowing a poet's heightened powers of perception: the viewer, like the tourist, is led to identify with the poet at the moment of looking. *Childe Harold,* again, was ripe for such treatment. Many travellers in the period used it as a guidebook on their own travels; John Murray even published a pocket edition for this purpose, and Byron quotations became *de rigueur* in tourist guidebooks to the Continent. Taken out of their challenging political contexts, these quotations offered tourists the means to stage themselves Byronically.[112] Early-Victorian touristic sensibility throve on this abstracted Romanticism, so the Childe pops up regularly in the literature of virtual tourism. An 1843 advertisement for a scale model of Venice (Fig. 7.13) claims that the astonished viewer, when "viewing this surprising object", will identify with Byron.[113] Its phrasing suggests that the model's verisimilitude prompts the viewer to "become" the gazing poet, registering the immediacy of the experience by altering Byron's "saw" to the present-tense "see".

Borrowing "the poet's eyes" also entailed framing the view in a par-

108. Anon. 1827, 3–4.
109. Wilcox 1988.
110. G. Wood 2001, 117.
111. See Miller 1847b, 351. On the art of "looking through another's eyes" as taught in didactic literature, see Keene 2007 forthcoming.
112. Buzard 1993, 114–30. For examples see Story 1863, I, 7; Wallach 1968, 377; Wilton 1990, 146.
113. The verse is from *Childe Harold* IV.1 (Byron 1986, 148). See also Buzard 1993, 115.

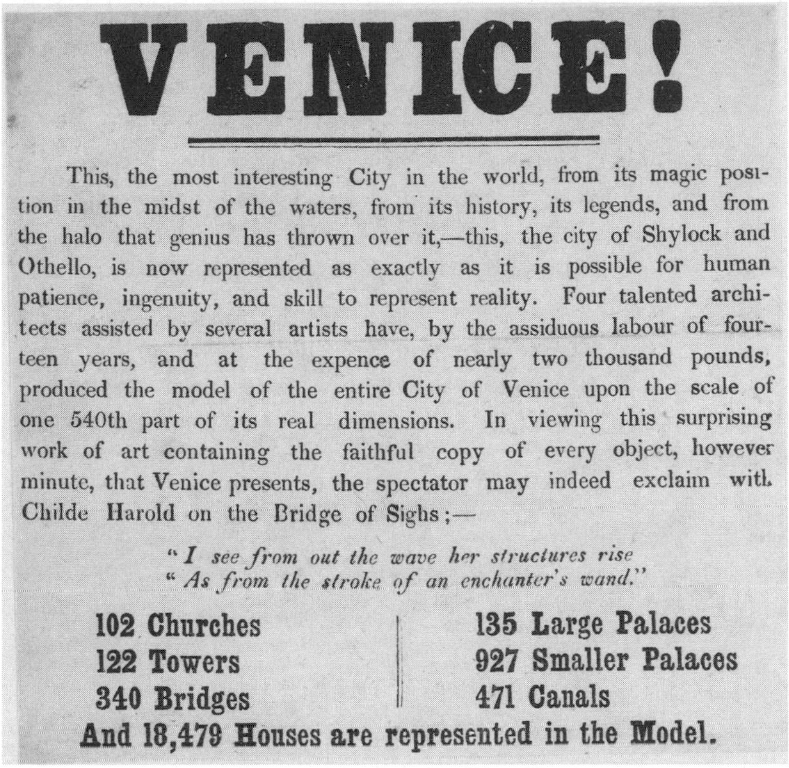

# VENICE !

This, the most interesting City in the world, from its magic position in the midst of the waters, from its history, its legends, and from the halo that genius has thrown over it,—this, the city of Shylock and Othello, is now represented as exactly as it is possible for human patience, ingenuity, and skill to represent reality. Four talented architects assisted by several artists have, by the assiduous labour of fourteen years, and at the expence of nearly two thousand pounds, produced the model of the entire City of Venice upon the scale of one 540th part of its real dimensions. In viewing this surprising work of art containing the faithful copy of every object, however minute, that Venice presents, the spectator may indeed exclaim with Childe Harold on the Bridge of Sighs;—

" *I see from out the wave her structures rise*
" *As from the stroke of an enchanter's wand.*"

| | |
|---|---|
| 102 Churches | 135 Large Palaces |
| 122 Towers | 927 Smaller Palaces |
| 340 Bridges | 471 Canals |

And 18,479 Houses are represented in the Model.

*Figure 7.13.* Advertisement for a scale model of Venice, situating the viewer as Childe Harold. Printed handbill (1843), Bodleian Library, John Johnson Collection, Dioramas Box 1 (78).

ticular way. The lens or frame analogies recall the Claude glass, a device enabling travellers to frame views in a picturesque manner: its convex mirror tinted the light, misted the distance, and made the trees in the foreground bend in, framing the prospect in the familiar manner of Claude Lorrain.[114] *Childe Harold* likewise offered familiar frameworks (visual, moral, emotional, and historical) for viewing landscapes. In the late 1840s, the Colosseum housed a pleasure-garden scattered with simulacra of Grand Tour locations. Its light-hearted guidebook is structured as an "imaginative tour" in which we "fly" or are "wafted" from one distant scene to the next.[115] Two kinds of illustration, both shown in Fig. 7.14, placed visitors *in situ*. Vignettes of fashionably-dressed spectators show the correct style for our stroll; but the views themselves are framed by quotations from Byron, with whom we are invited to identify and thus fancy ourselves actually present. *Childe Harold* predominated, except for the view of Mont Blanc where *Manfred* got a look

114. On the Claude glass see Clarke 1981, 33-5; Baugh 1987, 110–17.
115. Anon. 1845, 14, 16.

in.[116] One part of the gardens contained Roman, Greek, and Byzantine ruins—a *mélange* reflected in the guidebook, whose account of this area is dominated by a collage of twenty-three lines from different parts of *Childe Harold*. Byron had written these passages to describe individual sites, but here they have been carefully edited to remove specific topographical associations, transforming them into all-purpose meditations on antiquity. Commodification this may be, but Byron's work itself encouraged such collapsing of discrete identities:

> There is the moral of all human tales—
> 'Tis but the same rehearsal of the past:
> First freedom—and then glory; when that fails,

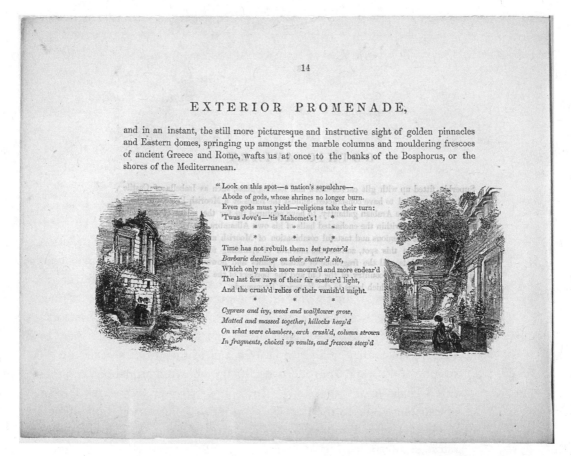

*Figure 7.14.* The virtual Grand Tour: a Byronic page from the guidebook to the Regent's Park Colosseum, describing the specially constructed ruins in the pleasure gardens. Anon. 1845, 14.

116. Anon. 1845, 15–17.

> Wealth, vice, corruption—barbarism at last;
> And history, with all her volumes vast,
> Hath but one page—'tis better written *here!*
> CHILDE HAROLD'S PILGRIMAGE.[117]

For Byron, the ruins of Rome represented history itself ("all human tales"). Historical specificity collapses into a pessimistic, cyclical vision of human endeavour, justifying and illuminating the acts of conflation perpetrated by the Colosseum's landscape gardener and guidebook author.[118]

Using well-known verse to frame a view could situate potentially unfamiliar scenes within a familiar aesthetic tradition, providing them with ready-made associations. According to idealist models of the imagination, the pleasure drawn from seeing an object arose not from the object itself, but from the trains of ideas and emotions generated in the observing mind. Such associations might be legendary, historical, personal, literary, or artistic;[119] when they seemed absent, as in the American wilderness, associations could be brought in from elsewhere by quotation or allusion.[120] The closest Leicester Square got to this territory was Niagara Falls, and in Burford's panorama guidebook (Fig. 7.15) we can see the Claude glass at work. Niagara was hardly unknown, having recently been represented dioramically at the Belgravia Pantechnicon; yet, by comparison with Grand Tour sites, it was unfamiliar even to many Americans.[121] So, although Burford claimed that these falls were "without parallel" and "altogether unique", defying "comparison", parallels and comparisons were precisely what was required in order to familiarize the view and trigger associations. Burford also asserted that words were inadequate to the task, and that "a Panorama alone" could convey the magnitude of the view; but his panorama was *not* alone.[122]

Niagara's textual frame—and at three Spenserian stanzas this was a large and ornate frame—was a passage from *Childe Harold* describing the Falls of Velino, near Florence. Only one alteration was made to Byron's original: for the name "Velino", a dash was substituted.

---

117. Anon. 1845, 15, quoting *Childe Harold* IV.108 (Byron 1986, 179).

118. The same passage had been used in 1836 to illustrate Thomas Cole's pan-historical landscape cycle *The Course of Empire* (Wallach 1968, 378).

119. Klonk 1996, 22–5.

120. On (Western) associations and the American wilderness see Boime 1991; Bedell 2001, 94–8; Lowenthal 2003, 24–7.

121. Altick 1978, 190; McKinsey 1985, 127–55.

122. My intepretation here differs sharply from that of G. Wood (2001, 117). On Burford see *ODNB*.

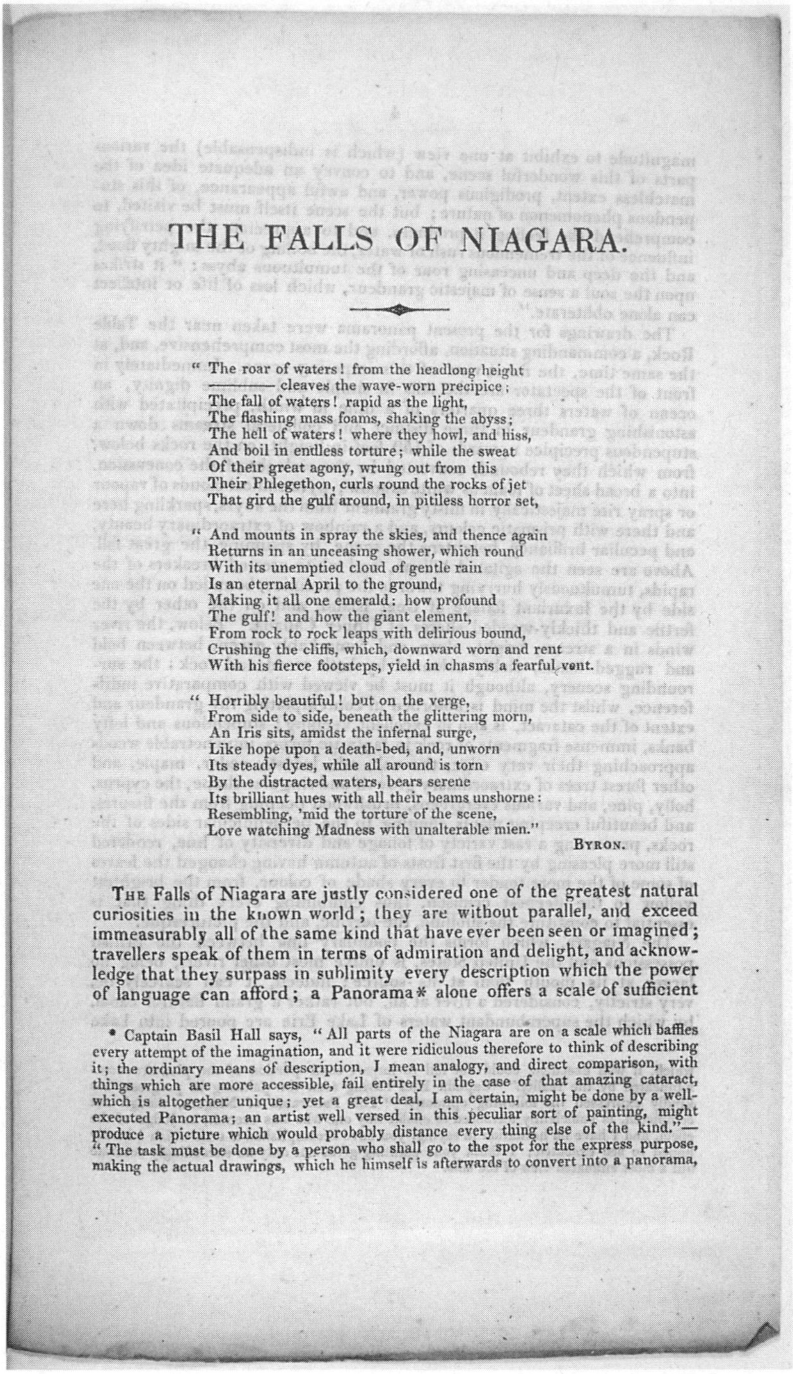

# THE FALLS OF NIAGARA.

" The roar of waters! from the headlong height
———— cleaves the wave-worn precipice ;
The fall of waters! rapid as the light,
The flashing mass foams, shaking the abyss;
The hell of waters! where they howl, and hiss,
And boil in endless torture ; while the sweat
Of their great agony, wrung out from this
Their Phlegethon, curls round the rocks of jet
That gird the gulf around, in pitiless horror set,

" And mounts in spray the skies, and thence again
Returns in an unceasing shower, which round
With its unemptied cloud of gentle rain
Is an eternal April to the ground,
Making it all one emerald : how profound
The gulf ! and how the giant element,
From rock to rock leaps with delirious bound,
Crushing the cliffs, which, downward worn and rent
With his fierce footsteps, yield in chasms a fearful vent.

" Horribly beautiful! but on the verge,
From side to side, beneath the glittering morn,
An Iris sits, amidst the infernal surge,
Like hope upon a death-bed, and, unworn
Its steady dyes, while all around is torn
By the distracted waters, bears serene
Its brilliant hues with all their beams unshorne :
Resembling, 'mid the torture of the scene,
Love watching Madness with unalterable mien."

BYRON.

The Falls of Niagara are justly considered one of the greatest natural
curiosities in the known world ; they are without parallel, and exceed
immeasurably all of the same kind that have ever been seen or imagined ;
travellers speak of them in terms of admiration and delight, and acknowledge that they surpass in sublimity every description which the power
of language can afford ; a Panorama* alone offers a scale of sufficient

* Captain Basil Hall says, " All parts of the Niagara are on a scale which baffles
every attempt of the imagination, and it were ridiculous therefore to think of describing
it ; the ordinary means of description, I mean analogy, and direct comparison, with
things which are more accessible, fail entirely in the case of that amazing cataract,
which is altogether unique ; yet a great deal, I am certain, might be done by a well-
executed Panorama ; an artist well versed in this peculiar sort of painting, might
produce a picture which would probably distance every thing else of the kind."—
" The task must be done by a person who shall go to the spot for the express purpose,
making the actual drawings, which he himself is afterwards to convert into a panorama,

*Figure 7.15.* A page from the booklet accompanying Burford's panorama of Niagara Falls.
Anon. 1833a, 3.

> "The roar of waters! from the headlong height
> ———— cleaves the wave-worn precipice [. . . .]" [123]

The reader was clearly expected to substitute the name "Niagara" for the dash.[124] The wild scene is framed, and hence tamed, by Byron's verse, with its Classical references, allegorical personages, clear visual trajectory, and authorial celebrity. Tourist guidebooks appropriated Byron in the same way, although separate guides to Niagara were published only after Burford had exhibited his panorama.[125] Horatio Parsons's guidebook of 1835 asserts Niagara's sublimity by borrowing a line from *Childe Harold*'s Alpine stanzas:

> The eye, unable to discover the mysterious phenomena at the bottom of the Falls [. . .] gives place to the imagination, and the mind is insti[n]ctively elevated and filled with majestic dread. Here is
>
> "All that expands the spirit, yet appals."—*Byron*.[126]

The implication is patriotic: Niagara, as much as the Alps, is the epitome of the sublime. Alpine associations are superimposed upon Niagara at the very point at which the body's eye is described yielding to "the imagination".

Originality was foreign to the poetics of virtual tourism. Using secondhand language, its guidebooks evoked preconceived meanings to elicit rehearsed responses. Seen in terms of their "service" to a poet's verse, they seem like debased currency; but this misses their point. They aimed to transport viewers to distant locations, not to serve other people's verse. They cut, pasted, altered, and recombined different poets' words to suit their own purposes. Yet the poetry did not lose all its authority in such transactions. Byron was a name with which to conjure—to invoke Alpine scenery, to restore ancient Nineveh, or to stamp an American waterfall with the seal of Western aesthetic approval. These appropriations, of course, flattened out his poetry's more challenging qualities, giving his imitators a head start: his stridently anticlerical *Heaven and Earth* provided a safer model for his conservative imitators like John Edmund Reade once it had been mediated by Martin's *Deluge*.[127] Imitation,

---

123. Anon. 1833a, 3, quoting *Childe Harold* IV.69 (Byron 1986, 168).

124. Horatio Parsons (1835, 32–3) did just this in his tourist guidebook to the Falls, inserting *"Niagara"* and quoting the same three stanzas.

125. McKinsey 1985, 133–4, 158–9.

126. H. Parsons 1835, 29–30, adapting *Childe Harold* III.62 (Byron 1986, 123).

127. On Reade see Chew 1924, 100–3.

plagiarism, and quotation diluted the pungency of the original by repeating it out of its initial context. This production of cliché was a positive benefit to the spectacle industry: by framing new landscapes in familiar verse, the panorama proprietors ironed out potential obstacles between spectator and spectacle. They made the view consumable, enabling the eye to devour every detail.

These techniques were clearly transferable to the uncouth objects and lost worlds of geology: similar framing devices were used to familiarize, moralize, and revivify rocks and fossils. Before we return to geology, however, it is necessary to look more closely at the prevalence of cliché in this whole enterprise, especially where time travel was involved. We are faced with a paradox. On the one hand, singularity and surprise were the prime attractions of these shows. Spectators wanted to be transported somewhere not merely distant, but different, and Byron's voice was used to ease this movement, making the viewer feel present at the site depicted. On the other hand, the imaginative layers (associations, frames) conferred by such quotation elided the particularities of specific sites into an aesthetic and historical continuum. This is especially clear in the Niagara and Colosseum guidebooks, where Byron's borrowed voice transforms unique objects into representatives of a general principle, whether this be "the sublime" or history itself.

This tension between the particular and the general was particularly evident when topographical spectacle took an apocalyptic turn, transporting viewers back in time to the scene of an ancient cataclysm, or outside time to the wilderness of hell.[128] Here the God's-eye view of the panorama became more than a visual convention. It contributed materially to the meaning of the scene displayed, drawing on traditional conceptions of the world as theatre. To witness the Deluge or the fall of Babylon was to glimpse divine forces operating on the threshold of the visible world. Restorations of prehuman worlds were rooted in the same tension between mystery and fact, which helps explain why Martin should have wished to bring them to life. These buried worlds, too, were poised on a threshold: they were fantastic otherworlds, inhabited by dragons, yet they were also part of the natural order. Fossils were both holy relics and natural objects. The geological scenes on show in the Colosseum (see Fig. 7.9) were displayed not with the optical technologies upstairs, but down in the "cavern" with the devilry of the phantasmagoria. Their location encapsulates the ambivalent cultural status of the ancient earth, hovering between fact and fiction, natural and unnatural, earth and hell.

128. On the vogue for apocalypse see Dahl 1953; Altick 1978, 186–7; Matteson 1981; Rupke 1983a; Esslin 1994; F. Stafford 1994; Paley 1986; 1999; Freeman 2004, 163–78.

## Apocalypse and the Scenery of Hell

Apocalyptic spectacle in the late eighteenth and early nineteenth centuries was informed by a marked shift in how the past was viewed. Convulsive demographic and socio-economic change, and the rapid transformation of the British landscape, sharpened the sense of the past as another country (Fig. 7.16).[129] Walter Scott's poems and novels, bringing various pasts to life through fiction, expressed this sense of historical change in its most popular form. Historiography became explicitly aetiological, seeking to understand the present by reconstructing the past and producing "Whiggish" narratives of causation and progress. Cautionary tales of decay and destruction built on a long-standing British fascination for the rise and fall of empires past. The idea of violently discontinuous historical change had gained a new edge with the French Revolution: violent physical causes like earthquakes and floods had gained new associations, and paintings of (super)natural disasters routinely invited political interpretation. The term "revolution" had once signified gradual, inexorable change; it now shifted, in both political and physical contexts, towards a sense of sudden and total transformation.[130]

Political resonances would not have been missed by visitors to the Cyclorama in 1848, another year of European revolution. Its reconstruction of the 1755 Lisbon earthquake was divided into four scenes like a play: sunrise, a journey upriver, the earthquake itself, and the aftermath. The real curtain rose only at the beginning, but verse quotations in the guidebook served as virtual curtain-raisers, realizing the image of the world as a theatre. The final scene was even ushered in by a quotation from Jacques's speech about the world as stage in Shakespeare's *As You Like It* ("Last scene of all").[131] The third scene was introduced as follows:

> The storms yet sleep, the clouds still keep their station,
>> The unborn earthquake yet is in the womb,
> The bloody chaos yet expects creation,
>> But all things are disposing for thy doom;
> The elements await but for the word,
>> "Let there be darkness!" and thou growest a tomb.—BYRON.[132]

---

129. Brooks 1998. For thoughtful accounts of historicism more generally see Lowenthal 1985 and Chandler 1998; on its influence on geology see Oldroyd 1979.

130. Rudwick 1972, 109; Feaver 1975, 44–7; Levere 1981, 170; Stott 1999; Constantine 2003.

131. Anon. 1848, 13, quoting Shakespeare 1993, 701 (II.vii.163).

132. Anon. 1848, 12, quoting *The Prophecy of Dante*, II.40–5 (Byron 1980–93, IV, 223).

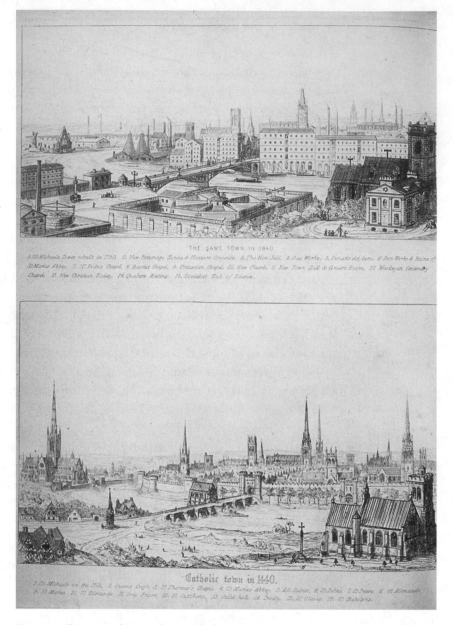

*Figure 7.16.* "Contrasted Towns". Diptych showing an imaginary English town in 1440 and 1840, drawn by the architect Augustus Pugin. This engraving encapsulates Pugin's view that the English urban landscape had changed for the worse since the Middle Ages. Pugin 1841, unnumbered plate.

These lines come from *The Prophecy of Dante* (1819), where they prophesy bloody political convulsions; at the Cyclorama the literal sense of the earthquake metaphor is restored. The role of Providence is underlined by the choice of quotations: the reversal of God's creative word in Byron's last line echoes the first scene, a sunrise illustrated by a quotation from

*Paradise Lost* on God's creation of light.[133] This combination of Milton and Byron stages the event as a *martinien* microcosm of sacred history, from Creation to Apocalypse.

These quotations point up the proximity of the general and the particular in apocalyptic spectacle. The doomed city was reconstructed with minute realism to transport the viewer to the place depicted, to the very point in its history which had, so to speak, made history. Yet the higher concerns associated with apocalypse—death, the end of the world, the purpose of history, theodicy, heaven, and hell—haunted the British public. Whether expressed in religious or secular terms, these concerns were more pressing than antiquarian details; on the other hand, specific details were precisely what restorations required in order to validate their claim to truth, allowing their larger meaning to be taken seriously. The paradox is only apparent: details and generalities support and validate each other. The point at which Lisbon made history was the very point at which it became a mythic site as well as an antiquarian one, a site around which archetypal meanings could cluster (particularly, in this case, theodicy) and new narratives bred within the ruins of the old. Equally, this transcendent significance impelled showmen and antiquaries to reconstruct specific events for a fact-hungry public. The same was true of Pompeii, Nineveh, and the "Country of the Iguanodon".

Poetry quotations in the guidebooks were often used to define the specific topography of these scenes. Martin's *Deluge* (see Fig. 3.4) shows this in action. William Feaver has drawn a contrast between Martin's intention to reconstruct a historical event with minute accuracy and his recourse to Byron's *Heaven and Earth* for "poetic licence", but the quotations in fact occluded the licentious parts of the original play and reinforced Martin's antiquarian project.[134] As in the Geneva panorama, the viewer is invited in the "Description" (see Fig. 7.11) to imagine the whole scene under very different atmospheric conditions, the day before the Deluge:

> Shall yon exulting peak,
> Whose glitt'ring top is like a distant star,
> Lie low beneath the boiling of the deep?
> No more to have the sun break forth,
> And scatter back the mists in floating folds
> From it's tremendous brow?[135]

Later, in the "Explanation", these lines are explicitly anchored to the mountain in the picture: Byron's generalized mountain-description con-

---

133. Anon. 1848, 9.
134. Feaver 1975, 92; compare Rudwick 1992, 22.
135. [J. Martin] 1828a, 4–5, adapting *Heaven and Earth* iii.22–7 (Byron 1980–93, VI, 355).

fers topographical specificity upon Martin's scene when it is given the label "Mount Caucasus, or Ararat" and a keyed number directing the eye towards the peak.[136] Such local detail paradoxically brought home the Flood's universality. *The Deluge*'s perspective is panoramic, and in depicting so realistically the imminent submergence of the Caucasus, Martin succeeded in portraying "the inundation of a world", just when Buckland was bringing the physical evidence for a global Deluge into the full glare of public attention.[137] Although Martin's guidebook displays no awareness of the new diluvialism, he highlights possible natural causes for the Deluge, such as "The sun, moon, and a comet in conjunction".[138] Such realistic displays gave new definition to the idea that God worked through natural causes.

But the materiality of these displays was not restricted to this world. Even heaven and hell were conventionally expected to fit into the physical universe. The theatrical, almost panoramic presentations of these realms in Milton's *Paradise Lost* had revitalized long-running debates on the nature (in both senses) of hell in particular.[139] This poem's unrivalled power of imprinting images of the invisible world onto the reader's mind made it, for many, the quintessence of the sublime, and by the mid-nineteenth century it was conventional to speak of Milton's "picture of the infernal world" or "brilliant phantasmagoria of contending angels".[140] Such images cried out for dramatic realization, which they duly received. British theatres were not allowed to stage dramas of sacred history, but individual episodes could be staged outside the theatres or sneaked into secular plays. Pandemonium was a particular favourite.

These realizations were strongly naturalistic. In the Pandemonium scene concluding the second season's programme of de Loutherbourg's Eidophusikon in 1781–2 (Fig. 7.17), the creation of Satan's naphtha-lit neoclassical city—a stylized fiction in Milton's epic—became part of a series of "Imitations of Natural Phenomena". Pandemonium might seem an odd conclusion to a programme comprising displays of terrestrial scenery, but surviving comments suggest that it impressed its audiences chiefly for its fidelity to nature. De Loutherbourg was praised for dramatizing "the effect of fire upon metal" and producing such realistic thunder and lightning. For many viewers, this was the essence of successful realiza-

136. [J. Martin] 1828a, 8.

137. Bulwer-Lytton 1970, 343. Evidence that either Martin or Buckland was aware of the other's work is lacking, despite the vague parallels and assertions made in Matteson 1981, Hopkins 2001, and Freeman 2004, 173.

138. [J. Martin] 1828a, 8. See Rudwick 1992, 24.

139. On the late-seventeenth-century geography of hell see Walker 1964, 39–40 and 100–1. On the theatrical roots of *Paradise Lost* see Demaray 1980.

140. C. Lyell 1830–3, III, 89; E. White 1876, 146.

*Figure 7.17.* Well-dressed spectators watching de Loutherbourg's "Eidophusikon" (1781–2 season), showing the scene in Milton's *Paradise Lost* where the city of Pandemonium is called into existence by the fallen angels in hell. The man on the right is scrutinizing the display through a spyglass. Watercolour by Edward Burney (*c.* 1782), British Museum Prints & Drawings, 1963-7-16-1.

tion, enabling de Loutherbourg to outdo Milton.[141] The identification of hell in terms of extreme natural phenomena is further seen in the placing and replacing of Pandemonium in the developing dramaturgy of the Eidophusikon and its imitators. The first season's sequence had concluded with a sea-storm and shipwreck. Pandemonium, the new finale for the second season, extended the storm's thunder, lightning, and anguish to preternatural proportions. Conversely, in some later imitations, violent nature stood in for de Loutherbourg's hell. The "New Eidophusikon" of 1799–1800 followed the usual sea-storm with an eruption of Mount Etna, and another "Eidophusikon" which Stanfield designed in 1824 featured a "terrific eruption" of Vesuvius and a *martinien* "Destruction of Babylon". The apocalyptic machinery of Bunn's *Manfred* made these links clear: its climactic Alpine avalanche dropped away to reveal Pandemonium itself amid the ruins of nature.[142]

141. Anon. 1782, 180–1; [Pyne] 1821, 217; Altick 1978, 121–5; Paley 1986, 54–5.
142. Altick 1978, 125–6, 214; Howell 1982, 103.

Martin's mezzotints for *Paradise Lost* (1825–6) took the naturalization of Milton's hell one stage further.[143] Martin presented Chaos, heaven, and hell in a clear spatial relation to the earth, poetically suggestive of the vast reaches of outer space, as if to underline hell's location within the physical universe (Fig. 7.18). This in itself was nothing new: it reflected the much-debated official dogma of most Christian denominations, and hell was usually located in the centre of the earth or on another planet. These debates maintained their vigour through the early Victorian period.[144] Milton himself had invited such speculation, and Martin's achievement in materially "embodying" these cosmic realms was widely celebrated. As one reviewer put it in 1825, "the wonders of that Heaven

*Figure 7.18.* The bridge over Chaos: one of John Martin's mezzotints for an illustrated edition of *Paradise Lost*. After Satan has facilitated the Fall of man, his offspring, Sin and Death, span the gulf between hell and earth with a triumphal bridge so that they and other evil beings can enter the world freely (Book X; Milton 1971, 519–29). The bridge's design recalls the civil-engineering projects of Isambard Kingdom Brunel and Martin himself. Milton 1853, facing p. 291.

143. For discussion see Pointon 1970, 90–166; Feaver 1975, 72–83; and, in particular, Treadwell 1993.

144. For examples see "Christian Philosopher" 1824, 4, 11; Christmas 1850, 289–91; Davy 1851, 18–65; Hontheim 1907–22; Crowe 1986. On later shifts from this theological naturalism see Rowell 1974; Wheeler 1990, 175–218.

and Hell which existed before earth was made, are magnificently embodied."[145] This conception of heaven and hell as a literally prehistoric realm of physical "wonders" hints at the continuity between Martin's Miltonic project and his later work on the Age of Reptiles. In the hands of some geological popularizers, as we shall see, the ancient earth was itself a hell, complete with fallen angels.

The naturalization of hell was sealed when Robert Burford and Henry Selous displayed Pandemonium and its surrounding scenery at Leicester Square in 1829 in the form of a panorama, the pre-eminent medium for accurate geographical representation. In his guidebook's "Description" Burford asserted that Milton's verse contained such "sublime and beautiful imagery" that it demanded to be "illustrate[d]" on a scale "worthy the elevation of soul and terrific conceptions of the Poet."[146] Some critics disapproved of the panorama's departure from terrestrial "fact",[147] but the guidebook treated Pandemonium with the same topographical precision as if it were another earthly city. Its information about individual demons and landmarks excerpted not only the relevant passages from Milton, but also biblical and Classical sources: Burford's panorama both illustrated, and was illustrated by, *Paradise Lost*. He defended his display by alluding to the poem's own naturalistic tendency to make hell "not less local than the habitations of man".[148] The panorama's minute realism was calculated to appeal to that large class of readers for whom *Paradise Lost* was no mere fiction, but a divinely-inspired prophetic utterance second only to Scripture itself.[149]

Burford and Selous clearly borrowed from Martin's *Paradise Lost* engravings, but their panorama's iconography was even more indebted to Martin's large-scale paintings of terrestrial apocalypses. As Burford's diagram reveals (Fig. 7.19), Pandemonium is set in a hellscape resembling the Caucasian scenery of Martin's *Deluge* (Fig. 3.4). The city itself conflates Martin's Pandemonium, Babylon, and Nineveh, its closest analogues being *Belshazzar's Feast* (plate 8) and *The Fall of Nineveh* (Fig. 7.20). Satan's "Hall of Audience" (no. 16, top right) adapts the exterior of Martin's Tower of Babel to fit the interior of Martin's Pandemonium. The vast temple-like complex of infinitely receding ziggurats, colonnades, and statues over which the hall towers, surrounded by swirling clouds and shivered by lightning, recombines the main features of Belshazzar's banqueting hall down to the last detail, even incorporating Martin's hissing serpent twined about the pillar in the form of "The Great Dragon"

145. Balston 1947, 99.
146. Anon. 1829a, 3.
147. Altick 1978, 182.
148. Anon. 1829a, 5.
149. Miller 1857, 258; see also Newlyn 1993, 54–6.

(no. 19, top right). The overall perspective reproduces that of *The Fall of Nineveh*, complete with the storm-clouds of divine wrath. Martin's twining serpent itself echoes those in de Loutherbourg's Pandemonium (Fig. 7.17), thus restoring the motif to its original infernal setting.[150]

It is no surprise that these twining serpents migrated from Pandemonium to Babylon and then back again. The three cities of sin, Babylon, Nineveh, and Pandemonium, were already barely distinguishable in Martin's hands, as hostile critics noted with glee.[151] Here again is that tense interdependence of the particular and the general so characteristic of apocalyptic spectacle. The events were depicted with minute realism, yet represented types of a larger pattern where they became as timeless as the architecture of hell. Their overarching lesson was that God rewards pride and luxury with destruction. Pandemonium supplied another variation on this theme: Milton had presented Satan as the prototype of oriental despotism, with Pandemonium explicitly prefig-

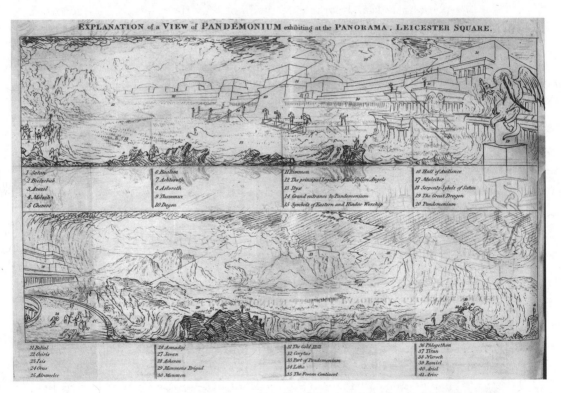

*Figure 7.19.* Keyed diagram of Burford and Selous's panorama of Milton's Pandemonium, with all the devils carefully labelled. This is a fold-out plate in the accompanying guidebook. Anon. 1829a, frontispiece.

150. Martin incorporated similar serpent designs into *The Fall of Nineveh*: on their significance see [J. Martin] 1828b, 14, and Monckton 1948.

151. Anon. 1834b, 457–8.

*Figure 7.20. The Fall of Nineveh* (1829), mezzotint engraving by John Martin, based on his painting of 1828. Nineveh, capital of the immense Assyrian empire, is sacked by the Medes and Chaldeans during the reign of the libertine king Sardanapalus. The king stages a sumptuous suicide: he and his concubines (seen at the top of the staircase) are about to burn to death on an enormous pyre with all his treasures. British Museum Prints & Drawings, Mm.10.5.

uring the Tower of Babel.[152] In Christian chronology, one ancient city after another re-enacted the fate of the antediluvians, and each cataclysm recalled and fulfilled its predecessors. Babylon re-enacted the fall of Sodom and Gomorrah, as the first Isaiah quotation in Martin's guidebook declaims: "Babylon [. . .] shall be as when God overthrew Sodom and Gomorrah."[153] The first version of Martin's *Crucifixion*, published in the 1830 *Amulet* annual, inspired the following lines by the poet and literalist pundit George Croly:

> This was the earth's consummate hour;
> For this had blazed the prophet's power;
> For this had swept the conqueror's sword,
> Had ravaged, raised, cast down, restored;
> Persepolis, Rome, Babylon,
> For this ye sank, for this ye shone.

152. *Paradise Lost* I.304–13, 692–7, 717–22, II.1–6, XII.24–62 (Milton 1971, 62–3, 83–6, 90–1, 610–13). See Demaray 1980, 106.

153. [J. Martin] 1832, 2.

This sense of past apocalypses rolled into one suited Martin's aims, and he incorporated Croly's poem as a textual illustration in his 1834 *Crucifixion* guidebook. Typology and allegory persisted even in matters of detail: alongside the "historically correct" reconstruction, a "Jewish Chief" falling from his horse is described in the guidebook as "a type of the Fall of Jerusalem".[154]

In the age of new approaches to biblical criticism, many Victorians maintained old ways of using the Bible, which included reading contemporary meanings into ancient catastrophes.[155] The spectre (or promise) of revolution and millenarianism, the fear (or hope) that society was changing faster than ever before, abandoning cherished (or hated) traditions, the conviction that supernatural phenomena periodically intruded into the natural order—all these feelings fuelled, and were fuelled by, the British fascination with apocalyptic spectacle.[156] Martin himself seems to have intended his early paintings partly as a critique of the hubristic

*Figure 7.21. The Fall of Babylon* (1831), mezzotint engraving by John Martin depicting the events immediately following those of *Belshazzar's Feast* (plate 8). The Tower of Babel, symbolizing the vanity of "science without God", is conspicuous in the distance as the Medes and Persians invade the corrupt Chaldean capital.

154. Booklet quoted in Feaver 1975, 135.

155. On the decline of typology in academic biblical criticism, however, see Frei 1974.

156. On millenarianism, see Harrison 1979.

excesses of Regency London, "the New Babylon". To a radically-oriented northerner, mistrustful of the establishment, the Prince Regent made an apt Belshazzar, ripe for a fall.[157] Both rulers presided over vast empires: Martin defended the architectural *mélange* of his doomed cities on the grounds that their despotic rulers would have commanded the services of architects from all over their subject territories. Much the same could be said of the ostentatious medley of architectural styles now adorning London; Martin himself produced plans for Pandemonium-like triumphal arches and riverside complexes which were intended to survive as colossal ruins to haunt the coming race when London was overwhelmed in its turn.[158]

On the other hand, the monarchs of the day (including George IV) were among Martin's most enthusiastic patrons,[159] and his political views did not dictate how his pictures were interpreted. From the conservative standpoint of the ancient English universities, his portentous biblical images could be seen to warn against the hubris of purely secular ambition. A gilt-framed engraving of Martin's *Fall of Babylon* (Fig. 7.21) hung in Sedgwick's rooms in Trinity College, Cambridge. James Secord has reflected on the significance of this image for Sedgwick in 1844 as he was struggling to refute what he saw as the pernicious materialism of *Vestiges of the Natural History of Creation*. Babylon was a city "corrupted by intellectual ambition and lascivious indulgence"; in the context of this treatise, disguised like a latter-day Whore of Babylon in the "modest garb of philosophy", the moral of the mezzotint was as clear as its Tower of Babel.[160] As Isaiah had put it, thundering from the pages of Martin's guidebook, "For thou hast said in thine heart, [. . .] I will exalt my throne above the stars of God."[161]

Martin's melodramatic images of apocalypse were thus open-ended enough to support diverse ideological commitments. That they all resembled each other was a corollary of their endlessly-repeatable moral; similar functions mould similar forms. But this reproducibility had its costs, making Martin's work vulnerable to piracy. Plagiarisms proliferated on canvas, glass, paper, porcelain, wood, steel, and copper.[162] Martin's images dominated the biblical end of Britain's virtual-tourism spectrum (Fig. 7.22); in France, where theatres were allowed to stage re-

157. On Martin's radicalism see Boime 2004, 562–3.

158. These ambitions reflected the earlier schemes of John Soane and Joseph Gandy: see Feaver 1975, 41–7.

159. Paley 1986, 135–6; Boime 2004, 558–9.

160. J. Secord 2000, 240, 235. Sedgwick returned to the image of the Tower of Babel when grappling with Darwin's *Origin of Species* in 1860: see Clark and Hughes 1890, II, 360.

161. [J. Martin] 1832, 2.

162. Balston 1947, 61–2; Feaver 1975, 111–12.

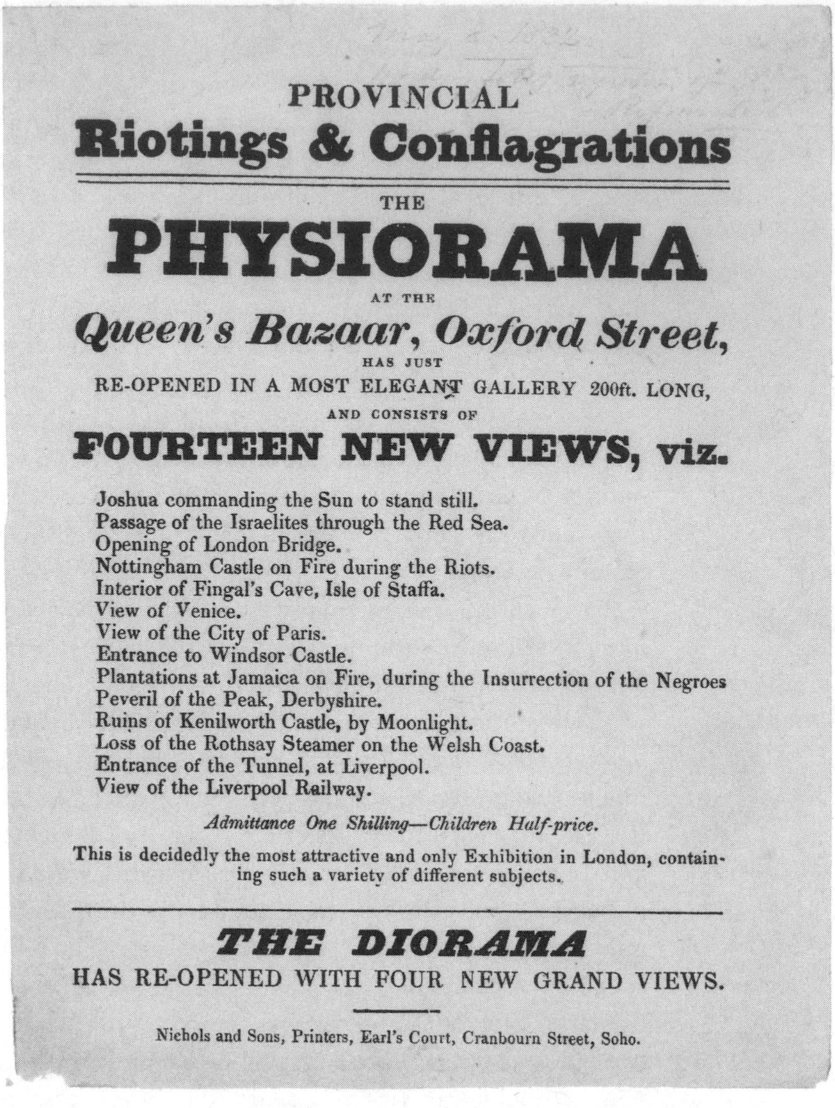

*Figure 7.22.* Advertisement for the Physiorama, a variant on the cosmorama, in which tiny backlit panoramic scenes were viewed through peepholes. The biblical scenes listed here—Joshua commanding the sun to stand still, the Israelites' passage through the Red Sea—almost certainly replicated John Martin's Bible illustrations and paintings. Printed handbill (1832), Bodleian Library, John Johnson Collection, Dioramas Box 2.

ligious dramas, his paintings were even turned into plays.[163] Time travel was the main aim of all these productions: the advertisement for Hippolyte Sébron's gigantic diorama of *Belshazzar's Feast* (Fig. 7.23), displayed in Oxford Street in 1833, claimed that "such is the extraordinary illusion with which it is painted, that the mind is led to contemplate it as a subject

163. Meisel 1983, 173.

*Figure 7.23.* Advertisement for Hippolyte Sébron's unauthorized dioramic version of Martin's *Belshazzar's Feast*. Printed handbill (1833), Bodleian Library, John Johnson Collection, Dioramas Box 2.

in reality." Martin tried to get a court order to close the exhibition, but failed; it was still drawing crowds six years later.[164]

Yet, as we have already seen, commodification did not imply levity or exclude serious reflection. Apocalyptic spectacle was generally designed to be taken seriously, and its commercial viability led to some remarkable unions of religious edification and marketing *savoir-faire*.[165] In October 1838 a series of "Pictorial Lectures on Sacred History", given by a man named Lewis, was shown at London's New Strand Theatre (Fig. 7.24). The advertisement first tells us that the lectures were "illustrated" by "dioramic views", but later that the lectures "illustrate" the dioramas. As the advertisement reveals, Lewis began his first lecture with a description of Babylon in ancient "splendour", accompanied by a large picture of the Tower of Babel and illustrated with reference to Herodotus. Recalling the guidebook to Martin's *Fall of Babylon*, Lewis then recited the Old Testament's "Prophecies denounced against Babylon", precipitating a change of scene to "the shattered and lightning-stricken Tower as it now exists". This show realized not only Martin's painting, but the very

164. Feaver 1975, 140.

165. Compare the Crucifixion panorama painted inside the apse of St Mary Moorfields, a Catholic chapel in London, in the late 1810s (Hyde 1988, 78).

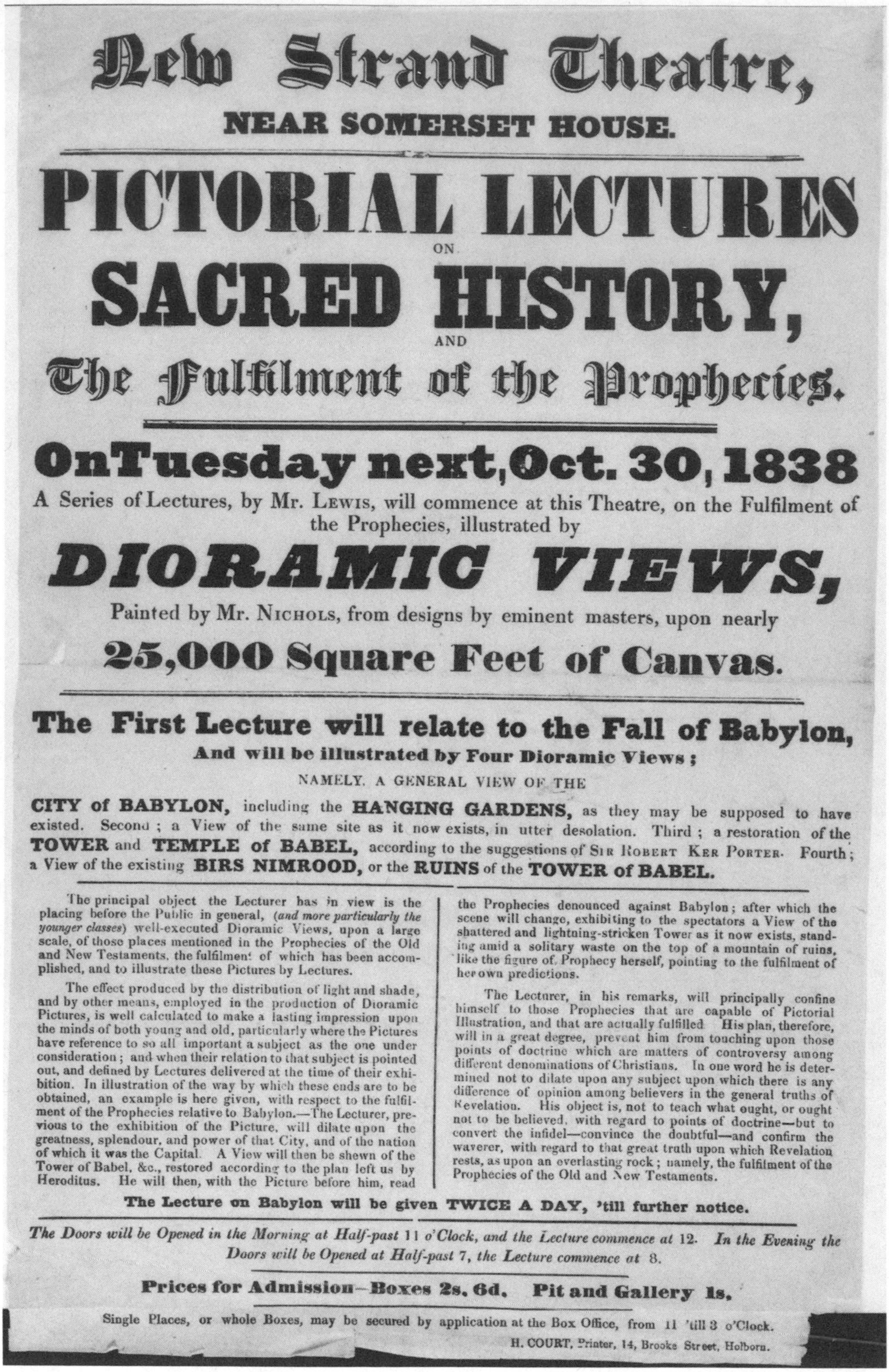

New Strand Theatre,

NEAR SOMERSET HOUSE.

PICTORIAL LECTURES

ON

SACRED HISTORY,

AND

The Fulfilment of the Prophecies.

On Tuesday next, Oct. 30, 1838

A Series of Lectures, by Mr. Lewis, will commence at this Theatre, on the Fulfilment of the Prophecies, illustrated by

DIORAMIC VIEWS,

Painted by Mr. Nichols, from designs by eminent masters, upon nearly

25,000 Square Feet of Canvas.

The First Lecture will relate to the Fall of Babylon,

And will be illustrated by Four Dioramic Views;

NAMELY, A GENERAL VIEW OF THE

**CITY of BABYLON,** including the **HANGING GARDENS,** as they may be supposed to have existed. Second; a View of the same site as it now exists, in utter desolation. Third; a restoration of the **TOWER** and **TEMPLE of BABEL,** according to the suggestions of Sir Robert Ker Porter. Fourth; a View of the existing **BIRS NIMROOD,** or the **RUINS** of the **TOWER of BABEL.**

The principal object the Lecturer has in view is the placing before the Public in general, (*and more particularly the younger classes*) well-executed Dioramic Views, upon a large scale, of those places mentioned in the Prophecies of the Old and New Testaments, the fulfilment of which has been accomplished, and to illustrate these Pictures by Lectures.

The effect produced by the distribution of light and shade, and by other means, employed in the production of Dioramic Pictures, is well calculated to make a lasting impression upon the minds of both young and old, particularly where the Pictures have reference to so all important a subject as the one under consideration; and when their relation to that subject is pointed out, and defined by Lectures delivered at the time of their exhibition. In illustration of the way by which these ends are to be obtained, an example is here given, with respect to the fulfilment of the Prophecies relative to Babylon.—The Lecturer, previous to the exhibition of the Picture, will dilate upon the greatness, splendour, and power of that City, and of the nation of which it was the Capital. A View will then be shewn of the Tower of Babel, &c., restored according to the plan left us by Heroditus. He will then, with the Picture before him, read the Prophecies denounced against Babylon; after which the scene will change, exhibiting to the spectators a View of the shattered and lightning-stricken Tower as it now exists, standing amid a solitary waste on the top of a mountain of ruins, like the figure of Prophecy herself, pointing to the fulfilment of her own predictions.

The Lecturer, in his remarks, will principally confine himself to those Prophecies that are capable of Pictorial Illustration, and that are actually fulfilled His plan, therefore, will in a great degree, prevent him from touching upon those points of doctrine which are matters of controversy among different denominations of Christians. In one word he is determined not to dilate upon any subject upon which there is any difference of opinion among believers in the general truths of Revelation. His object is, not to teach what ought, or ought not to be believed, with regard to points of doctrine—but to convert the infidel—convince the doubtful—and confirm the waverer, with regard to that great truth upon which Revelation rests, as upon an everlasting rock; namely, the fulfilment of the Prophecies of the Old and New Testaments.

**The Lecture on Babylon will be given TWICE A DAY, 'till further notice.**

*The Doors will be Opened in the Morning at Half-past 11 o'Clock, and the Lecture commence at 12. In the Evening the Doors will be Opened at Half-past 7, the Lecture commence at 8.*

**Prices for Admission—Boxes 2s. 6d. Pit and Gallery 1s.**

Single Places, or whole Boxes, may be secured by application at the Box Office, from 11 'till 3 o'Clock.

H. COURT, Printer, 14, Brooke Street, Holborn.

*Figure 7.24.* Advertisement for Lewis's dioramic "Pictorial Lectures on Sacred History". The first lecture showed the fall of Babylon. Printed poster (1838), Bodleian Library, John Johnson Collection, Education Box 27.

concept of prophecy: the shattered tower was compared to "the figure of Prophecy herself, pointing to the fulfilment of her own predictions".

Allegory and antiquarianism here worked together. Lewis's aim was to prove "the fulfilment of the Prophecies of the Old and New Testaments." In his first lecture, this "fulfilment" was enacted in Babylon's simulated transformation. The Bible's prophetic veracity was conveyed via the reality effect of a diorama, which was "calculated to make a lasting impression on the minds of both young and old, particularly where the Pictures have reference to so all important a subject as the one under consideration". Of equal importance was the fact that these dioramic views occupied "nearly 25,000 Square Feet of Canvas". As with Martin's canvases, a deluge of specific sensory stimuli was supposed to enable the spectator's imagination to receive "a lasting impression" of general truths beyond the reach of the senses.

## Conclusion: The Poetics of Time Travel

The virtual-tourist industry, especially its time-travel department, offered science writers practical techniques for promoting geology. The iconography of apocalypse, as defined by Martin, was itself a potent device. Poetry quotations and pictures could work together to stimulate or reassure the reader's imagination. By drawing on the conventions of the guidebook, writers could direct readers to look with a new pair of eyes. Deep time thus became a consumable spectacle, a status it has retained.[166] But before we examine how its popularizers used these various techniques, we must briefly consider how the gaze invited by these various panoramic displays, especially in apocalyptic spectacle, mapped onto the geologist's-eye view. These matters lie at the heart of the science's much-touted poetic status, over and above the individual poetic techniques used by its popularizers.

To summarize our findings so far: panoramic displays in the early nineteenth century often called for two distinct kinds of double vision, each inviting both specific and general views. The first operated at the level of scale. On the one hand, the viewer's attention was drawn to the truthful representation of minute details. Guidebooks drew the eye towards the details with keyed diagrams, facts, and illustrations. The implied spectator is at this point craning forward, straining his or her eyes to test the illusion of reality. On the other hand, the viewer was invited to stand back and allow the eye to immerse itself in the whole scene. In both ways the aim was to make viewers feel they were really at the site

166. S. Montgomery 1991; W. Mitchell 1998.

depicted—an end to which all the mechanical ingenuity of the age was directed.

The second level of double vision operated at the level of meaning, between the specific site and more general patterns. Much of the appeal of virtual tourism, as with the real thing, was the change of scene it provided. Yet the scene was often framed by texts which invited spectators to view it not just as a particular locality, but also as representative of some larger pattern, which might be aesthetic, historical, moral, or theological. These purposes usually merged: to mark out a landscape as a type of the sublime, as in Burford's Niagara guidebook, also invited reflections on the Creator's power. Furthermore, apocalyptic spectacle combined not only both kinds of virtual tourism (travel in space and time), but also all these different broader associations, reconstructing a great city or civilization at the very moment of its morally legible fall. This layering of associative meaning was achieved by a combination of iconography and textual accompaniment.

In sum, the viewer's gaze seems to have been required to oscillate between general and specific objects—whether spatially (between details and the whole), imaginatively (between familiar and unfamiliar), or semiotically (between a specific site and a timeless truth).[167] This oscillation was inscribed into the panorama guidebook itself. First, its "Explanation" referred to numbers on a diagram of the panorama: the viewer's eye was directed to assimilate the view by moving back and forth between panorama and diagram, and between diagram and explanation. Second, the guidebook's text swerved between fact and affect, or prose and verse (although these two pairs were not identical). Third, verse quotations often encouraged a "mental diorama", with body's eye yielding momentarily to mind's eye.

In view of the close link between knowledge and vision in Enlightenment epistemology, the oscillating movement invited by panoramic displays bears comparison with the processes of vision and thought involved in nineteenth-century science. Natural-history texts inscribed a similar movement at the level of scale, between the minutiae of specific objects and their place in a surrounding environment or history. The naturalist's mind voyaged in and out of substance.[168] At the level of meaning, too, the view or understanding of a specific locality or process provided a mental springboard for more elevated reflections. Lyell's prose exemplifies this

167. My interpretation here differs from Galperin (1988, 46), who has suggested that panoramas could never ultimately provide their much-touted "larger" view, only an aggregation of details. On Alexander von Humboldt's precarious balancing-act between general view and particular detail, see Leask 2002, 243–98.

168. Merrill 1989, 107–37; Bedell 2001, 103–5. On the "voyage into substance" see B. Stafford 1984.

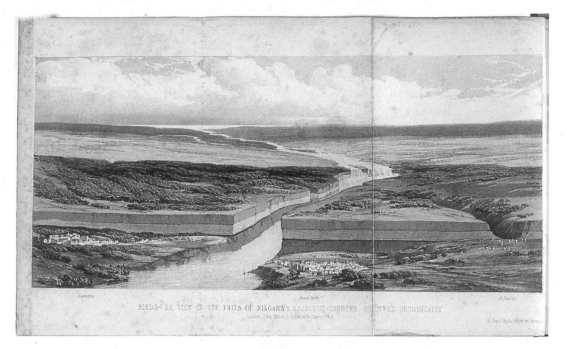

BIRDS-EYE VIEW OF THE FALLS OF NIAGARA & ADJACENT COUNTRY, COLOURED GEOLOGICALLY.

*Figure 7.25.* "Bird's-eye View of the Falls of Niagara & Adjacent Country, Coloured Geologically", a fold-out lithograph used as the frontispiece of volume 1 of Lyell's *Travels in North America* (1845).

upward flight: in *Principles of Geology,* his numerous geological anecdotes and examples resolve into a God's-eye view of the fluctuating economy of nature. His later observations on Niagara Falls took this imaginative movement closer to outright mysticism. Impressive as the falls are in themselves, the geologist looks with a different eye, tracing the events which have, over long ages, eroded the gorge to its present position. He becomes rapt to higher spheres:

> The geologist may muse and speculate on these events until, filled with awe and admiration, he forgets the presence of the mighty cataract itself, and no longer sees the rapid motion of its waters, nor hears their sound, as they fall into the deep abyss. But whenever his thoughts are recalled to the present, the tone of his mind,—the sensations awakened in his soul, will be found to be in perfect harmony with the grandeur and beauty of the glorious scene which surrounds him.[169]

Lyell opens the door to ever larger associations, soaring beyond even the panoramic bird's-eye viewpoint of his frontispiece (Fig. 7.25) to contem-

169. C. Lyell 1845, I, 53. See McKinsey 1985, 97–8; L. Wilson 1998, 22–5.

plate time itself. The specific scene, astonishing in itself, dissolves in the depths of eternity.

Lyell's numinous lyrical flight, with its emphasis on the "perfect harmony" between man and nature, recalls the tone and philosophy of certain landscape-descriptions in Wordsworth's 1805 *Prelude,* in particular the Alpine chasm described in Book VI. Here the particular features of the landscape are seen locked in mutual struggle, yet from this desolate sublimity the poet intuits a higher unity beyond time and space:

> The immeasurable height
> Of woods decaying, never to be decayed,
> The stationary blasts of waterfalls,
> And everywhere along the hollow rent
> Winds thwarting winds, bewildered and forlorn,
> The torrents shooting from the clear blue sky,
> The rocks that muttered close upon our ears,
> Black drizzling crags that spake by the wayside
> As if a voice were in them, the sick sight
> And giddy prospect of the raving stream,
> The unfettered clouds and region of the Heavens,
> Tumult and peace, the darkness and the light—
> Were all like workings of one mind, the features
> Of the same face, blooms upon one tree;
> Characters of the great Apocalypse,
> The types and symbols of Eternity,
> Of first, and last, and midst, and without end.[170]

Wordsworth's stance here is unambiguously prophetic, figuring the supreme cultural authority of the poet. This was the authority he was so concerned to defend from the encroachments of visual commodity-culture, maintaining the primacy of the poetic word in exercising the mind's eye. In 1802, in his programmatic preface to *Lyrical Ballads,* he had asserted that "Poetry is the first and last of all knowledge", using the apocalyptic "Alpha and Omega" trope as part of a bold redefinition of the poet's status.[171]

For many people, however, the concept of poetry was already broadening its semantic range, valorizing the visual associations evoked over the merely metrical nature of the text. For Coleridge, Burnet's *Sacred Theory of the Earth* proved that "poetry of the highest kind may exist with-

170. W. Wordsworth 1972, 238–40. For commentary see Abrams 1971, 105–7; J. Wordsworth 1982, 192–4; Wyatt 1995, 160–1. On Wordsworth see also *ODNB.*
171. W. Wordsworth 1944–72, II, 396.

out metre", and Wordsworth himself argued forcefully for the inclusion of prose composition under the term "poetry".[172] Increasingly, the true poet was conceived of as someone with unusual "insight", prone to "visionary" experiences, and preferably (but not necessarily) capable of communicating these experiences to the imaginations of others, whether in verse, prose, art, architecture, or music.[173] Despite the protests of an idealizing minority (the so-called Romantics, from Wordsworth to Baudelaire), representational literalism was the key to the new "poetry".[174] This was true even of symphonic music, which developed along strongly pictorial lines, as seen in the scenic overtures of Felix Mendelssohn and the dramatic symphonies of Hector Berlioz. For the first time, detailed programme-notes guided audiences to "read" the music, and in the 1840s the term "symphonic poem" (*sinfonische Dichtung*) was coined by Franz Liszt to reflect this new literalism.

The painterly purveyors of apocalyptic spectacle were clear contenders for the new title of "poet". Like *Paradise Lost,* seen as the crown of post-Shakespearian poetry, this genre sent the mind flying back and forth: into the deep past, along the punctuated time-line of sacred history, out to the present-day political or religious situation, and on to the end of time. Individual scenes could embody the entire narrative ("first, and last, and midst, and without end") or the process of history. Such multi-specific visions, dealing in absorbing particularities and sublime generalities with equal intensity, were intoxicating; those who succeeded in evoking them were extolled not only as poets, but as prophets. Martin's many admirers saw him in both terms ("That second Milton", "Mighty prophet −seer sublime!").[175] Bulwer-Lytton claimed to speak for his age when he delivered this encomium of Martin in 1833:

> I see in him [. . .] the presence of a spirit which is not of the world—the divine intoxication of a great soul lapped in majestic and unearthly dreams. He has [. . . .] gone back into the drear Antique; [. . . .] He has looked upon "the ebon throne of Eld," and imbued a mind destined to reproduce what it has surveyed, with
>
> A mighty darkness
> Filling the Seat of Power—as rays of gloom
> Dart round.

172. Coleridge 1985, 318; W. Wordsworth 1944–72, II, 391–2 and nn.

173. See *OED,* s.v. *poetry* II 3c, 3d, 5; Anon. 1834c, 117–18; Whewell 1852, 5–6. Compare the late-eighteenth-century use of the word "painter" (Barrell 1972, 5–6).

174. On "Romantic" protests against a purely imitative art, see G. Wood 2001. On representational literalism in sculpture, see J. Secord 2004, 147.

175. R. Montgomery 1839, 128; [Reade] 1829, 248. See Balston 1947, 119.

[. . . .] Alone and guideless, he has penetrated the remotest caverns of the past, and gazed on the primæval shapes of the gone world.[176]

Note that Martin's genius is said to consist in his ability to "reproduce what [his mind] has surveyed", words which recall the then-emerging medium of photography.[177]

Bulwer-Lytton's verse quotation evokes the cave of Demogorgon in Shelley's lyrical drama *Prometheus Unbound*, framing Martin himself as a latter-day Prometheus. Prometheus was more commonly associated with science than poetry, and Bulwer-Lytton's conception of genius also applied to men of science with a comprehensive vision: "the mooned loveliness and divinity of Nature reveals itself only to the rapt dreamer upon lofty and remote places."[178] The persistent image of the natural philosopher as solitary genius maintained strong links between science and poetry which new Kantian ideals of science as essentially communal had not yet dislodged.[179] Nature itself was the greatest poem of all, written by God the *poiētes* (creator).[180] The duty of the science writer was to translate this poetry for his or her readers, and to teach them to read nature themselves.

Geology's associations with the act of Creation gave it a special place among the sciences. Its practitioners were literally "penetrat[ing] the remotest caverns of the past", unearthing evidence of spectacular cataclysms and creative wisdom.[181] In the 1820s, when Martin was at the peak of his reputation, geologists too were beginning to offer time travel into alien landscapes. The antediluvian scenery of the literalists was no less enthralling than the succession of prehuman worlds presented by old-earth geologists: as one reviewer put it, "Everything ante-diluvian is poetical. The flood washed away a world from life into imagination."[182] Geology recalled the series of catastrophes punctuating sacred history, enabling the mind to travel towards the very beginning of that timeline to witness the formation and destruction of the landscape itself. This mental panorama was James Rennie's chief reason for calling geology "romantic".[183]

176. Bulwer-Lytton 1970, 343. Bulwer is quoting Shelley's *Prometheus Unbound*, II.iv.1, 2–4 (Shelley 1989–2000, II, 171). On Bulwer see *ODNB*.

177. On early photography in its art-historical context see Stevenson 2002; Boime 2004, 413–38.

178. Bulwer-Lytton 1970, 329.

179. On these Kantian concepts see Daston 2001, 81–2; on the image's practical limitations, see Golinski 1992, 1–10.

180. Miller 1859, 87.

181. On this analogy see Leask 1998b and Sommer 2003.

182. Anon. 1823, 63 (noted by Paley 1999, 219).

183. [Rennie] 1828, 7–8; see above, pp. 153–5.

The Deluge's typological role as precursor of the final Apocalypse, reinforced by Martin's iconography, brought geology close to prophecy, the highest form of (human) poetry. These associations go back to the previous century's "theories of the earth", sacred or secular: Burnet's final conflagration and Buffon's universal Ice Age offered contrasting visions of the end of the world.[184] In the early nineteenth century, both Granville Penn's eschatological claim that geology leads our thoughts "by an indissoluble chain from that which '*was*' and '*is*,' to that which '*is to come*,' [. . . .] in contemplation of *another earth*", and Robert Bakewell's philosophical remark that in geology "We are led almost irresistibly to speculate on the past and future condition of our planet, and on man, its present inhabitant" harmonized with a poetic ideal best expressed in Byron's pithy aphorism, "What is Poetry?—The feeling of a Former world and Future."[185]

As we have seen, Buckland and his colleagues were self-consciously assuming the mantle of the prophet in their private communications—a stance which enabled them to contest the authority of the Old Testament and the prophetic author of Genesis.[186] But the same stance was used by literalists like Andrew Ure for very different reasons:

> In exhuming from their earthy beds, or spar-bespangled vaults, the relics of that primeval world, we seem to evoke spirits of darkness, crime, and perdition; we feel transported along with them to the judgment-seat of the Eternal, and hear the voice of many waters coming to execute the sentence of just condemnation, on an "earth corrupt and filled with violence." The powers of prophecy overshadow us. The bony fossil starts to life, and conjures us in mysterious mutterings, to flee from the wrath to come. How solemn, to walk through this valley of death! Methinks the very stones cry out, "The Lord reigneth; righteousness and judgment are the habitation of his throne."[187]

As Marianne Sommer notes, this passage was later excerpted in the second edition of Octavian Blewitt's tourist guide *The Panorama of Torquay* (1832), where it functioned as "a dramatic vision with moral concern" to illustrate Buckland's cave-geology and introduce the wonders of Kent's Cavern.[188] In its original setting in Ure's chapter on the Deluge, however, it serves a specifically religious function, combating the vogue for

184. On Burnet's prophet-persona see F. Stafford 1994, 34–7.

185. Penn 1825, I, 269–71; Bakewell 1828, 481; Marchand 1973–94, VIII, 37 (Byron's Ravenna Journal, 28 January 1821).

186. Sommer 2003, 196–7.

187. Ure 1829, 505.

188. Blewitt 1832, 108; Sommer 2003, 191.

natural theology among gentlemen of science by indicating the literally prophetic message of diluvial remains. The fossil's "mysterious mutterings" recall Wordsworth's "rocks that muttered", and both are "characters of the great Apocalypse". Rather than Wordsworth's mystical union of mind and deity, however, the voices of Ure's rocks point to a literal apocalypse. It is a time-traveller's vision of the Deluge as Judgement Day, with the geologist as representative of sinful humanity.

That such imaginative transport could function at a purely secular level, on the other hand, is best substantiated not by the intense mysticism of Wordsworth or the ambiguous exegetical manoeuvres of Buckland, but by the motif's emergence in polite fiction. In Benjamin Disraeli's first novel, *Vivian Grey* (1826–7), geology is portrayed as a form of time travel and secular prophecy in the words of a wise German naturalist (with no apparent ironic intent):[189]

"See now," said Mr. Sievers, picking up a stone, "to what associations does this little piece of quartz give rise! I am already an antediluvian, and instead of a stag bounding by that wood, I witness the moving mass of a mammoth. I live in other worlds which, at the same time, I have the advantage of comparing with the present. Geology is indeed a magnificent study! What excites more the imagination? What exercises more the mind? Can you conceive any thing sublimer than the gigantic shadows, and the grim wreck of an antediluvian world? Can you devise any plan which will more brace our powers and develope our mental energies, than the formation of a perfect chain of inductive reasoning to account for these phenomena? What is the boasted communion which the vain poet holds with Nature, compared with the conversation which the geologist perpetually carries on with the elemental world? Gazing on the strata of the earth, he reads the fate of his species. In the undulations of the mountains is revealed to him the history of the past: [. . . .] the geologist is the most satisfactory of antiquarians, the most interesting of philosophers, and the most inspired of prophets [. . . .]"[190]

Geology's poetic nature is here defended using the same tropes as Bulwer-Lytton would use of Martin a few years later. These links between earth history, poetry, prophecy, apocalypse, and simulation come across most clearly in a panegyric on Cuvier in *La Peau de chagrin* (1831), a novel by Honoré de Balzac. Cuvier is praised for having "restored worlds from whitened bones" by applying his "retrospective gaze" in "a kind of re-

189. In its lack of apparent irony the passage contrasts with Disraeli's later portrait of a geological bore, the air-headed *Vestiges* reader Lady Constance, in *Tancred* (1847). See Klaver 1997, 158–60; J. Secord 2000a, 188–90.

190. Disraeli 1826–7, IV, 128–30, discussed in Klaver 1997, 159–60. On Disraeli see *ODNB*.

verse Apocalypse", an achievement which made him, rather than Byron, "the greatest poet of our century".[191]

All these effusions fulfilled Wordsworth's own prophecies about the relations of science, poetry, and truth, though perhaps not in the way he had imagined. In his 1802 Preface, Wordsworth had distinguished between "the Man of Science", who "seeks truth as a remote and unknown benefactor", and "the Poet", who "rejoices in the presence of truth as our visible friend and hourly companion". He had predicted that if scientific discoveries should ever create a "material revolution" in the way ordinary people perceived their world, the poet would be on hand to "lend his divine spirit" to the "remotest discoveries of the Chemist, the Botanist, or Mineralogist", "carrying sensation into the midst of the objects of the Science itself."[192] As Roger Sharrock has noted, the "first stirrings" of such a revolution in world-view were brought about by geology and its revelations of former worlds.[193]

The geological literature we have been examining did indeed represent "science [. . .] familiarised to men"; but, rather than requiring the assistance of "the Poet" to ennoble these ideas, men of science were themselves now staking a double claim to the poet's status. Works like Robert Hunt's *Poetry of Science* (1848) made these claims explicit, building on Wordsworth's own statement that "Poetry [. . .] is the impassioned expression which is in the countenance of all Science".[194] Furthermore, the literature of science was itself becoming increasingly poetic in order to "familiarize" the middle classes with the new knowledge. As Wordsworth's idealizing image of "truth" yielded to a new reification of "fact", panoramic visions of former worlds could be looked on as the highest form of poetry. Sensation—in both senses of the word—was brought into the midst of the science.

191. Balzac 1964, 24–6 ("reconstruit des mondes avec des os blanchis", "regard rétrospectif", "une sorte d'Apocalypse rétrograde", "le plus grand poète de notre siècle"). On Balzac and Cuvier see d'Alsó 1934.

192. W. Wordsworth 1944–72, II, 396–7.

193. Sharrock 1962, 72. Wordsworth's views have been reassessed by Yeo 1993, 65–71; J. Smith 1994, 48–62; Wyatt 1995, 7, 189–92. See also Herbert 2005, 135–6.

194. W. Wordsworth 1944–72, II, 396–7.

# Literary Monsters

8

Writing to his friend Edward Cowell in July 1847, the poet Edward Fitzgerald called the unfolding of earth history "a greater Epic than the Iliad". He exclaimed that the "vision of Time" expressed in Lyell's Wordsworthian description of Niagara was "more wonderful than all the conceptions of Dante and Milton".[1] Cowell was not convinced, so a week later Fitzgerald gesticulated in italics and capitals:

> Geology which has certainly discovered to us so much of the Past—and the Being of this Earth when *we were not;* is a more wonderful, grand, and awful, and therefore *Poetical,* idea than any we can find in our Poetry. For it is a FACT![2]

Fitzgerald's confidence in geology's factual truth was not shared by everyone. Geology presented its popularizers with a difficult challenge: how to represent landscapes and animals no human eye had seen, elicit-

---

1. Terhune and Terhune 1980, I, 566, discussed in J. Secord 1997, xxxix.
2. Terhune and Terhune 1980, I, 569. On Fitzgerald see *ODNB*.

ing wonder without compromising the "factuality" of their claims.[3] Geo-logical restorations necessarily belonged more to the world of fancy and imagination than did visions of the fall of Nineveh, whose history could at least be claimed to go back to ancient eyewitness-accounts. "[T]he past is always a fiction", but especially the prehuman past.[4]

Science was supposed to deal in literal truths, not fiction. This gave rise to some self-consciousness when authors moved from the exposi-tion of specimens or discussion of theories to providing restorations of a former world. Typically, restorations were demarcated from the sur-rounding prose in style and textual location, so that they would not be received as "fact" on quite the same level as other parts of the text. Yet they played a pivotal role in producing the text's illusion of factuality and the visionary authority of science. The truth Fitzgerald was referring to—a panoramic vision of time, "FACT" in capitals—was constituted by adding doses of fiction to a fact-packed discourse. The gap between the two created a certain tension.

The beginning of David Ansted's *The Ancient World; or, Picturesque Sketches of Creation* (1847) presents an extreme example. Cautious by in-clination, Ansted nevertheless wished to catch his readers' imaginations with verbal and visual "sketches". His frontispiece (Fig. 8.1) claims to be an unmediated view of "The Vegetation of England during the Coal Pe-riod". Turn the page, however, and a brace of caveats greets the read-er's eye:

> THE Frontispiece is *intended* to give an *idea* of what *may probably* have been the aspect of Vegetation in England during the Coal period. The trees introduced are chiefly those living forms which *seem* most analo-gous to extinct species. [. . . .] on the right [. . .] are the tops of Araucarias, coniferous trees *nearly allied* to which have been found in the coal-mea-sures. The remaining trees and plants are inhabitants of Norfolk Island or Eastern Australia, but *seem* to have had representatives in ancient times.[5]

The necessary suspension of disbelief is here broken under the sheer weight of provisionality. What is more, we are told that the picture is largely a *mélange* of present-day plants from around the world. Elsewhere Ansted vividly describes the desolate landscape of the primary epoch, then hastily adds, "It must, however, be distinctly understood that this

---

3. On this conundrum see Lowenthal 1985, 188–9.
4. Fortey 1997, 253.
5. Ansted 1847, ii (my italics). On Ansted see *ODNB; DNBS*.

THE VEGETATION OF ENGLAND DURING THE COAL PERIOD —Page xii.

*Figure 8.1.* A Carboniferous composite: "The Vegetation of England during the Coal Period", the frontispiece of David Ansted's *Ancient World* (1847).

view is strictly hypothetical, and is, after all, only one means of explaining certain phenomena."[6] Too much caution damages visibility.

Yet Ansted made creative use of the fact-fiction tension when it suited him. A dramatic saurian scene is introduced by what seems at first like a caveat: "and yet, knowing with absolute certainty these points, we hardly dare draw the conclusions which are suggested."[7] In fact this is not scientific caution but showmanship, like Mantell's claiming he had initially been "anxious" to reduce the dimensions of his reconstructed Iguanodon because the "enormous monster" appalled him.[8] Ansted relates the subsequent battle-scene with gusto, and after a repeat performance in the Cretaceous he exclaims, "There is scarcely any freak of the imagination,

6. Ansted 1847, 21.
7. Ansted 1847, 179.
8. Mantell 1833, 315.

however wild or vague, that does not seem surpassed by some reptilian reality during this remarkable period."[9] The rhetoric of caution, carefully built up in earlier chapters, enhances the theatrical effect.

Scientific inquiry in general was promoted alongside a fantasy of total, panoramic visibility, with its associated metaphors of revelation, illumination and enlightenment. These aims were inseparable from the theatrical culture examined in chapter 7. For the geologist, rendering the invisible visible—Peter Brook's definition of good theatre[10]—meant filling the gulf between fact and fiction with literally theatrical performances: pageants, panoramas, pantomime villains in ichthyosaur outfits, even a comedy monkey.[11] Geological literature's stylistic instability was a structural corollary of the epistemological demands its authors faced, sewing together fact and fiction to produce truth.

There are analogies to be drawn here with palaeontological practice. Cuvier had promoted the image of the geologist as a latter-day Ezekiel in the Valley of Dry Bones, reassembling disconnected skeletons under the guidance of the spirit of comparative anatomy. Mantell reworked this image in the Gothic mode, presenting himself as a reluctant Victor Frankenstein, driven by science to bring forth appalling monsters.[12] The literalist Samuel Best therefore warned: "let us not be too sure that in putting together the bones of extinct species [. . .] we are not out of collected fossil remains creating to ourselves a monster."[13] Best's worries were understandable. In their descriptions of these strange beasts, geological writers were familiarizing them for their readers by taking a known animal and adding other animals' features—the very technique used by Renaissance taxidermists to create monsters. Thus William Conybeare's Plesiosaurus was turtle-like and swan-like; Mantell's Iguanodon had affinities with the lizard, crocodile, and rhinoceros; Buckland's pterodactyle was a lizard in the form of a fiendish vampire. Monstrosity was inscribed into their names: Ichthyosaurus ("fish-lizard"), Pterodactyle ("wing-finger"), Pterichthys ("winged fish").

The practical business of reassembling a skeleton had its own pitfalls. While the dividing-line between reconstruction (or restoration) and fabrication was recognized as all-important, not everyone drew it in the same place. Thomas Hawkins, another *Frankenstein* enthusiast who exulted in his role as the Promethean "creator" of his fossil dragons,[14]

9. Ansted 1847, 223.

10. Brook 1990, 47.

11. For the monkey, see Mill [c. 1855], 215–17.

12. Cuvier 1812, III, 3–4 (see above, p. 115); Mantell 1833, 315.

13. Best 1837, 28–9.

14. Hawkins 1834a, 27. For his reading of *Frankenstein* see Hawkins 1834a, 29.

excited the suspicions of the British Museum staff when he was found to be creating mock-up bones and even limbs to make his specimens look more complete, without colouring these bones to mark their "fictional" status in the museum's favoured manner. Hawkins was enraged by the criticism he received, and a lengthy battle ensued.[15] The self-consciously scientific discipline of palaeontology, far from ousting an old-fashioned fascination with fabricated "monsters", was now opening up new possibilities for spectacular hoaxes, such as Albert Koch's thirty-foot-long elephantine "leviathan" *Missourium theristrocaulodon* and his 114-foot-long "sea serpent" *Hydrarchos sillimani*. Koch assembled these in the 1840s from the bones of several mastodons and Zeuglodons respectively, and they drew large crowds in America and Europe.[16] The tension between fact and fiction inherent in geological reconstruction often crystallized around issues of monstrosity, of parts joined together (and resurrected) that should have remained separate (and dead).

Curiously enough, compilation—the creation of a literary "monster", with all the stitches showing—was a favourite means of navigating this tension. Ansted's Carboniferous scene, assembled from present-day species, is a pictorial example. On a larger scale, James Secord has shown that *Vestiges* was seen by contemporaries as a kind of Frankenstein's monster, an animated aggregation of the "scattered limbs" of its component sciences;[17] but *Vestiges* was itself drawing on the generically hybrid tradition of geological literature. By the 1830s cross-referencing had become widespread in these texts, presenting the appearance of a connected body of literature whose authors had closed ranks. In the more speculative sections, cross-referencing often gives way to quotation, sometimes amounting to a patchwork of established authorities. Restorations of former worlds needed this kind of authoritative backing.

As a result, extinct animals were frequently brought to life by quoting from the more vivid descriptions of authors originally associated with each beast's identification and restoration (i.e. creation). Conybeare, acknowledged master of the Plesiosaurus, had speculated before the Geological Society about its appearance when hunting for fish,[18] and Mantell had painted a brief scenic restoration of his Iguanodon country in his 1826 fossil treatise.[19] Neither passage had reached a wide audience in its origi-

---

15. On this row see McGowan 2001, 124–48; Taylor in *ODNB*. Compare Torrens (1995, 264) on the Plesiosaurus controversy of 1824.

16. Altick 1978, 289; Semonin 2000, 382–6. The Zeuglodon had recently been identified as a fossil cetacean.

17. J. Secord 2000, 41.

18. Conybeare 1824, 388–9.

19. Mantell 1827, 82–4.

nal form, but both made their appearance in many cheaper publications thereafter, whether as quotation, plagiarism, or close paraphrase.[20] This process accelerated from the 1840s on, when a growing body of popular books by well-known geologists added to the pool of verbal restorations. Buckland's description of pterodactyles as Miltonic demons was likewise reproduced time and again.[21] Compendia like Thomas Milner's *Gallery of Nature* were, in large part, patchworks of quotations and paraphrases from older writings (or older patchworks).

These compilatory practices have important implications for geology's reception history. One might deduce from the poor sale and relatively low profile of Mantell's *Illustrations of the Geology of Sussex* that its concluding restoration of the country of the Iguanodon had a low impact. But this would be to overestimate the importance of the physical object, the book in which these words were first published. The book did not entomb the text, but gave birth to it: Mantell's concluding paragraphs went on to paint an image of the Wealden delta in the minds of all those Victorian children who read Milner's *Gallery of Nature*. Periodicals reproduced these texts still further. The rise of the cheap miscellany and its scientific spinoffs, as charted by Jonathan Topham, meant that the circulation of particular texts became progressively less dependent on their original vehicles. To reverse Topham's own dictum, texts are far too important to be treated merely as books.[22]

The same applies to the pictures incorporated into these texts. In the next chapter we shall examine the nuggets of vivid description which brought former worlds before the mind's eye; but, as we have already seen, pictures themselves played an increasingly important role in stimulating the mind's eye. Drawing on the insights gleaned from our exploration of the virtual-tourist guidebooks, we now examine how texts and printed images worked together to feed the geological imagination.

## Pictures

Pictures had long been indispensable to the communication of geological knowledge, and Martin Rudwick has charted the development of the

20. Conybeare's description recurs in [C. Lyell] 1826b, 521–3; Pidgeon 1830, 376–7; W. Buckland 1836, I, 211–12; Goodrich 1849, 172–3; Milner 1846, 722–3; Broderip 1848, 347–8; Owen 1854, 33–4; Nicholson 1877, 244–5; Northrop 1887, 36–8. It was closely paraphrased in Ure 1829, 242–3 and Hack 1835, 304–5, and (at one remove) in an advertisement for an exhibition in Whitby in 1841 (Osborne 1998, 180). Louis Figuier (1863, 149–50) paraphrased it in French without acknowledgement; his English translator, Ormerod, turned this back into direct quotation (Figuier 1865, 201–2). Mantell's description is reproduced (besides his own recyclings) in Bakewell 1828, 283–4; Hawkins 1834a, 8; Milner 1846, 730–1, and reprints up to 1880.

21. See below, pp. 422–3.

22. Topham 1998, 262. On the miscellanies see Topham 2004a.

now-familiar "visual languages" of strata sections, geological maps, and restorations in the eighteenth and nineteenth centuries.[23] Less attention has been given to the precise dynamics of the text-to-image relation in geological literature.

We often take a picture's relation to reality for granted, accustomed as we are to the visual syntax of science books, and practised in balancing sensory and rational responses to such images. Photography has revolutionized the extent to which we view such pictorial evidence as a truthful representation of the object depicted, as a "visual fact".[24] In the early nineteenth century, however, this truth-value was hotly debated.[25] Some educationalists were suspicious of pictures as a means of training the eye, partly because they carried the moral danger of sensory overstimulation common to spectacle in general, but also because they were felt to be representationally crude by comparison with verbal suggestion. As with the panoramas and the Crystal Palace monsters, texts were used to guide readers' interpretations of printed images—although, with pictures and texts bound together into the same object, the visual syntax was somewhat different. Let us examine how the reader was invited to look from text to picture and back again.

Engravings and woodcuts of fossils frequently stood in for the real objects. To take a typical example, most of the woodcuts in Mantell's *Medals of Creation* are printed on the same page as the text referring to them. They are introduced laconically as "visual facts", often as part of the title of the relevant section: "Lepidotus.* *Lign*. 132, 133."[26] This formulation implicitly equates picture and object; only occasionally does Mantell spell out that an object "is represented" or "figured in" a picture.[27] Even when he refers to pictures published in other books, the act of looking seems taken for granted, resulting in a rather jerky reading experience if his directions are consistently followed.

Buckland's technique in *Geology and Mineralogy* resembles Mantell's. His pictures are all in the second volume, up to twenty being crammed into a single plate. Each figure is keyed in at the relevant point in the text, for example: "This part of the Belemnite is [. . .] made up of a pile of cones, placed one within another, having a common axis, and the largest enclosing all the rest. (See Pl. 44, Fig. 17.)."[28] "Plate 44" is reproduced in Fig. 8.2, of which Buckland's "Fig. 17" is at the bottom right-hand corner.

23. Rudwick 1976; 1992. See also Rupke 1998.

24. A. Secord 2002, 32.

25. These debates are examined in A. Secord 2002.

26. Mantell 1844, 637. The asterisk directs readers to a footnote referring rather cryptically to Louis Agassiz's treatise on fossil fishes: "Poiss. Foss. Tom. II. p. 233."

27. Mantell 1844, 640–1.

28. W. Buckland 1836, I, 372 n.

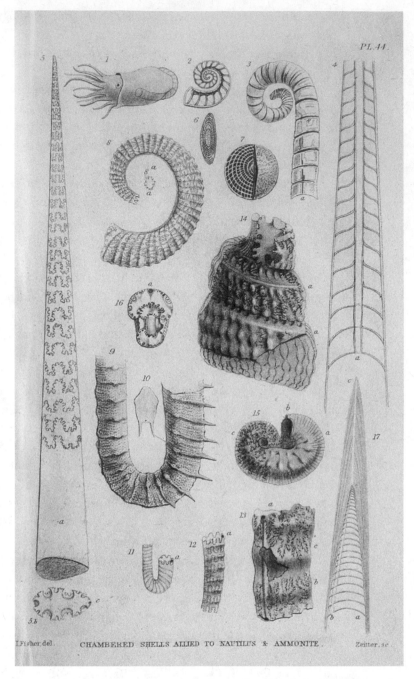

*Figure 8.2.* A typical plate in the second volume of Buckland's *Geology and Mineralogy*, showing "Chambered shells allied to nautilus & ammonite". This volume was entirely taken up with such plates, each one stuffed with specimens. W. Buckland 1836, II, plate 44.

A description of "Plate 44" is given at the beginning of the second volume, in a separate section covering all the plates and adding further details: "Fig. 17. Longitudinal section of the calcareous Sheath and Alveolus of a Belemnite." [29] The reader has to do a lot of thumbing around, again making for a spasmodic reading process. To return to our analogy, the stitches between text and image are obvious. Yet the composite result is an admirably clear idea of how belemnite cones were constructed—one more example to build up a sense of the Deity's ingenuity. The image-reality relation appears transparent; but we are very far here from Pestalozzi's theories of a purely visual education. Both Buckland and Mantell used pictures in tandem with, rather than instead of, descriptions. The combination of media and the profusion of details guide the reader's imagination to home in on the animal in a kind of mental pincer-movement.

This movement was taken to extremes by Miller, whose *Old Red Sandstone* contains only a few pictures. His description of the *Pterichthys* begins by alluding to his own specimen, which he first asks us to imagine rather than simply look at:

> I have placed one of the specimens before me. Imagine the figure of a man rudely drawn in black on a gray ground, the head cut off by the shoulders, the arms spread at full, as in the attitude of swimming [. . . .] [30]

This verbal "sketch" concludes apologetically two pages later, inviting us at last to see a drawing of the specimen (Fig. 8.3):

> Such is a brief, and, I am afraid, imperfect sketch of a creature whose very type seems no longer to exist. But for the purposes of the geologist the descriptions of the graver far exceed those of the pen, and the accompanying prints will serve to supply all that may be found wanting in the text. [31]

This book was originally written as articles for the general reader, without pictures. In the second passage Miller seems to imply that the pictures were added to the book in order to serve "the purposes of the geologist", thus appearing to assign the two forms of representation to two distinct readerships (the general reader being satisfied with verbal evocations alone). But, like Miller's rhetoric of modesty, this hint may be disingenuous: most of the woodcuts in *The Testimony of the Rocks* are simply sprinkled amid the prose without any connecting links at all, as a kind of textual conglomerate (Fig. 8.4). Miller clearly intended his pic-

29. W. Buckland 1836, II, 66.
30. Miller 1841, 49.
31. Miller 1841, 51.

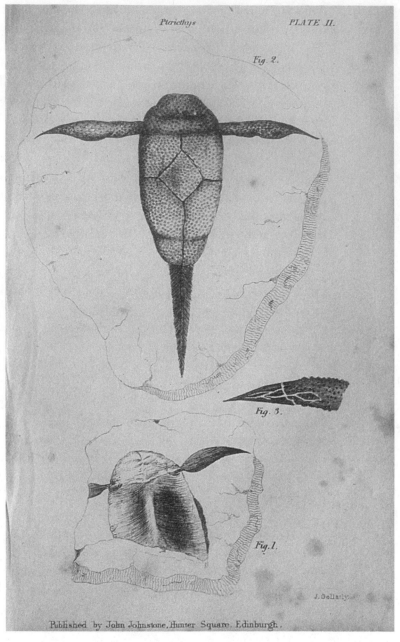

*Figure 8.3.* The fossil fish *Pterichthys* (or *Ptericthys*), drawn by Hugh Miller for his first geology book *The Old Red Sandstone*. Miller 1841, plate 2.

tures for unlearned eyes as well as learned ones, and those in *The Old Red Sandstone* are no exception. Indeed, their vividness is enhanced by the attention Miller draws to their status as artefacts, specifically mentioning "the graver". Words and pictures become interchangeable: his description is a verbal "sketch", while the plates are visual "descriptions". Image and text illustrate each other.

Oolitic Systems, all fishes, though apparently as numerous individually as they are now, were comprised in the
ganoidal and placoidal orders. The period of these
orders seems to have been nearly correspondent with
the reign, in the vegetable kingdom, of the Acrogens
and Gymnogens, with the intermediate classes, their
allies. At length, during the ages of the Chalk, the
Cycloids and Ctenoids were ushered in, and were

Fig. 55.

LEBIAS CEPHALOTES.
Cycloids of Aix. *(Miocene.)*

gradually developed in creation until the human period, in which they seem to have reached their culminating point, and now many times exceed in number and importance all other fishes. We do not
see a sturgeon (our British representative of the ganoids) once in a twelvemonth; and though the skate
and dog-fish (our representatives of the placoids) are
greatly less rare, their number bears but a small proportion to that of the fishes belonging to the two pre-

Fig. 56.

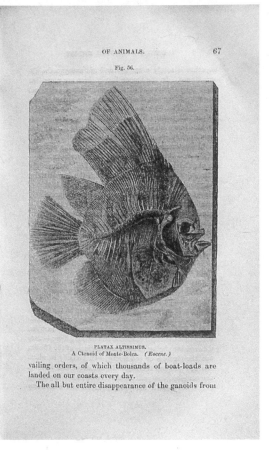

PLATAX ALTISSIMUS.
A Ctenoid of Monte-Bolca. *(Eocene.)*

vailing orders, of which thousands of boat-loads are
landed on our coasts every day.

The all but entire disappearance of the ganoids from

*Figure 8.4.* Illustrated two-page spread from Miller's *Testimony of the Rocks*. Miller 1857,
66–7.

Miller mentions that the top drawing in Fig. 8.3, showing the whole
fish, is a reconstruction:[32] it does not represent a specific fossil, but is
constructed ideally from several incomplete specimens. But more than
just a fish is figured here. Look again at the drawing. Rather than showing the fish in outline, like the reconstructed ichthyosaur and plesiosaur
skeletons depicted in countless texts of this period, Miller presents it on
a stylized slab of sandstone, giving the reader the impression that he or
she is looking at a real fossil. The slab is pure white, and every detail
of the fish is beautifully clear and black. Miller's purpose is illuminated
towards the end of the book, where he describes the marvellous variety
of hues seen in living fish, and bemoans the fact that the geologist is co-
lour-blind. He describes his pictorial and verbal restorations alike as "a
mere profile drawn in black,—an outline without colour".[33] Again, this

32. Confusingly, Miller uses the word "restorations" (Miller 1841, 51).
33. Miller 1841, 250.

is a rhetoric of modesty. Three pages later, he describes with present-tense immediacy how, in the "green depths" of the Cornstone Formation seas, "Shoals of *Cephalaspides* [. . .] sweep past like clouds of cross-bow bolts in an ancient battle".[34] It is difficult to believe that Miller's readers would have imagined a green sea populated only by black-and-white fish. Rather, his longing remarks about the hectic colours of modern-day fish a few pages before serve to prime the reader's visual imagination, so that he or she imagines these *Cephalaspides* in all their "hues of splendour",[35] filling in the gaps unknown to the geologist with modern-day analogies. The many admirers who praised Miller's "colourful" descriptions were responding as he meant them to, using his words and pictures as a springboard for the mind's eye. The engraving in Fig. 8.3 may not be coloured, but Miller's self-deprecating term "outline" belies the beauty of this finely textured image, which invites us to revivify it in imagination. In combination with his word-painting, such an image may well have seemed a window upon a former world, and an emblem of geology's imaginative power.

In terms of sensuous appeal, few engravings match those executed by George Scharf for Hawkins's two treatises. Many of the melodramatic verbal restorations in *The Book of the Great Sea-Dragons* are presented as if they sprang directly from the engravings. Figure 8.5 shows the twentieth plate in the book:

> the head is instinct with life, the eye glowers in his socket, as if in the last agony, the spine twists to and fro, as though the nervous filaments still tortured its extremest parts, one foot digs into the ground, and the lashing tail writhes under the general throe, agitating for Death. His heart-strings were wrenched asunder so quickly, so rudely, that Death failed to stamp his Effigies upon the resisting bones, and so left them there at the bottom of the Seas, lifeful, despite himself, as we behold them here.[36]

This is the royal "we": in his earlier tome Hawkins had been (mostly) a mere "I", but now he has passed to higher spheres. Yet, unlike most geological writers, who were all too apt to forget the various wives, sisters, daughters, friends, and professional artists who produced pictures for their books, Hawkins is full of praise for Scharf, calling him the equal of Raphael and Titian:

> Who but Scharf could so portray the naked bones of these Taninim, and seize their metaphysical aspect, so tenuous and shy! With an eye to the out-

34. Miller 1841, 256–7.
35. Miller 1841, 251.
36. Hawkins 1840a, 18.

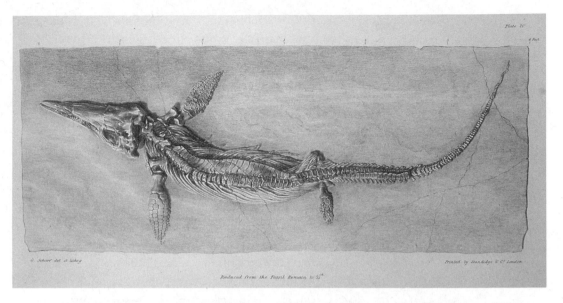

*Figure 8.5.* George Scharf's exquisitely detailed engraving of an ichthyosaur or "Dragon, in Stone, from [the village of] Street", for Thomas Hawkins's *Book of the Great Sea-Dragons*. Scharf was much in demand as a geological illustrator, but found this work tedious and sketched scenes of London life to cheer himself up. Hawkins 1840a, plate 20.

ward form, certes, and a sense known only to Genius of artistical dexterity manifested by a stroke, Scharf shall multiply these Sea-Dragons throughout Christendom, and embellish our Chronicles beyond all others.

Indeed, Scharf is praised more effusively even than Martin, who had provided the book's only saurian restoration in the frontispiece (see Fig. 10.3). Martin's "stupendous Powers" are acknowledged, but Hawkins suggests that "Martin has barely attained, with all his stupendous Powers, the utter hideousness their own", whereas the "Genius" of Scharf has "seize[d] their metaphysical aspect".[37] Visualizing the prehistoric world today, we rely on pictorial or modelled restorations of flesh-and-blood scenes, but Victorian imaginations worked harder: Martin's restoration was a single snapshot in a whole gallery of scenes.[38] The raw material comprised (mainly) fossil engravings and verbal restorations: readers, guided by Hawkins, had to stitch these media together in their imaginations, resulting in a series of baroque visions. As in virtual-tourist guidebooks, the text frames the way the image is read.

Framing was particularly significant for frontispieces, whose syntactic relation to the text was relatively distant. Juxtaposed with the title-page, the frontispiece is the first object that meets the reader's eye as he or she

37. Hawkins 1840a, 18.
38. This argument is developed in R. O'Connor 2003b.

opens the book. As the title of a book stands in a synecdochal relation to the book itself, so the combination of frontispiece and title-page serves as an emblem of what the book has to offer. Several frontispieces were infused with a strong sense of romance, to draw the reader in and reveal that such things may indeed be true. In some books the frontispiece is given a strong syntactic relation to the text once it has made this initial visual statement to the reader, but in Hawkins's two treatises the frontispieces remain aloof. Their titles do little more than paraphrasing the title or subtitle of the book in which they appear, reinforcing their emblematic role *vis-à-vis* the title-page: "The Sea-Dragons as they Lived" and, in *Memoirs of Ichthyosauri*, "Extinct Monsters of the Ancient Earth" (see Fig. E.1). This isolation helps to sustain their dreamlike, slightly unreal atmosphere.

In *Memoirs of Ichthyosauri* Hawkins often addresses the reader directly. Before telling the story of how he saved a fossil plesiosaur from being destroyed by two quarrymen, he inserts a paragraph consisting solely of the two words "Listen reader" (strictly an oxymoron, but one which tends to heighten the immediacy by suggesting that Hawkins is right there in front of you, ordering you to pay attention).[39] When presenting his plates he, like Ashe, can sound like an auctioneer:

> Turn, reader to the thirteenth plate; I am sure you will understand how my heart fluttered when that gem of price was placed before my flashing eyes [. . . .] That head [. . .] possesses two hundred and sixty long sharp teeth [. . . .][40]

Such directness and deictic language ("that gem", "that head") was commoner in children's books. Chapter 2 of *Peter Parley's Wonders of the Earth, Sea, and Sky,* by Samuel Clark, begins:

> I will show you a picture of what creatures were once living where the town of Lyme Regis, in Dorsetshire, now stands, and tell you something about their habits of living. You may perhaps be ready to think that a great deal of what we profess to know concerning them, is the work of fancy, but I can assure you it is not [. . .][41]

The page layout is reproduced in Fig. 8.6: opposite the text is a picture labelled "Extinct Animals", showing three saurians.[42] The fact-fiction ten-

---

39. Hawkins 1834a, 41.
40. Hawkins 1834a, 31.
41. [Clark] 1837, 5.
42. The picture is discussed in Rudwick 1992, 73–5.

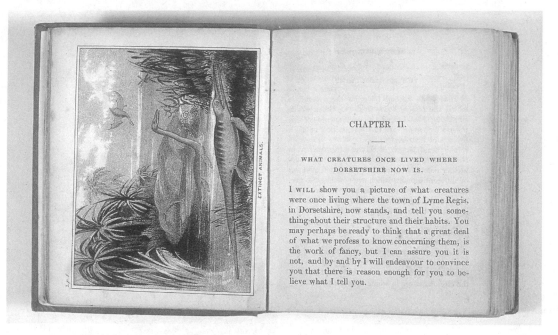

*Figure* 8.6. Introducing children to the Age of Reptiles: two pages from *Peter Parley's Wonders of the Earth, Sea, and Sky.* [Clark] 1837, 4–5.

sion characteristic of geological restoration erupts in pictures like this, and Clark here anticipates his young readers' scepticism. In this chapter, sometimes with the help of smaller pictures inserted directly into the text, he proceeds to explain the habits and anatomy of each animal in turn, just as they appear in the picture: "That odd looking creature which is flying in the air [. . .] has been called the Pterodactyle".[43] The educational value of pictures is likewise inscribed into the narrative of Hack's *Geological Sketches*. Harry, and hence the reader, is given pictures which come directly from the portfolios of his knowledgeable Mother and Father; by directing his viewing, these adults ensure that young Harry does not get overexcited.

Hack's method of stitching images into the text gives a much smoother read than (say) Buckland's or Mantell's books. For the smoothest read of all, however, we must return to John Mill's *Fossil Spirit*. Most of Mill's pictures are small outline drawings of fossils or restored animals, little monsters stitched into a composite text. Some of these he presents directly to the reader (Fig. 8.7); some he incorporates in mid-sentence as pictograms, but with labels just to make sure (Fig. 8.8); others are sprinkled without connecting links. Most strikingly, each "Evening" (like the book's title itself) begins emblematically with initial capitals

43. [Clark] 1837, 17.

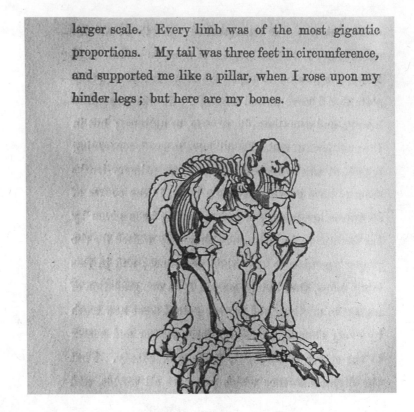

larger scale. Every limb was of the most gigantic proportions. My tail was three feet in circumference, and supported me like a pillar, when I rose upon my hinder legs; but here are my bones.

*Figure 8.7.* "Here are my bones". John Mill used a variety of ways of weaving pictures into the text of *The Fossil Spirit*. Here the fossil is presented directly. Mill [*c.* 1855], 78, detail.

I soon after became a sea reptile of the most strange and wonderful kind that ever existed. Here is my likeness. I was a

Plesiosaurus.

I had a small head and a very long neck. The neck

*Figure 8.8.* "I was a Plesiosaurus". Mill's restorations of extinct animals were usually inserted as pictograms, a common device in educational literature for young children, but—since the names were unfamiliar—adding labels as well. Mill [*c.* 1855], 34, detail.

made up from bits of fossils and other natural objects (Fig. 8.9). This striking literalization of the "book of nature" analogy is also embodied in the book's title-page (Fig. 8.10): pictures are not only closely woven into the text, but also become texts themselves. Nevertheless, even such an intensive and inventive use of pictures as this remains a long way from Pestalozzi's "education through the eye". Mill's pictogram technique may suggest that he felt words on their own were insufficient, but his pictures are no less dependent on the words, both in the overarching narrative and in the little labels underneath each pictogram.

That said, Mill's frontispiece, drawn by H. Melville, remains for a long time textually aloof. Unlike most frontispieces, it is unlabelled. In this connection we may also note that the book's front cover and spine give only the subtitle, *A Boy's Dream of Geology* (plate 5); the "main" title does not make its appearance until the book is opened. For a young reader unacquainted with the wonders of geology, the subtitle may have seemed amply justified by the experience of opening the book and seeing this Gothic-Oriental dragon-scene. Melville's picture is dreamlike and poetic, suggestive of the *Arabian Nights;* without a label it, like the word "dream", can make no factual claims. Only in the fourth "Evening" is the picture at last hauled into the realms of fact, insofar as the fakir's story of transmigration can be called "fact". Having described the land he lived in as an Iguanodon, the fakir muses:

> The world of the Iguanodon has perished, and is buried deep under foot, but it can be restored to the imaginative eye by the power of scientific analysis: the brain of the poet and the pencil of the artist.*

The asterisk directs us to a footnote which tells us to "see our Frontispiece, by H. Melville, Esq."[44] The syntactic link is made: the dragon-haunted frontispiece becomes a scene of the English Wealden period, and fantasy resolves into fact.

Yet, in its emblematic isolation and structural prominence, Melville's engraving retains its dreamlike flavour. Both the picture and its textual referent derive their iconography and rhetoric from Mantell's *Wonders of Geology*, where Martin's *Country of the Iguanodon* provides the frontispiece (see Figs. 7.6 and 8.11). Martin's picture has clearer and firmer syntactic links to its host text than Melville's. Mantell refers to it at the end of volume 1, in terms very similar to Mill's meditation;[45] a later section on the pterodactyle refers the reader again to Martin's

44. Mill [c. 1855], 46.
45. Mantell 1838, 369.

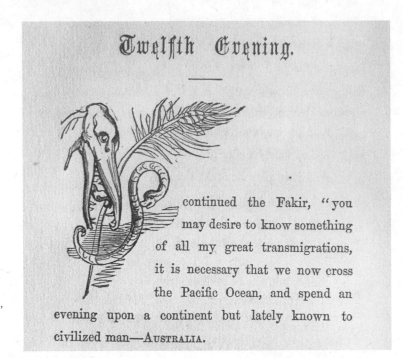

Twelfth Evening.

continued the Fakir, "you may desire to know something of all my great transmigrations, it is necessary that we now cross the Pacific Ocean, and spend an evening upon a continent but lately known to civilized man—AUSTRALIA.

*Figure 8.9.* Decorated initial letters like those of this "As", constructed from bits of fossils, were the most unusual aspect of Mill's pictorial procedure. Mill [*c.* 1855], 130, detail.

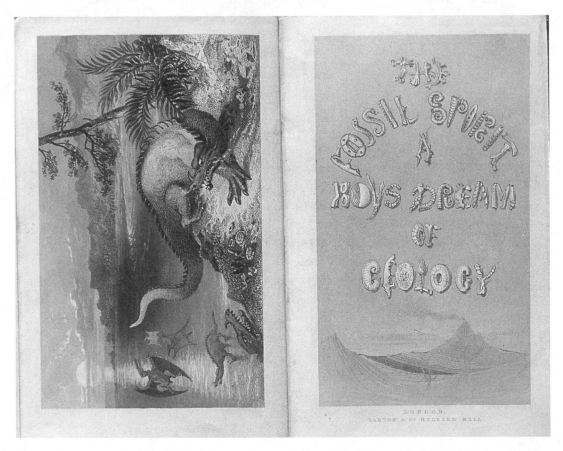

*Figure 8.10.* Frontispiece (by H. Melville) and title-page of Mill's *Fossil Spirit*. The frontispiece shows two Iguanodons about to engage in battle at sunset, as described later in the book; the title-page combines fossil letters with a section of the earth's crust. Mill [*c.* 1855].

picture;[46] and the picture itself is labelled as having been based on Mantell's "Geological Discoveries". In the third and later editions, a page-long "Description of the Frontispiece" was inserted after the title-page, keyed into Mantell's text: "The reptiles comprise the iguanodon (p. 389), hylæosaurus (p. 401), megalosaurus (p. 389), crocodiles (p. 385), and turtles (p. 384). An iguanodon attacked by a megalosaurus and crocodile, constitutes the principal group".[47] Such referencing recalls Martin's other engravings, furnished with explanatory descriptions and keys, such as *Belshazzar's Feast* (which Mantell specifically mentions here). The reader is thus invited to "read" Martin's picture—though here, by contrast with

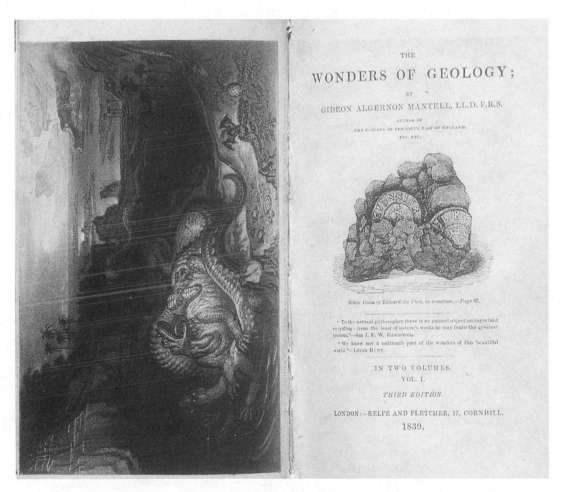

*Figure 8.11.* Frontispiece, by John Martin, and title-page of the third edition of Mantell's *Wonders of Geology* (1839). The legend concealed beneath the frontispiece begins "The Country of the Iguanodon, Restored by John Martin".

46. Mantell 1838, 437.
47. Mantell 1839, v.

Martin's own guidebooks, a great deal of page-thumbing is necessary to follow all the references.

Despite all these direct syntactic links between text and image, Martin's scene has a still more dreamlike effect than Melville's.[48] It sustains the sense of antique romance which Mantell, for all his factual emphasis, was keen to plant in his reader's imagination, using words like "wondrous" and "dragon-forms" in his main description of the Iguanodon country.[49] As Fig. 8.11 shows, the picture continues to serve an emblematic function alongside the title-page, which also contains a vignette of thirteenth-century English coins embedded in ironstone—an apt image of the "medals of creation", blending human and geological antiquity in a discreetly patriotic manner. The picture's title, the book's title, and the epigraphs combine with the frontispiece itself to project a sense of "wonder".[50]

The importance of pictures in science writing is increasingly acknowledged by historians, even if pictorial reproductions are relatively thin on the ground in most current scholarship.[51] But it is equally important to be aware that a printed image exists in one or more specific relations to its host text, just as Martin's paintings and the panoramas were designed to be viewed in conjunction with their guidebooks. The emblematic function of frontispieces may be taken for granted today, but the "meaning" of such a picture cannot be fully understood simply by reproducing it and analysing its iconography in isolation from its textual accompaniment. The same applies to vignettes and plates, and to the diagrams and maps we have not even touched on here. However vivid all these pictures may be, they were not intended to bear sole responsibility for awakening the imagination. It is no coincidence that Mantell, immediately after referring his readers to Martin's Iguanodon-picture, launches into the most detailed piece of literary spectacle in the whole book,[52] or that Mill's reference to Melville's picture is followed by the most vivid of all the fakir's life-stories: a sunset battle-scene with a Megalosaurus in which the imagery of Martin's own picture is transformed into saurian

48. Its "nightmarish" quality is analysed in Rudwick 1992, 78–81.

49. Mantell 1838, 369.

50. Fig. 8.11 is taken from Mantell's third edition of 1839, whose title-page and frontispiece are identical to that of the first edition, except that the subtitle acknowledging Richardson's contribution has been omitted and a new epigraph by John Herschel added to Leigh Hunt's.

51. This neglect has been ascribed to a "non-visual (or even anti-visual?) tradition that dominates social and historical studies of science" (Rudwick 1992, 262 n. 18); but it surely has as much to do with the often exorbitant cost of reproducing pictures and gaining permission to do so, whereas quoting from early-nineteenth-century published texts costs nothing.

52. Mantell 1838, 369–76.

autobiography (albeit with more emphasis on "the pleasure of fighting" than in Martin's gloomily pointless warfare).[53] Text and image combined to mount their "geologic drama" upon the reader's mental stage. And, as in the panorama guidebooks, poetry quotation helped to frame and stage these visions of the past.

## Poetry Quotation: Byron and the Geologists

Since the eighteenth century, quotations from well-known poems had helped writers to communicate practical and scientific information by appealing to a shared literary culture.[54] The choice of poets reflected the nature of the target audience. In the 1800s and 1810s most poetry quotations in geology books were from Classical verse, often in the original language, serving to introduce a suspect science to a Classically educated establishment. Such quotation could help reinforce what Gillian Beer has called "the accretive and benign power of scientific enquiry", familiarizing it by gesturing towards the gentlemanly cohesion of author and reader.[55] Latin tags reflected the process of "ingrafting [. . .] the new and curious sciences of Geology and Mineralogy, on that ancient and venerable stock of classical literature".[56] Poetry was still a language of authority, and geology's status could be buttressed by borrowing its voice.

Quotations from English poetry, let alone contemporary poetry, were sporadic in geological literature before the 1830s. The rising middle classes were, however, increasingly bombarded with extracts from modern poetry in giftbooks, guidebooks, and romance fiction. Science writers wishing to extend their catchment area beyond the gentry soon followed suit. In geology, Lyell made the first real plunge, having tested the waters in his periodical writing.[57] He courted the wealthy readers of his first edition by ostentatiously quoting poetry and prose in the original Latin, Greek, French, and mediaeval Italian (without translation).[58] But these were far outnumbered by his many English quotations: the readers of the less expensive 1834 edition did not need a Classical education to understand and enjoy Dante, Ovid, Horace, and Aristophanes

---

53. Mill [c. 1855], 46–52.

54. This was particularly true of the literature of agricultural improvement: see Barrell 1972, 61, and Heringman 2004.

55. Beer 1996, 210.

56. W. Buckland 1820, 2–3.

57. [C. Lyell] 1826b, 509, 518, 524, 538.

58. For example, C. Lyell 1830–3, I, 166, 185, 277, 376, 409; II, 279. Some of these were translated in later editions.

in English translation, old favourites like Shakespeare and Milton, and new favourites like Scott and Byron.[59] Turner's *Sacred History of the World* contained a similar profusion of verse in its compendious and anecdotal footnotes.[60] Both books were very successful, and from then on modern poetry was a familiar presence in popular geological prose.

The study of these acts of quotation, as with the giftbook anthology or virtual-tourist guidebook, has been hindered by the strongly negative value-judgement traditionally placed on them. This value-judgement is rooted in today's post-Wordsworthian perception that artistic integrity (imagined as a reified "originality") was replaced in the nineteenth century by a commodity culture favouring the simulacrum.[61] In an otherwise superb study of British naturalists, David Allen has offered a discouraging evaluation of popular natural-history writing:

> By the 1830s the true Romantic inspiration, its roots increasingly obscured, was growing less and less familiar, the so-called Romanticism that usurped it more and more patently a pose. The tasteful exercise of sentiment, so long and generally accepted as the hallmark of a cultivated gentility, tended to lie beyond the reach of minds dulled by industrial routines or by the no less stunting effects of a too literal fundamentalism. Increasingly, it turned instead into a feigned emotion: into Sentimentality, the mere sop to fashion of those who could not or would not commit themselves in the fuller way required—a debased substitute which by reason of its very shallowness was able to travel much faster and much farther.

This tendency, Allen argues, led to an "infection" of poetry quotation in natural-history books, so that "even sober works of science" were "for the sake of an easier sale [. . . .] liberally spangled with snippets of assorted verse", making them into "ludicrous half-breeds".[62] Class determinism is one problem here—were factory workers and biblical literalists really unable to experience a genuine emotional response to nature?—but the focus on sincerity is misplaced. Certainly, writing in this period was rarely lucrative, and authors and publishers needed to try every trick in the book to attract readers; but the manipulation of rhetoric does not necessarily imply insincerity. Commodification and sentiment have always been inherent in public science. The result may seem "debased" from the scientific viewpoint of a practising naturalist today; but the multi-layered imaginative commodities they offered their audiences helped to foster

59. For example, C. Lyell 1830–3, I, 11–14, 38, 273, 459; III, 380.
60. For example, S. Turner 1832, 136–46.
61. For a critique of this complex of attitudes, see Ruthven 2001.
62. D. Allen 1976, 75.

the rise of natural history in the first place, and they deserve sympathetic historical attention.

Pulled out of its pejorative context, Allen's term for these poetry-strewn science books, "half-breeds", is well chosen. The stitch-marks in these textual monsters often do show: an indented poetry quotation tends to draw attention to itself as much as does the insertion of a picture. As with the virtual-tourist guidebooks, verse quotation offered what Beer calls "*stories* by whose means to imagine the world",[63] vivid mental images through which the object could be viewed. The effect depended on the reader's already knowing the poem, or at least its style—not only in order to project a sense of community with a middle-class sentimental readership (though this was important), but also to situate a novel object within a familiar aesthetic tradition, supplying it with associations and intelligible meanings. The landscapes of the remote past were stranger and wilder than any American wilderness.

Here we may apply the same cluster of analogies as in the previous chapter: looking through a poet's "eyes" can be conceived in terms of framing, superimposing, or looking through a lens. With this science, however, poetry became charged with an additional layer of possible significance: it drew attention to geology's strange status, poised uneasily yet enticingly between fact and fiction. The epistemological brinkmanship implicit in geology's claim to present "marvels strange beyond even the conceptions of the poet" was both eased and signalled by sprinkling fragments of poetry within the geological text.[64] These quotations were visually marginal to the text, occurring as indented interruptions, epigraphs, or chapter endings. As such, they had the power to transform a text's margin into a frame, giving shape to the whole. In examining how they worked we focus again primarily on Byron, partly because he was the most commonly-quoted poet in this context, but also because we have already built up a clear picture of how his work was used in the wider literary and visual culture.

Beginnings and endings were natural locations for locating a science in a larger moral framework. The two-page spread consisting of frontispiece and title-page was a likely site for moral generalization, with the epigraph a favourite slot for gnomic quotation in poetry or prose. The title-page epigraph in Mill's *Fossil Spirit,* for instance, is taken from Nicholas Wiseman's lectures about science and religion: "'The science of Nature's antiquities.'—WISEMAN."[65] The two pages of "select Sentences"

63. Beer 1996, 210.
64. Miller 1859, 148.
65. Mill [*c.* 1855], iii, quoting from N. Wiseman 1836, I, 277.

which follow the title-page represent a miniature anthology of moral sentiments, not unlike an extended epigraph. All but one of these excerpts are in calm prose, invoking a distant philosophical perspective on the vicissitudes of time. The odd one out is a quotation from Byron's most notorious work, *Don Juan:*

"The very generations of the dead
Are swept away, and tomb inhabits tomb,
Until the memory of an age is fled,
And buried sinks beneath its offspring's doom."—Byron.[66]

As the virtual-tourist guidebooks have shown us, Byron's verse was often used to connote a sentiment of *sic transit gloria mundi,* the vanity of worldly pomp. His words here add a tinge of melancholy to the enlarged but emotionless views of Pythagoras and Aristotle. The fakir's story, which at one level allegorizes the providential legibility of earth history, subsequently shows how the apparently forgotten "memory of an age" may be unearthed and restored by modern science.

Conversely, a verse quotation might be placed at the end of a book or chapter to guide readers to respond to the preceding prose by joining in with an exclamation.[67] Here the touristic use of *Childe Harold* served a number of purposes. Consider the imaginary excursions in the Peak District incorporated into Mantell's *Medals of Creation,* complete with practical advice about inns and roads. This section of the book opens with a quotation from *Childe Harold* advertising the area's scenic charms in Grecian terms. One excursion ends with Mantell admitting that not all his readers will have understood the geological phenomena he was trying to explain, but that, all the same,

the hours we have passed together [. . .] will not have been spent in vain; for in the beautiful language of the noble bard;—

"To sit on rocks, to muse o'er flood and fell,
To slowly trace the forest's shady scene,
Where things that own not man's dominion dwell,
And mortal foot hath ne'er or rarely been:
To climb the trackless mountain all unseen,
With the wild flock that never needs a fold;

---

66.1 Mill [*c.* 1855], v, quoting from *Don Juan,* IV.102 (Byron 1986, 544). The original reads "inherits" in the second line.

67. Compare the examples from Mantell and Rennie examined above, pp. 123, 150.

Alone o'er steeps and foaming falls to lean;
This is not solitude; 'tis but to hold
Converse with Nature's charms, and view her stores unroll'd."
CHILDE HAROLD, Canto II. xxv.[68]

Ostensibly supplying these lines for the benefit of his more confused readers, letting them off the philosophical hook, Mantell at the same time emphasizes the picturesque appeal of a geological excursion for all his readers. Geology becomes a new form of pleasure.

A less conventional projection of the touristic Byron occurs in the last chapter of William Broderip's *Zoological Recreations,* entitled "Ancient Flying Dragons", which draws to a close by meditating on the vicissitudes of time. Broderip then pulls our perspective out still further:

MAN has the dominion over all. In future ages *his* remains will fill the bosom of the earth; and the traveller in some far distant century will feel the full force of Byron's lines wherever he sets his foot:—

Stop!—for thy tread is on an Empire's dust!
An Earthquake's spoil is sepulchred below![69]

In other words, civilization as we now know it may become extinct, but Byron's verse will survive long enough for tourists of the future to continue using *Childe Harold* as a mouthpiece for their emotions. At the same time, this imaginary future maps onto the reader's present, and the "full force of Byron's lines" is reflected back over the preceding descriptions of fossil zoology, prompting us to "feel" that the whole earth is "An Earthquake's spoil".

In Mill's "sentences", the Byron quotation stands aloof from the main text, like Hawkins's frontispieces, contributing tone and colour without being closely woven into the discourse. In the two examples just quoted, the poet gains more authority, speaking on behalf of the philosopher at a rhetorical pressure-point. Lyell, however, gave Byron still more authority at the end of the climactic chapter in volume 1 of the *Principles.* As we saw, Lyell introduced a passage from *Childe Harold* for its "philosophical" approach to marine erosion, using this verse to give his readers a panoramic viewpoint on the earth's history, "boundless, endless, and sublime" like Byron's "dark-heaving" ocean.[70] The quotation bridges the gap between scientific prose and poetic fiction: the poet momentarily be-

68. Mantell 1844, 933 and 967, quoting *Childe Harold* II.25 (Byron 1986, 60).
69. Broderip 1848, 380, quoting *Childe Harold* III.17 (Byron 1986, 109).
70. C. Lyell 1830–3, I, 459, discussed above, pp. 175–6.

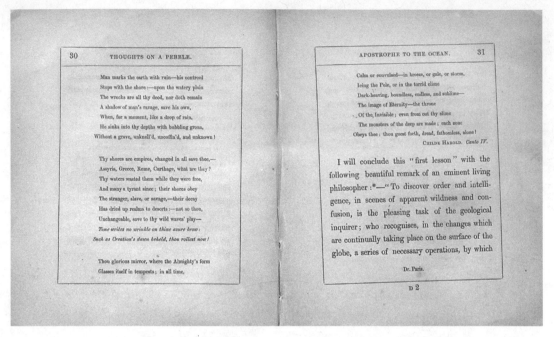

*Figure 8.12.* Quotations from Byron's *Childe Harold* in the middle of Mantell's *Thoughts on a Pebble*. Mantell 1849, 30–1.

comes a geologist. Lyell's friend Mantell was much taken with this use of his favourite poet, excerpting the same verse quotation in his own *Geology of the South-East of England* (1833) and calling it "as philosophical, as it is highly poetical and sublime".[71] In later books he used the entire stanza as an embodiment of the poetry of "modern geological researches", while in the eighth edition (1849) of *Thoughts on a Pebble,* he added two more stanzas (Fig. 8.12).[72]

*Thoughts on a Pebble* contains a good example of a widespread feature of the excerpting process, a feature often mistakenly termed "misquotation":

If we stroll along the sea-shore, we shall find an answer to some of these enquiries; for

There is a *language* by the lonely shore,—
There is society where none intrudes,
By the deep Sea, and music in its roar.[73]

71. Mantell 1833, xiv.
72. Mantell 1836a, 16–17; 1838, 100; 1842, 19; 1849, 29–31, quoting from *Childe Harold* IV.179, 182, 183 (Byron 1986, 199–200).
73. Mantell 1836a, 13; reprinted with variations in Mantell 1842, 15; 1849, 25–6 (adapting *Childe Harold* IV.178; Byron 1986, 199).

Byron's original reads "There is a rapture on the lonely shore", but Mantell has altered it to make Byron speak of the "language" of geological evidence. Such alterations were frequent among writers who loved poetry and knew a large amount of it by heart, such as Mantell, Broderip, and Miller. In many cases, they were intentional, as Mantell's italics suggest.[74] But the question of intentionality is not particularly important. More to the point, these adaptations (a more appropriate term than "misquotations") enabled verse to be woven into prose more smoothly and to give it new meaning in its new context.

Besides these general meditations, quoting poetry could also help, as at the panorama, to focus the mind's eye upon particular objects, bringing them more vividly before the eye while at the same time colouring and associating them with particular moods. Mountains were particularly ripe for this treatment, since descriptions of the Alps were now regularly assuming the language of the Byronic tourist-guide. Mont Blanc shows attracted huge audiences, from Alfred Bunn's production of *Manfred* in 1834 to Albert Smith's dioramic lectures about the ascent of Mont Blanc in the 1850s,[75] and both virtual and real tourist-guidebooks pillaged "beauties" from *Manfred* and *Childe Harold*. It soon became difficult to mention a glacier in any context without alluding to its "cold and restless mass" moving onward day by day.[76]

Mont Blanc itself forms the climax of an imaginary Grand Tour in Edward Hitchcock's *Religion of Geology* (1851), a fine piece of virtual tourism designed to demonstrate the aesthetic pleasure that could be derived from geological formations The scene is brought before the mind's eye by quoting from *Manfred*:

should we reach this summit of the Alps, we should [. . .] behold a scene which but few eyes ever have, or ever will, rest upon. We should

"breathe
The difficult air of the iced mountain's top,
Where the birds dare not build, nor insect's wing
Flit o'er the herbless granite."

We should, in fact, have reached the climax of the sublime in natural scenery.[77]

74. Mantell's alteration is bolder than the minor grammatical adjustments typically found in the embedded verses of romance fiction (on which see Price 2000, 93). On Miller's quoting practices, see Mackenzie 1905, 11. Compare Walter Scott's practice when quoting verses in his novels (Price 2000, 170 n. 105).

75. On Bunn's *Manfred* see above, pp. 283, 305; on Smith's lectures see Curwen 1940, 290; Altick 1978, 475–8; Hyde 1988, 146–9.

76. Laurance 1835, 12; J. Forbes 1843, 124; see also *ibid.*, 90.

77. Hitchcock 1851, 166, quoting *Manfred* II.ii.62–5 (Byron 1986, 290).

As in the Niagara panoramas, quoting Byron becomes a shorthand for "the sublime", allowing Hitchcock to make his point without having to go into detail. By 1851, for Hitchcock's readers on both sides of the Atlantic, the metaphysical anguish and unorthodox theological views of Byron's Manfred had been thickly pasted over with scenic effects: it is difficult to imagine a contemporary reader feeling any discordance between Hitchcock's natural-theological message and his poetic vehicle. The purely descriptive purpose of his Byron-quotation is reflected in the syntactic immediacy with which it is woven into the surrounding prose. Rather than introducing the quotation with "We should feel the full force of Byron's lines" (after Broderip) or "We should, in the language of the noble bard" (after Mantell), Hitchcock moves smoothly in and out of Manfred's voice without even mentioning Byron's name.

On a few occasions, the fact-fiction tension was dealt with by disparaging the poem's inferior status. In his manual *Elements of Geology*, Lyell treated Byron in a much less flattering way than we have hitherto seen, quoting (and adding an exclamation mark to) Byron's line, "The dust we tread upon was once alive!" Far from praising its geological truthfulness, however, he exclaims, "How faint an idea does this exclamation [. . .] convey of the real wonders of nature!" [78] Poetry is here held up for disparagement in order to show geology's superiority to the vague utterances of verse—substituting, as Miller put it, "for the smaller poetry of fiction, the great poetry of truth." [79] The traditional equation of poetry with falsehood could help to debunk a false theory, as in Wiseman's mocking account of William Whiston's cometary theory of the earth. Here he quotes from *Manfred*:

> Whiston was still more poetical. He supposed our earth to have roamed, for ages, through space,
>
> > "A wandering mass of shapeless flame;
> > A pathless comet;"—BYRON.
>
> till, at the period of the Mosaic creation, its course was bridled in, and it was reclaimed from its vagrant state [. . . .] [80]

Wiseman's book was published not long after *Manfred* itself had been staged by Alfred Bunn. As many of his London readers would recall, the section in which the above lines appear had been accompanied by

<hr>

78. C. Lyell 1841, I, 58, quoting *Sardanapalus*, IV.i.65 (Byron 1980–93, VI, 89). Compare Francis 1839, 67.

79. Miller 1859, 80. Compare Mantell 1846, 1–7.

80. N. Wiseman 1836, I, 279, quoting *Manfred* I.i.117–18 (Byron 1986, 278). This whole passage was later excerpted in Mill [*c.* 1855], xii–xiii. On Wiseman see *ODNB*.

a spectacular Eidophusikon-like display of a comet in an orrery. Byron himself had already made fun of cometary deluge-theories in his *Vision of Judgment*; Wiseman's act of quoting here amounts to a knowing wink, playing up the melodrama of an exploded theory.

All the objects so far given Byronic treatment were themselves theoretically visible to the human eye. It is now time to turn to the techniques by which antediluvian scenes and creatures were themselves restored in poetic form. *Cain* was the only one of Byron's dramas to present the reader directly with scenes of this kind, in the phantasmagoric voyage to Hades conducted by Lucifer in Act II. As we saw in chapter 2, Byron's demonization of Cuvier's geology had threatened to disrupt Buckland's careful dovetailing of his science with Anglican learning. But now, in 1836, Buckland's friends and colleagues Broderip and George Poulett Scrope used *Cain* to puff Buckland's *Geology and Mineralogy* in the *Quarterly Review*. Buckland appears here as a new Lucifer:

> Those who have listened spell-bound to that conversational eloquence with which the Professor is so peculiarly gifted—an eloquence which, when dilating on such subjects, absolutely calls up before his audience—
>
> > "The monstrous shapes that one time walk'd the earth,
> > Of which ours is the wreck,"
>
> will, however, imagine the vivid and fascinating manner in which he brings out [. . .] illustrations of the great truths of Natural religion [. . . .][81]

In Byron's drama, Cain speaks these lines as he remembers how Lucifer had called up the "shapes" in question; Lucifer's (not Byron's) didactic aim had been anti-theological, presenting mammoths as evidence for extinction and, therefore, for a capricious and tyrannical Jehovah.

It seems remarkable that Broderip should have puffed Buckland's book in this way for the *Quarterly*'s conservative readership. Evidently, by this time the *Cain* furore had simmered down. In his satire *The Professor's Descent*, Buckland had dramatized his credentials by casting himself as a latter-day Odin outwitting Byron's Lucifer; so perhaps it was natural for him to take on Lucifer's own mantle and put his powers to better use. Lucifer, like the Byron invoked in Lyell's *Elements,* had produced only vague, ghostly shapes, whereas the palaeontologist could call up extinct monsters in all their living detail, not to mention a more correct view of the Deity. All the same, in 1836 *Cain* still seemed challenging enough that the poet P. J. Bailey felt it necessary to defend his nascent

81. [Scrope and Broderip] 1836, 43, adapting *Cain* II.ii.359–60 (Byron 1986, 918). Mantell (1844, 874) quoted the same lines *verbatim*.

Byronic-Goethean drama *Festus* by insisting that "there is nothing of-fensive in it; nothing cainish".[82] For geological popularizers, this contin-ued sensitivity allowed *Cain,* if quoted judiciously, to provide a fresher *frisson* than the thoroughly sanitized *Childe Harold.* The act of quotation pulled Byron's visions of wrecked worlds out of their immediate polem-ical context, putting them to pious uses just as Martin had tamed the Deluge scenes from *Heaven and Earth.* In an edited form, even Shelley's overtly atheistic dream-vision *Queen Mab* was able to be used by Mantell two years later to support the natural theology of Thomas Chalmers, doyen of Evangelical Scotland, with its diaphanous vision of microscopic animalcules.[83]

Broderip's metrically smoother alteration of Byron's original first line from "Mighty Pre-Adamites" to "The monstrous shapes" suggests that he too was writing from memory. *Cain* certainly seems to have been a favourite work of his: four pages later he describes how "We are next car-ried back" (like Cain) to the *"age of reptiles",* when England was

> peopled by monsters [. . .] which stalked amid marshy forests of a luxuriant tropical vegetation, or floated huge on the genial waters,—
>
> "Their earth is gone for ever." [84]

Immediately after this line, Broderip addresses the fact-fiction tension which these quotations so often highlight, claiming that the facts of ge-ology may well seem at first "much more like the dreams of fiction and romance than the sober results of calm and deliberate investigation".[85] Broderip returned to *Cain* in his miscellany *Zoological Recreations* (1847), using it in two chapter-epigraphs to allude to extinct animals. The last chapter is on pterodactyles and is textually framed by poetry. It ends with the *Childe Harold* quotation discussed above ("Stop!—for thy tread is on an Empire's dust"), and begins with the following extract from *Cain* as its epigraph:

> "Their earth is gone for ever—
> So changed by its convulsion, they would not
> Be conscious to a single present spot."
>
> BYRON.[86]

---

82. Philip Bailey to Thomas Bailey, 26 April 1836, quoted in McKillop 1925, 753.
83. Mantell 1838, 517–19.
84. [Scrope and Broderip] 1836, 47, quoting *Cain* II.ii.120 (Byron 1986, 911).
85. [Scrope and Broderip] 1836, 47.
86. Broderip 1848, 380 and 369; compare 311 and 325.

This is a darker Byron, dwelling on loss and destruction. The philosophic melancholy of *Childe Harold*'s vision of empire is suggestively yoked to its larger-scale progeny, the existential melancholy of Lucifer's cosmos.

Yet Broderip's treatment of the fact-fiction tension was more often playful, as befits his "recreational" purpose. His last four chapters all contain the word "Dragons" in their titles: the first one begins by claiming that the earth really was once "a place of dragons"; it ends by asserting Byronically that our earth "is one vast grave of cities, of nations, of creations".[87] Within this geological framework is a breezy account (or rather anthology) of "fabulous" dragon legends, ballads, and romances in verse (see Fig. 6.5). Broderip rattles through story after story, deluging us with poetic fictions and unleashing a wilderness of dragons. The three remaining chapters focus determinedly on real, fossil "dragons", as if to show how dragon-lore has moved on from the false to the true. Even here, fable regularly intrudes in the form of verse quotations and Old Norse mythology.

In Broderip's delightfully piebald prose, poems could function as magic doors between the realms of past and present, or fact and fiction, enabling him to leap between worlds—a device we shall now explore in more detail. In his preface, Broderip had expressed a hope that the book "might cherish, or even awaken a love for Natural History." In these geological chapters, he made the lesson of Byron's *Don Juan* explicit: "Truth *is* stranger than fiction".[88] Yet such visions of "the fairy-land of science" had their own complications. Their ostensible aim may have been to set up a simple hierarchy of truth, attracting the reader with the entertaining deceptions of romance only to assert that the sober truths of science are still more wonderful. Yet, to a considerable extent, this "scientific" wonder was constructed from poetry itself.

87. Broderip 1848, 326, 338.
88. Broderip 1848, viii, 364, adapting *Don Juan* XIV.101: "Truth is always strange, / Stranger than Fiction" (Byron 1986, 818). The same quotation was used by the radical journalist William Chilton with reference to the transmutation of species (C[hilton] 1845, 9).

# Scenes and Legends from Deep Time

9

Poetry quotation was only part of the reason geology was so often described as "poetical". In an age which often seemed to collapse the boundary between the poetic and the visual, typically along theatrical lines, geology's main claim to poetry consisted in its ability to call up dramatic images of vanished worlds. The nineteenth-century development of pictorial images of the ancient earth has been charted by Martin Rudwick, in his stimulating book *Scenes from Deep Time*.[1] In the early-Victorian period, however, restorations of the distant past were painted far more frequently and confidently in words than in pictures.

Why was this? First, most geologists had to rely on (and pay for) professional lithographers to execute their "visions" for publication, whereas they could compose verbal restorations themselves and, as Rudwick puts it, retain "complete control over their material".[2] But, in another sense, pictures of extinct animals called for *more* authorial control than words did, since they demanded the representation of unknown or unknowable

1. Rudwick 1992.
2. Rudwick 1992, 235.

details. Modern examples are legion: the question of skin colour is the most obvious, but anything beyond the normally-fossilized bones and teeth, such as skin and sense-organs, causes problems for the illustrator (let alone the sculptor or animator). Whether *Tyrannosaurus* was feathered—a logical possibility given its close relationship to birds—cannot be finally settled until palaeontologists unearth a feathered specimen;[3] until then, pictorial restorations must necessarily come down on one side or the other. Verbal descriptions do not need such systematically complete detail: awkward gaps in the fossil reconstruction can simply be omitted from the description, or circumvented by various narrative devices. A full-scale pictorial restoration serves up all the visual ingredients, leaving the artist open to criticism from all sides; but a verbal restoration recalls the case of Schrödinger's cat, with the image of the dinosaur floating in a kind of superimposed quantum state, compelling the reader to complete the picture in the mind's eye.[4]

But Victorian verbocentrism was not simply a matter of scientific caution. We have already seen that stimulating the mind's eye was central to contemporary theatrical entertainment, and some early popularizers of geology were happy to commit themselves to a high level of speculative detail in print. No less than novelists, scientific writers recognized and wished to exploit the dramatic power of the written word to stage the buried past on the reader's "mental retina".[5] How they did so is the subject of this chapter and the next. Having seen how pictures and poems were incorporated into the prose matrix of a geological text, we turn to the prose itself, to the textual equivalents of Rudwick's "scenes from deep time". The first two sections of this chapter examine how single scenes were restored, beginning with straightforward descriptions and ending with visionary apparitions. We then explore how these scenes were tied together into theatrical sequences. Chapter 10 focuses initially on Miller, who developed the new forms of landscape pageantry to their fullest; his self-consciously Miltonic creations point up with special clarity the religious, poetic, and scientific ideologies underpinning these acts of simulated vision.

## Description, Anecdote, and Diptych

Verbal restorations of the past may be examined in terms of two variables: time-frame and narratorial stance. In terms of time-frame, a text

---

3. On this debate see Gauthier 1999.

4. I owe the concept of "Schrödinger's dinosaur", and much of this analysis, to Michael Taylor (personal communication).

5. Ainsworth 1836, xii. On theatrical narrative techniques see Meisel 1983; Buzard 1993, 172–92; Carroll 2004, 49–50.

might show a single snapshot, a few minutes of action, or a whole sequence of scenes. The second variable determines how the restored scene
is mediated to the reader: the narrator might describe having "seen" the
past landscape in his or her imagination, or else take the reader with him
or her on a time-travelling guided tour. Alternatively, the narrator might
simply describe the view impersonally, not as something "seen" but as
something which simply "was", perhaps conjecturally restored on the
basis of geological evidence. Let us begin, for the sake of clarity, by fixing
the narrative stance at this impersonal level, in order to explore the different possibilities of the time variable.

With respect to time, the simplest technique is more or less to ignore
it, restoring an animal, plant, or landscape in a "timeless" verbal vignette
or scene. These restorations are not timeless in geological terms, being
set in a definite period; but they represent that period in general, rather
than a specific moment in time. They display a typical scene, usually
involving animals standing around and eating each other. The amount
of the stratigraphic column concentrated into such a scene varies, from
high-resolution scenes representing a single formation such as the Lias,
through lower-resolution scenes of the whole "Age of Reptiles", to very
low-resolution scenes of the entire "antediluvian world". We find a similar range reflected in the pictorial restorations analysed by Rudwick,
most of which have a parallel function. The period, however broadly
defined, is always specified, often with a vague locality (the "Weald
of Sussex", "Dorset", "where Paris now is"); but these descriptions are
not localized in chronological terms beyond vague indications of day
or night. As such, the short-term passage of time is not foregrounded.
The past tense used is generally the continuous past, often hedged about
by modal verbs ("must have", "may have"), participles, and conditional
tenses.

Yet dramatic potential was not lacking. In *Memoirs of Mammoth*,
Thomas Ashe's restoration of the Napoleonic Megalonyx had been conducted entirely in modal verbs and conditionals, right up to the resounding last sentence: "he would retreat fighting, always keeping his face to
the enemy, looking proud, great, and ferocious."[6] Thomas Hawkins enthusiastically deployed the continuous past on the final page of *Memoirs
of Ichthyosauri,* compensating for the lack of rhetorical immediacy with
memorably lurid imagery:

> The sometime terran, sometime oceanic pterodactyles—those more than
> vampire monsters, which had solitary occupation of the wastes of sand
> when black night fell down upon them—were an after-thought [. . . .]

6. Ashe 1806, 60.

How did they gloat over the million million Medusæ—the boneless zoophites of an element wide as the world, and all their own: innumerable swarmed they, like Milton's cloud of locust angels, and the sauri amongst them as Satan, Molock and Abaddon.[7]

These Miltonic demon-similes strongly recall Buckland's colourful descriptions of the pterodactyles' world in his Geological Society paper of 1829.[8] In his treatise *Geology and Mineralogy* Buckland brought these two descriptions together for his wider public:

Thus, like Milton's fiend, all qualified for all services and all elements, the creature was a fit companion for the kindred reptiles that swarmed in the seas, or crawled on the shores of a turbulent planet.

"The Fiend,
O'er bog, or steep, through strait, rough, dense, or rare,
With head, hands, wings, or feet, pursues his way,
And swims, or sinks, or wades, or creeps, or flies."
Paradise Lost, Book II. line 947.

With flocks of such-like creatures flying in the air, and shoals of no less monstrous Ichthyosauri and Plesiosauri swarming in the ocean, and gigantic Crocodiles, and Tortoises crawling on the shores of the primæval lakes and rivers, air, sea, and land must have been strangely tenanted in these early periods of our infant world.[9]

This passage, complemented by a miniature pictorial restoration (see Fig. 9.1, bottom left), shows how shifts in tense could add immediacy to ordinary descriptions. Buckland's own verbal forms are participial (*swarming*), simple past (*swarmed*), or modal (*must have been tenanted*)— the latter construction usefully combining caution with conviction. But the embedded Milton quotation marks a departure from this past-tense description, dramatizing Buckland's simile (*like Milton's fiend*). As Buckland's target readership knew, Milton's original passage describes Satan journeying through the realm of Chaos on his way to earth—a realm also depicted by John Martin (see Fig. 7.18). This quotation, already rich in pictorial associations, provides a Miltonic lens through which this extinct world may be viewed. Moreover, the verse is in the historic *present* tense, a shift in narrative stance which enhances the passage's immediacy. Buckland's concluding sentence returns to conventional description,

7. Hawkins 1834a, 51.
8. See above, p. 160.
9. W. Buckland 1836, I, 224–5, quoting from W. Buckland 1835, 219 and 218.

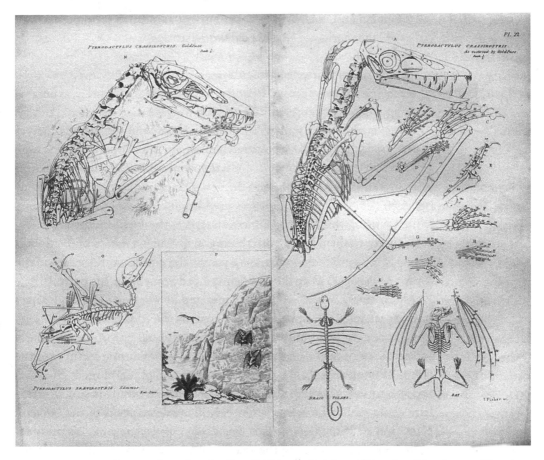

*Figure 9.1.* Pterodactylon, from Buckland's *Geology and Mineralogy.* The vignette (P) at the bottom left is entitled "Imaginary restoration of Pterodactyles, with a contemporary Libellula, and Cycadites" (i.e. dragonfly and tree-fern). This is the only restoration of a scene from deep time in the whole book. W. Buckland 1836, II, plate 22.

its main verb being the modal *must have*; but, since this verb does not appear until the end of the sentence, from the reader's perspective the three present participles (*flying, swarming, crawling*) are left temporarily hanging in syntactic limbo. They maintain both the immediacy of Milton's historic present and the imagery of his last line (*flies, swims, creeps*). One invisible world delicately folds onto another; the poetry works as a magic portal, momentarily transporting us back in time.

Neither of these vivid passages can really be described as a "story". They are straightforward descriptions. But descriptions could in turn generate narrative of three kinds: anecdotes, diptychs, and sequences. First, the writer might transform a description into an anecdote by focusing like a novelist on the scene's local time-frame and on the individual rather than the species. Such localized narrative was often prompted by unusually striking fossils, which stood out for their individual qualities. For instance, New Red Sandstone slabs containing the footprints of a fos-

*Figure 9.2.* The progress of civilization, as seen in one of William Tegg's pirated London editions of Nathaniel Hawthorne's bestselling *Universal History,* designed to teach history to children using the fictional persona Peter Parley. The bottom scene ("The highest state of Civilization") features ocean-going liners, a train, cranes, bridges, a hot-air balloon, and a crate of books from Tegg & Son. Peter Parley's original creator, Samuel Goodrich, had commissioned this book from Hawthorne in 1837, and tried unsuccessfully to prevent Tegg from reprinting it in London. [Hawthorne] 1860, frontispiece.

Condition of the terraqueous surface, as to Vegetation and Vertebrated Animals,
during four remarkable Geological Epochs.

*Figure 9.3.* "Condition of the terraqueous surface [. . .] during four remarkable Geological Epochs", John Whichelo's engraving for Joshua Trimmer's textbook *Practical Geology.* Here the chronological order runs from bottom to top, as in the strata. The scenes represent the "carboniferous era" (including sauroid fish), the "saurian age" (including Iguanodon), the "early tertiary age" (including Anoplotherium and Palaeotherium), and "the elephantoid age" (including modern mammalian genera). To the right, in the background of the last scene, a hyaena is seen entering his den—perhaps an acknowledgement of William Buckland's importance in stimulating these visions two decades earlier. Trimmer 1841, frontispiece.

There rolls the deep where grew the tree.
O earth, what changes hast thou seen!
There where the long street roars, hath been
The stillness of the central sea.[14]

Here, too, a double vision is communicated by musical means. "Tree" rhymes with "sea" (and partly with "deep"), joining two images other-

14. A. Tennyson 1969, 973 (first published 1850). For discussion see Dean 1985, 12–13.

The word "picture" points up the close relation between these imaginary visions and the Classical literary device of ekphrasis, where a work of art is described in vivid detail.

This congruence with Classical rhetoric also emerges in the second form of narratorial immediacy. Moving to present-tense narrative, a hypothetical "spectator" might be introduced to mediate between author-narrator and reader, again paradoxically enhancing the immediacy. As in much travel narrative, readers were decorously invited to identify themselves with such a "spectator", who stands in rhetorically for the author. For example, the introduction to fossil vegetables in Buckland's *Geology and Mineralogy* concludes with a marvellous vision of a Bohemian coal-mine:

> The most elaborate imitations of living foliage upon the painted ceilings of Italian palaces, bear no comparison with the beauteous profusion of extinct vegetable forms, with which the galleries of these instructive coal mines are overhung. The roof is covered as with a canopy of gorgeous tapestry, enriched with festoons of most graceful foliage [. . . .] The spectator feels himself transported, as if by enchantment, into the forests of another world; he beholds Trees, of forms and characters now unknown upon the surface of the earth, presented to his senses almost in the beauty and vigour of their primeval life; their scaly stems, and bending branches, with their delicate apparatus of foliage, are all spread forth before him; little impaired by the lapse of countless Ages, and bearing faithful records of extinct systems of vegetation, which began and terminated in times of which these relics are the infallible Historians.[18]

Unusually for this lavishly illustrated treatise, no engraving of these fossils is supplied in volume 2, leaving the words alone to convey the image. Buckland's emphasis on the scene's wild "beauty" recalls his indulgently Gothic description in *Reliquiæ Diluvianæ* of a German stalagmite cave which, though it lacked fossils, had aesthetic appeal.[19] Now Buckland imports the same romantic enthusiasm for cave scenery into a book where, up to this point, the words "beauty" and "beautiful" have signified the structural perfection of a providential design rather than a purely aesthetic quality.[20] Buckland makes no natural-theological gestures in this passage; he seems caught up in the wonder of the scene. The surface of the rock becomes a "graceful" and "gorgeous tapestry", superior to Italian painting and displayed in its own "gallery" (a technical mining term,

18. W. Buckland 1836, I, 458.
19. W. Buckland 1823, 127–8, quoted above, pp. 106–7.
20. For example, W. Buckland 1836, I, 222.

here given a double meaning). It has the added "enchantment" of being literally true—"instructive", "faithful", and "infallible" (this last word had a sting in the tail for biblical literalists)—while retaining all the sensory seductions of human works of art. As with Miller's "Wren's Nest" dipytch, the "transport[ing]" effect of these fossils on the spectator's "senses" is transferred to the reader by the music of Buckland's prose, in the long sentence beginning "The spectator". Alliteration (*scaly stems, bending branches*), discreet emphasis (*Trees, Ages, Historians*), and the division of this long sentence into four balancing cola separated by semicolons, confer a stately elegance upon this vision.

Third, the readers themselves could be invited to "imagine" the vanished object, picturing it directly in their mind's eye. This object might be a landscape, as in Lyell's 1827 review of Scrope's *Geology of Central France*.[21] More commonly, this technique summoned up the forms of extinct animals, as in the climactic passage of Mantell's 1827 and 1833 treatises:

> Imagine an animal of the lizard tribe, three or four times as large as the largest crocodile; having jaws, with teeth equal in size to the incisors of the rhinoceros; and crested with horns; such a creature must have been the Iguanodon![22]

The modal verb "must" here sheds all provisionality and takes on a purely theatrical force. Whereas the techniques of visionary immediacy outlined above propelled the observer into the past, Mantell's monster-raising owed more to the concept of the phantasmagoria, where spirits were recalled from the dead to terrify spectators in this world. This metaphor was taken up by his friend, the poet Horace Smith, in two dream-poems about Mantell's museum. In a poem for Mary Ann Mantell's album, "To M[rs.] Mantell of Lewes" (1833), Smith imagined that

> each inanimate reptile & beast
> Bursts it's case, and its flesh-cover'd bones are increased
> To their former dimensions gigantic [. . . .]

In this "phantasmagoric confusion" the revivified monsters, Godzilla-like, wreak havoc in the town of Lewes.[23] The second poem, "A Vision" (1838), was written to commemorate the removal of Mantell's collection to London, and imagines another frightening resurrection. This time the

21. [C. Lyell] 1827, 472–3, quoted above, pp. 167–8.

22. Mantell 1827, 83; 1833, 286.

23. Poem pasted into Charles Daubeny's "Literary Common-Place Book" (Magdalen College, Oxford, MS 377, I, 157–9), quoted by permission of the President and Fellows of Magdalen College, Oxford. On Smith see *ODNB*.

"antediluvian horde" (including "a huge Peterodactyle") want to avenge themselves upon Mantell for having made a public spectacle of them. Mistaking the poet for the geologist, they gore him to death, whereupon he wakes from his "phantasmagorical dream".[24] Edward Hitchcock used similar theatrical shock-tactics in his *Religion of Geology*. Of the Labyrinthodon, conceived as "a frog as large as an ox, and perhaps as large as an elephant", Hitchcock exclaims, "Think of such animals swarming in our morasses at the present day!" Of the Iguanodon he writes, "What an alarm would it now produce, to have such a monster start into life in the forests of England [. . .]!" The Dinotherium he places directly before the reader's astonished eyes:

> imagine the bones of the dinotherium to start out of the soil, and become clothed with flesh and instinct with life. You have before you a quadruped eighteen feet in length, and of proportional height, much larger than the elephant [. . .][25]

The fourth and the most fertile means of enhancing immediacy was to bring readers face-to-face with the restored scene by including them within the narrator's own visionary experience. This inclusive device frames the narrator as guide, bringing us close to the rhetoric of the virtual-tourist guidebooks. The visual impact of Martin's *Deluge* is powerfully evoked, for instance, in the vision with which Sharon Turner concluded his self-styled "Panorama of Creation",[26] *The Sacred History of the World:*

> We can but faintly conceive the appalling scene. Mankind were surprised [. . .] by the sudden alarm of portentous danger rapidly rushing on them from the blackening and howling sky. The Sun was seen no more—midnight darkness usurped the day—lightnings dreadfully illuminated—thunder rolled with increasing fury—all that was natural, ceased; and in its stead, whirlwind and desolation—Earth rending—cities falling—the roar of tumultuous waters—shrieks and groans of human despair—overwhelming ruin—Universal silence!—and the awful quiet of executed and subsiding retribution![27]

24. Printed broadsheet, Alexander Turnbull Library, Wellington, New Zealand, MS-1956 (stanza 11).

25. Hitchcock 1851, 227–9. Compare Anon. 1837b, 30.

26. S. Turner 1832, 463.

27. S. Turner 1832, 520. Compare Fairholme 1833, 285, on diluvial drifting: "The whole scene now presents itself to the imagination".

*Plate 1.* Two tiny textbooks by Samuel Clark: *The Little Geologist* (*c*. 1840) and *The Little Mineralogist* (*c*. 1838), 7 × 9½ cm, formerly in the collection of John Thackray. Though brightly packaged in coloured wrappings and attractively illustrated, the texts themselves are a little dry. There was also third volume, *The Little Conchologist*.

*Plate 5*. Another geological giftbook. The red cloth case-binding of Mill's book *The Fossil Spirit: A Boy's Dream of Geology* featured gilt-embossed fossil lettering.

OPPOSITE

*Plate 6*. Iguanodon in colour for the first time: George Scharf's watercolour sketch (316 × 225 mm) for his three-yard Mantellian painting *Reptiles Restored* (1833). A marginal note mentions that the long-tailed Iguanodon was "calculated from the remains to have been 100 Feet long". The spiny creature on the opposite bank is labelled "Megalosaurus" but closely resembles subsequent representations of the Hylaeosaurus, first described by Mantell in 1833. Alexander Turnbull Library, Wellington, New Zealand, E-330-f-001.

*Plate 7*. Two rhinocerine Iguanodons in the Crystal Palace Gardens, Sydenham. The jungle surroundings which add authenticity to the scene are the result of neglect rather than design: this photograph was taken by John O'Connor in 1980, before the overgrown foliage was pruned and the sculptures restored.

*Reptiles restored, the remains of which are to be found in a fossil state in Tilgate Forest, Sussex*

*Plate 8.* John Martin, *Belshazzar's Feast*, oil on canvas (1821). The prophet Daniel interprets the divine judgement pronounced upon King Belshazzar of Babylon. What happens next is seen in Fig. 7.21.

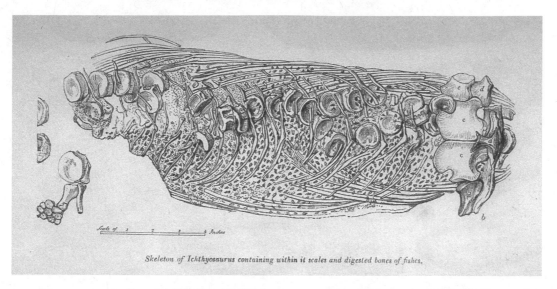

Skeleton of Ichthyosaurus containing within it scales and digested bones of fishes.

*Figure 9.4.* "Skeleton of Ichthyosaurus containing within it scales and digested bones of fishes", from Buckland's *Geology and Mineralogy*. W. Buckland 1836, II, plate 14.

The first sentence frames this narrative as a shared vision, which then lapses into past-tense description. Immediacy, however, is maintained by the dashes, especially after "all that was natural, ceased": here past-tense verbs are submerged in a boiling cascade of participial and substantive phrases. The dashes break up the "appalling scene" into its constituent narrative details, like the eye straining to take in one of Martin's mezzotints.

A more visceral approach was offered by Buckland's meditation on an ichthyosaur's stomach. Here Buckland refers to a particular skeleton depicted in volume 2 (see Fig. 9.4), which he brings to life before "us" to bring his ichthyosaur section to a resounding close:

> When we see the body of an Ichthyosaurus, still containing the food it had eaten just before its death, and its ribs still surrounding the remains of fishes, that were swallowed ten thousand, or more than ten times ten thousand years ago, all these vast intervals seem annihilated, time altogether disappears, and we are almost brought into as immediate contact with events of immeasurably distant periods, as with the affairs of yesterday.[28]

These last words recall the claims of the panoramists. The use of a fossil to trigger a vision was taken further by Broderip in the conclusion to

28. W. Buckland 1836, I, 201–2. Compare the gastric vision in Hawkins 1840a, 19.

*Zoological Recreations,* commanding the reader to "look" with him at an imaginary fossil:

> Look at the reptilian relic in the stone which helps to form that cottage wall. As we gaze, the wall disappears; and, to the mind's eye, its place is occupied by a vast sea, which, when circulation animated that bone, covered its site. Through the waters of this sea, Ichthyosaurs, Plesiosaurs, Mosasaurs, and Cetiosaurs dart, swim, and gambol. If we turn landward, the sluggish river, the marshy jungle and the dreary plain seem peopled by ancient Crocodilians, Iguanodons and Megalosaurs, while Pterodactyles appear to hover in the murky atmosphere of the old dragon times.
>
> Now, how changed the scene! [. . . .] Where the herbivorous Iguanodon revelled, the ox, the deer, and the sheep, quietly crop the fragrant herbage [. . . .][29]

The end of the first paragraph marks the sudden end of the present-tense vision, and we revert to the "then-and-now" narrative of a diptych. The foetid scenery into which we were momentarily plunged reappears, safely buried in the past tense beneath a providentially "fragrant" present-day landscape.

Fossils, then, could function like poetry quotations, becoming enchanted portals into a former world. If the narrator's guiding stance was maintained, this rhetoric could lead into full-scale virtual tourism, where narrator and reader stepped together *through* the portal and out onto the other side for a journey into the deep past. This drew not only on the conventions of the virtual-tourist guidebook or museum guide, but more fundamentally on two ancient and linked genres, the dream-vision and the fantastic voyage. Both were predicated on the idea of transport into another world, whether mental or bodily. Since ancient times, both genres had served as literary clothing for social satire, philosophical speculation, and didactic purposes. In the late-Victorian period, voyages into deep space or time would become subsumed into the technology-conscious realm of scientific romance or science fiction, but this should not lead us to label Davy's "Vision" and Byron's cosmic voyages simply as happy "anticipations" of science fiction.[30] They reflected a long-established pattern in Western literature by which a protagonist enters another world, often in a dream: the astronomer Johannes Kepler's dream-journey to the moon, Dante's spiritual quest through hell, purga-

---

29. Broderip 1848, 379.

30. On the development of Renaissance cosmic voyages into twentieth-century science fiction see Nicolson 1960; on the early history of science fiction see Stableford 2003.

tory, and paradise, the Icelander Star-Oddi's voyage into a Viking past, and the otherworldly sea-journeys of mediaeval Irish sagas and saints' lives.[31]

One finds monsters aplenty in these otherworlds, as in the "grey, dream-encircled past" of their geological descendants.[32] Until the end of the century, however, visitors to the ancient earth were usually imagined as disembodied spirits with extremely good eyesight, a useful fiction carried over from the dream-vision. These walks on the wild side were characteristically offered as part of a larger historical sequence of scenes. Such a narrative frame, moving across aeons in a single stride, could not be sustained in a conventional "tour-guide" format. Some form of time travel was required to transport reader and narrator from one scene to the next. Dreaming and voyaging provided the necessary overarching metaphors, often structured around theatrical models—theatrical spectacle being the chief commercial supplier of virtual dreams and virtual tourism.

Such narratives drew on potent Enlightenment images of the man of science as both visionary sage and explorer of unknown regions.[33] This second image suggested a range of different travel-metaphors. We might sail, like "an old fabulous navigator" or "Columbus", across the "seas of Time" to an unexplored country[34]—or to an unknown world, an alien planet in the "sea of space".[35] We might "enter" a particular stratigraphic formation;[36] travel up a metaphorical mountain, "rising" into a high antiquity;[37] or "descend" the strata, entering the underworld.[38] The entomologist William Kirby took this last metaphor literally, speculating in his Bridgewater Treatise that the giant saurians were not extinct but lived in a huge subterranean cavern, the Bible's "habitation of dragons".[39] Metaphors became mixed, much to their poetic advantage: Byron's Lucifer and Cain fly into outer space to reach Hades (compare Fig. 9.5), while Miller and his readers "sail upwards into the high geologic zones".[40] The idea of being "carried back" into antiquity was already a commonplace

---

31. See, respectively, Lear 1965; Alighieri 1981; R. O'Connor 2006b, 93–113; Wooding 2000. On the early history of the closely related genre of exotic travel-writing, see M. Campbell 1988; Romm 1992.

32. Miller 1854, 128; 1993, 127.

33. Knight 1967; Sommer 2003.

34. Miller 1859, 85; 1841, 270; Fullom 1854, 103; Hawkins 1840a, 20.

35. Byron 1986, 906–7; Hawkins 1834a, 33; Miller 1847b, 63.

36. Miller 1841, 269.

37. Mantell 1838, 604–5; 1844, 874; Miller 1859, 118.

38. Miller 1859, 148 and 239; 1847b, 226–7.

39. Kirby 1835, I, 20–42.

40. *Cain* II.i.143–205 (Byron 1986, 905–7; see also Fig. 2.11); Miller 1849, 201–2.

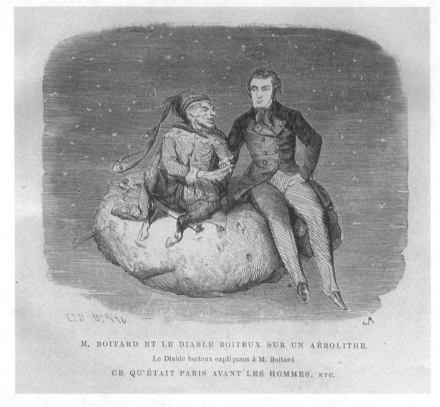

*Figure 9.5.* Engraving from *Paris avant les hommes* ("Paris before Man"), a didactic novel written in the 1830s by the radical French naturalist Pierre Boitard and published post-humously. The novel's narrator dreams that a droll demon conducts him into the deep past (via deep space, as in *Cain*). The words in the picture read: "M. Boitard and the *diable boiteux* on an asteroid. The *diable boiteux* explaining to M. Boitard what Paris was like before man". Boitard 1861, facing p. 10.

among geologists;[41] it was now ripe for literary development. It was only a short step to the scientific romances of Jules Verne, H. G. Wells, Arthur Conan Doyle, and Edgar Rice Burroughs, and thereafter to the dinosaurs of outer space which litter the early-twentieth-century science-fiction magazines.[42]

The visions of earth history we have been examining combined theatrical devices with epistemologically teasing allusions to dreaming. The dream-vision mode itself highlighted the potentially imaginary nature of these scenes, yet simultaneously provided a legitimate space for such fiction—a double function which the dream-vision had performed ever

41. Playfair 1802, 476; Sedgwick 1826–33b, 299.
42. On Burroughs and the magazines see Glut 1980; for a late-nineteenth-century example of dinosaurs in space see Hebblethwaite 2006.

since the Middle Ages.[43] What better model for the geologist's fragmentary and disconnected glimpses of the buried past?

## Earth as Archive, Earth as Theatre

So far, we have been analysing the various narratorial stances schematically, using isolated quotations. In the texts themselves, of course, these stances were not stable, but slipped from one to another. Past-tense description could snap into the language of the tour guide; the utterance of verse could, like a charm, transform a discussion of fossils into a time-travel adventure. The ease with which this slippage occurred is itself of interest. Before we look at how longer sequences of scenes were constructed, let us briefly examine these fluid movements between description and dramatization.

These two modes of presentation reflected different ways of looking at the earth geologically: as an archive and as a theatre. In the first view, the strata could be seen as pages of earth history. The earth's crust might resemble a palimpsest, a manuscript containing several superimposed inscriptions, "'written within and without' with wonderful narratives of animal life".[44] Read by the geologist, the more fossiliferous rocks became "illustrated books" or a "geologic library of nature" to rival that of Alexandria.[45] Sometimes, as the mouldering quarto speaks to Geoffrey Crayon in Washington Irving's story, these stony archives might "speak" to the geologist.[46] Geological expertise provided the necessary key or charm commanding the silent stones to speak in this way.[47] Miller, an assiduous folklore-collector, used this metaphor with particular feeling.[48] He presented geological inquiry in the social framework of a friendly conversation: some rock formations responded well to his inquiries, proving "frank and communicative" and providing him with "many a curious little anecdote".

> The boulder-clay, on the contrary, remained for years invincibly silent and sullen. [. . . .] It was a morose and taciturn companion [. . . .] I had no *"Open Sesame"* to form vistas through [its boulders] into the recesses of the past. And even now, when I have, I think, begun to understand the boulder-clay

43. Spearing 1976, 74–5; R. O'Connor 2005b, 165.

44. Miller 1841, 33, quoting Ezekiel 2.10. On the archive metaphor see Heringman 2004, 161–90.

45. [Dickens] 1851, 217; Miller 1859, 156–8.

46. Irving 1996, 112–20; Hawkins 1840a, 6.

47. Sedgwick 1833, 9–10.

48. On Miller's folklore collecting see Alston 1996.

a little, and it has become sociable enough to indulge me with occasional glimpses of its early history in the old glacial period [. . .] there are obscure recesses within its precincts into which I have failed to penetrate.[49]

Miller's language hovers precariously between metaphors of speaking and showing. Indeed, once the writer's interest focuses on these "vistas" themselves rather than the stony text in which the story is embedded, the earth suddenly shifts its aspect from archive to theatre, a stage on which geological processes and individual lives are played out.

While the archive metaphor celebrates the legibility of geological evidence, theatrical metaphors celebrate the visibility of the original scenes (and seek to reinforce the author's control over his or her reader). The shift between the two recalls contemporary models for the human mind, which was itself sometimes viewed as a "mighty palimpsest" on which successive sense-impressions fall "softly as light".[50] According to Thomas De Quincey, extreme circumstances could cause these "endless strata" to resolve themselves into a theatrical display. The drowning girl sees scenes from her life as in a "mighty theatre [. . .] within her brain"; inside De Quincey's own opium-lit brain, the "pall" concealing the past "draws up, and the whole depths of the theatre are exposed".[51] The transformation of these stratified deposits of the mind into theatrical sequences may reflect the influence of geology on De Quincey's thought; but antiquarian metaphors of decipherment had long underlain both earth science and philosophies of mind,[52] and textual critics now placed more emphasis on the fragmentary nature of manuscript evidence. The palimpsest embodied the interpretative problems involved: to Thomas Carlyle, history itself seemed a "complex Manuscript, covered over with formless inextricably-entangled characters".[53]

To bring its objects before the imagination, geology, the science of scenery, drew on archival metaphors of theatrical scene-changing which were simultaneously being used to conceptualize the imagination itself. The resulting texts swerve between the analogies of archive and theatre as frequently as they do between ruin and restoration, and writers like Miller sometimes seemed uncertain as to whether they wanted to tell a story or display a drama. In fact, the earth-as-theatre analogy had an extended pedigree as a means of shaping historical *narrative,* long before

49. Miller 1859, 45–6.

50. De Quincey 1985, 144 (from *Suspiria de profundis*). On this model see McDonagh 1987.

51. De Quincey 1985, 145–6; see *ODNB*. Compare Cobbe 1872, 351–2: "whole panoramas of beauty and horror".

52. Rudwick 1979, 68–73; Rappaport 1982; McDonagh 1987.

53. Carlyle 1888, II, 258.

and long after its emergence as the dominant metaphor of the Elizabethan stage. Its career is worth a brief glance.

In Britain, the analogy of human life as a drama was first fully developed by mediaeval clerical scholars like John of Salisbury and John Wyclif. God was seen to stage-manage and, with his angels, watch the pageant of human history or an individual's career.[54] In the Renaissance this metaphor became a commonplace of dramatic as well as narrative literature: in Shakespeare's hands it took on more sophisticated forms, serving as a bridge between stage and spectators and defining the unique and fragile actor-audience relation of Elizabethan drama.[55] The unusual formality of Jaques's set-piece speech on the seven ages of man in *As You Like It,* beginning "All the world's a stage", may be read as a playful gesture towards the moralizing pessimism characteristic of both the mediaeval topos and the self-dramatizing Jaques.[56] But when part of this passage was quoted in the guidebook to the Cyclorama's "Lisbon Earthquake" show in 1848, introducing the final tableau with the indented quotation "Last scene of all", all these layers of irony and ambiguity melted into thin air, and the old analogy re-emerged in its mediaeval form to illustrate the workings of "a Divine and inscrutable Providence".[57]

In a more optimistic vein, Buckland alluded to the same part of Jaques's speech to emphasize the providential arrangement of coal seams in the earth's crust. When, after aeons of preparation, the coal is at last used by modern man, it has reached the "seventh stage of its long eventful history".[58] Shakespeare's theatrical language has here been assimilated back into narrative. In fact this use of the topos had not experienced any hiatus since its mediaeval inception: the Elizabethan dramatists' teasing and sophisticated employment of the play metaphor was specialized, unorthodox, and (as Anne Barton has shown) short-lived.[59] By the mid-eighteenth century, "theatre" often denoted simply a place where something happens. Yet this metaphor's conventional familiarity did not iron out its dramatic force or theological connotations. These remained powerful, particularly after the French Revolution: Edmund Burke's first *Letter on the Proposals for Peace with the Regicide Directory of France* (1795) alluded with feeling to "the awful drama of Providence, now acting on the moral

54. Curtius 1953, 138–44; Righter 1962, 65.

55. Righter 1962.

56. Shakespeare 1993, 701 (II.vii.139–66).

57. Anon. 1848, 13.

58. W. Buckland 1836, I, 483. Shakespeare's original (1993, 701) reads "strange, eventful history". For similar allusions see Mantell 1838, 271; Hutchinson 1892, x (on the "exits and entrances" of the dinosaurs).

59. Righter 1962.

*Figure 9.7.* "Creation", G. F. Sargent's engraving for Mary Roberts's literalist treatise *The Progress of Creation* (1837), showing God creating the earth while angels look on. The epigraph, from John's Gospel, reads: "All things were made by him, and without him was not any thing made, that was made." Mary Roberts 1837, frontispiece.

earth as if seen from orbit.[66] But maps cannot reproduce living creatures, so Roberts next constructs an imaginary panorama, enabling us to see people, animals, and even tiny microscopic animalcules from our angelic height:

> Look again, but with the mental eye, for the visual organ can no longer follow it: dissimilar races of men are conspicuous in various portions of the globe. One part is crowded with fair men, in another are seen clear olive faces, in another black. [. . . .] Each continent [. . .] has also its own peculiar kinds of quadrupeds [. . . .] Myriads of insects, and creeping things innumerable are seen walking in the green savannah, to them forests of interminable length [. . . .] And more than even these, every leaf that quivers in the sunbeam, and every flower that drinks the dew of heaven, is in itself a world of animated life.

Roberts then shows us whose viewpoint we have been unwittingly sharing: "Over the mighty whole, watches One who never slumbers". The rhetoric of the panorama shifts easily into that of Psalm 139:

> He is our Father, his eye is perpetually upon us, the darkness of the night cannot hide from him, he spieth out all our ways. He will not overlook us in the thronged city; nor need we fear to be forgotten in the most solitary place.[67]

The new geology, too, maintained its links with Burnet's angelic perspective. Burnet's theatre of the earth, a fragile backdrop for human history, became Lyell's "theatre of changes" or Buckland's "drama of universal life", shifting without immediate reference to mankind.[68] The viewpoint remained the same: in an age when a dividing-line between extraterrestrials and angels was not often conceived of, let alone clearly drawn, the "beings of superhuman intelligence" invoked by Mantell and Chambers seem little different from the angels in Burnet's frontispiece. The imagined gaze of an extraterrestrial was never far away when other planets were depicted, from the theatrical frontispiece of Fontenelle's *Conversations on the Plurality of Worlds* to the verdant lunar landscape depicted in Roberts's *Progress of Creation* (see Fig. 9.8).[69] This theologi-

---

66. Hyde 1988, 146; Oettermann 1997, 90–3.

67. Mary Roberts 1837, 284–5.

68. W. Buckland 1836, I, 134.

69. Fontenelle's frontispiece is reproduced in Terrall 2000, 242. On the extraterrestrial-life debate, see Crowe 1986.

Or, as Tennyson expressed it:

> The hills are shadows, and they flow
>   From form to form, and nothing stands;
>     They melt like mist, the solid lands,
>   Like clouds they shape themselves and go.[73]

This angelic viewpoint gave geology much of its intoxicating appeal. It removed readers and geologists from worldly concerns, separating them from the earth as spectators from a spectacle, and allowed them to gaze down on it as a drama. Let us now examine how the drama itself was put together.

## Sequences of Scenes

Mantell's work was of seminal importance in the transformation of virtual guided tours into geological dramas. However, the overall structure of *Wonders of Geology* and the retrospect in *Medals of Creation* presented the scenes of "the earth's physical drama" in reverse.[74] Their narrative framework was that of a voyage back in time, descending the strata, fulfilling Lyell's prediction of the future geologist leading his followers as Virgil had led Dante:

> We have entered upon the confines of the past, and already we find ourselves surrounded by an innumerable population of unknown types of being—not as dim and shadowy phantoms of the imagination,—but in all the reality of form and structure [. . . .][75]

In other words, this is more than a mere phantasmagoria, and all the more impressive for being "real". But in the third lecture Mantell pauses in this downward journey to give his audience a higher-resolution "Retrospect" of the Tertiary period in central France, as a straightforward past-tense narrative in chronological order. Its content echoes Lyell's 1827 review-article, but whereas Lyell's description was hypothetical, framed within a very Lyellian rhetorical question, Mantell's is presented as pure fact:

> a change came over the scene—violent eruptions burst forth from craters long silent—the whole country was laid desolate—its living population

---

73. Poem 123 (A. Tennyson 1969, 973).
74. Mantell 1844, 873.
75. Mantell 1838, 167.

swept away—all was one vast waste, and sterility succeeded to the former luxuriance of life and beauty. Ages rolled by—the mists of the mountains and the rains, produced new springs, torrents, and rivers—a fertile soil gradually accumulated over the cooled lava currents [. . . .] Another vegetation sprang up—the mammoth, mastodon, and enormous deer and oxen now quietly browsed in the verdant plains—other changes succeeded—these colossal forms of life in their turn passed away [. . . .][76]

The dramatic profusion of dashes recalls Sharon Turner's vision of the Deluge, but here it also enhances the dreamlike quality of these "awful changes", which—along with Mantell's first six words—suggests the refrain of Byron's poem *The Dream*.[77] In terms of Victorian theatrical entertainment, one might call this description a set of dissolving views, each one melting into the next in clauses spanning millions of years each.

In the fourth lecture Mantell takes us back to the Age of Reptiles. At the end he provides another past-tense narrative "retrospect", this time including all the previous epochs as well. Then comes a surprise. Having finished his retrospect, Mantell repeats it—only this time, to heighten the immediacy, he imagines a "higher intelligence from another sphere" who had visited the Brighton area at various points in its history, and who now reports on the changes he sees:

Countless ages ere man was created, he might say, I visited these regions of the earth, and beheld a beautiful country of vast extent, diversified by hill and dale, with its rivulets, streams, and mighty rivers, flowing through fertile plains. Groves of palms and ferns, and forests of coniferous trees, clothed its surface; and I saw monsters of the reptile tribe, so huge that nothing among the existing races can compare with them, basking on the banks of its rivers and roaming through its forests [. . . .] And after the lapse of many ages I again visited the earth; and the country, with its innumerable dragon-forms, and its tropical forests, all had disappeared, and an ocean had usurped their place. And its waters teemed with [. . .] innumerable fishes and marine reptiles. And countless centuries rolled by, and I returned, and lo! the ocean was gone, and dry land again appeared [. . . .] And I beheld, quietly browsing, herds of deer of enormous size [. . . .] and I heard the roar of the lion and the tiger, and the yell of the hyena and the bear. And another epoch passed away, and [. . .] the face of the country no longer presented the same aspect; it was broken into islands, and the bottom of the sea had become dry land [. . . .] Herds of deer were still to be seen on the plains [. . . .] And I beheld human beings, clad in the skins

76. Mantell 1838, 262–3.
77. See above, p. 181; compare also [C. Lyell] 1827, 472–3.

Imagine then one of these monstrous animals, a *Plesiosaurus*, some sixteen or twenty feet long [. . .] with four large and powerful paddles, almost developed into hands; an animal not covered with brilliant scales, but with a black slimy skin. Imagine for a moment this creature slowly emerging from the muddy banks, and half walking, half creeping along [. . . .][87]

388

389

The slimy skin is again speculative, recalling Hawkins's sea-dragons as depicted by Martin (see Fig. 10.3). Like Martin's picture, and like the word "monstrous" itself, Ansted's vivid details serve a dramatic rather than a strictly philosophical purpose. As his restorations gain in immediacy, he approaches ever nearer to the stance of a tour guide. The sketch continues with the tempting invitation, "now let us see what goes on in the deeper abysses of the ocean". Down there is an ichthyosaur "Prowling about at a great depth", and "we may fancy we see this strange animal, with its enormous eyes directed upwards, and glaring like globes of fire". This touch of melodrama tips the balance at last, and for the first time Ansted launches into a splendid present-tense description:

Suddenly [. . .] the monster darts through the water at a rate which the eye can scarcely follow towards the surface. The vast jaws, lined with formidable rows of teeth, soon open wide to their full extent; the object of attack is approached—is overtaken. With a motion quicker than thought the jaws are snapped together, and the work is done. The monster, becoming gorged, floats languidly near the surface, with a portion of the top of its head and its nostrils visible, like an island covered with black mud, above the water.[88]

The Miltonic final image, recalling the slimy skin and muddy haunts of Ansted's plesiosaur, alludes to a commonly-drawn analogy between extinct reptiles and Milton's simile of Satan lolling like Leviathan in the infernal lake (see Fig. 10.4).[89] But this time, instead of finishing with a cautious retreat, Ansted returns to the theatrical analogy and declares that similar "scenes" must have been regularly "enacted" during this epoch.[90] Perhaps understandably, one of Ansted's literalist critics not only found his cautious stance disingenuous, but found a discordance between these scenes of saurian savagery and the underlying claim that God had created these creatures to "enjoy" life: "And this, it seems, was enjoyment!"[91]

87. Ansted 1847, 179.
88. Ansted 1847, 180–1.
89. See Milner 1846, 722; *Paradise Lost* I.192–220 (Milton 1971, 55–7).
90. Ansted 1847, 181.
91. Anon. 1847, 396–7 and 422–3.

Describing the country of the Iguanodon, Ansted at last assumes the mantle of the time-travelling tour-guide. He begins, however, by fore-grounding the effort of imagination required to do so:

> Let us imagine ourselves placed upon a projecting headland [. . .] command-ing a view of the open sea, which then covered the greater part of our is-land. Placed in imagination in this commanding position, let us endeavour to recall the scenes once enacted near some tract of low flat land—a sandy shore of the oolitic period—on which, at a distance, a few solitary palm trees stand out against the blue sky [. . . .]

The first page is vivid enough, but conditional, referring back to "our supposed position".[92] Then Ansted slips into the present tense, and we slip into the landscape:

> Presently a noise is heard, and a huge animal advances [. . . .] it is far longer and also taller than the largest elephant; its body hangs down near the ground, but its legs are like the trunks of great forest trees [. . . .] The monster approaches, and trodden down with one of its feet, armed with powerful claws, or caught between its long and narrow jaws, our crocodile is devoured in an instant.[93]

Here, too, the rhetoric of hypothesis remains: the crocodile is offered two alternative death-scenes.

By the 1850s, visionary narratives connecting the whole of earth his tory were becoming bolder and more widely used. Mill's *Fossil Spirit* has already been discussed as an unusual book-length example, though with-out the tour-guide mode. One particularly lively tour into the past was directly linked to the panoramic displays of the time: Henry Morley's four-page story "Our Phantom Ship on an Antediluvian Cruise" was published in *Household Words* in 1851 as part of a series of imaginary voy-ages to distant places by the "Phantom Ship". In 1851 Wyld's Great Globe had appeared in Leicester Square, and Morley begins his piece by joking that his ship is obsolete "now that we can visit any portion of the globe by taking a cab [. . .] to Leicester Square", and it is time for her to "retire decently, recede into the past with a becoming dignity". This "ghostly voyage back into the past" is conducted by sea and narrated in the first person plural,[94] and Morley makes constant reference to the Great Globe and other shows of London for the purposes of analogy. But the pace

92. Ansted 1847, 219–20.
93. Ansted 1847, 221–2.
94. [Morley] 1851, 492. This story is briefly discussed in Shatto 1976.

*Figure 10.1.* Hugh Miller the stonemason, holding his mallet and resting against a gravestone. This calotype was made by David Octavius Hill and Robert Adamson in 1844, many years after Miller had stopped working as a mason, but as editor of the *Witness* he maintained his working-class image. Scottish National Photography Collection, PGP HA 283.

in a chaste yet exuberant Augustan prose style modelled on Addison, Goldsmith, and Scott.[2]

As editor of the *Witness,* Miller was one of Britain's most talented and prolific men of letters. If he has been largely forgotten by the literary world today, the current official status of non-fiction as non-literature must take much of the blame. Moreover, as a determined and deeply religious opponent of evolutionary theory just before the publication of Darwin's *Origin of Species,* Miller is sometimes seen as a man fighting a disastrous rearguard action which could only end in suicide—a fine example of the *post hoc ergo propter hoc* fallacy, but one which sadly makes his scientific writing unacceptable to many present-day commentators.

For Miller, bringing landscapes before the mind's eye was one of the primary functions of a man of letters, exemplified by the prospect views of his favourite poets, James Thomson and William Cowper.[3] This pri-

2. Mackenzie 1905, 25–50. On Scott's influence see Robertson 2002.
3. Knell and Taylor 2006, 94.

ority was reflected in his science writing: "Much of the interest of a science such as geology must consist in the ability of making dead deposits represent living scenes".[4] When putting this into practice, Miller often resorted to imaginary guided tours. This inherently sociable mode was well suited to a journalist, and still more so to an inveterate autobiographer whose sense of self and community were founded on exalted notions of friendship.[5] The "imaginary ramble" allowed him to project an ideal companion with whom he could share his delight in the natural world. This principle underpins much of his geological popularization, whether "we" are standing at a commanding viewpoint, exploring a coastline, gazing at the strata, or sailing the seas of time.[6] When Miller moves from shared vision into autobiographical reminiscence, the friendly rapport he has established with his reader makes these fragments of autobiography seem engagingly inclusive, and they naturally gravitate back towards the first person plural.

Miller's descriptions of the ancient earth are numerous, varied, and often extensive. They are of exceptional literary quality, and this chapter provides little more than a sketch of how their more theatrical techniques developed.[7] Although some commentators found his prose overwrought, it enthralled the middle-class reading public on both sides of the Atlantic, exerting a lasting influence on writers as varied as Alfred Tennyson, John Ruskin, and Thomas Carlyle. Nor is this surprising: on opening his books we find prose whose poetic density, vigour, and rhythmic balance often surpasses that of his more famous contemporaries. It is a central thesis of this book that nineteenth-century scientific writing demands to be taken seriously as imaginative literature, but with Miller in particular this argument becomes more than merely historical. Besides its value for the historian, Miller's writing transcends its context on a purely aesthetic level, throwing Mantell, Buckland, and even Lyell into the shade. In part, this is because aesthetic concerns were closer to Miller's heart: he devoted more energy than most to distilling the poetry of the science, imposing poetic form on his mythopoeic visions of the ancient earth.[8]

This determination in turn makes Miller the most significant figure in our historical inquiry. It is for this reason, not simply because his books

4. Miller 1854, 445; 1993, 443.

5. On Miller's sense of self see Paradis 1996 and A. Secord 1996 (and compare Miller 1995, 115–16, 155–6). On his journalistic style see J. Secord 2003a, 331–3. On Miller as tour guide see Knell and Taylor 2006, 94.

6. Merrill 1989, 245.

7. In a subsequent book I hope to offer a fuller treatment.

8. On literary techniques in Miller's geological prose, see Merrill 1989, 236–54; Paradis 1996; R. O'Connor 2003a; Knell and Taylor 2006.

"fleets of *nautili*" recalls the present-day prospect described at the beginning of his chapter, with its "glitter of sails" and the "fleet riding at anchor" in the bay.[16] Journeying through thick fog is, moreover, a common motif in Scottish folklore (Highland and Lowland alike), where it signifies the traveller's unwitting passage into the Otherworld.

Back in the present-day landscape, Miller next conducts us on an imaginary field-trip among the "ruins" of the landscape he had just restored.[17] The "data" having been "collected", he embarks on a second act of restoration, this time a sequence of scenes leading up to the cataclysm which had raised the great hill on which we now stand. This narrative begins with the scene of his earlier vision:

> In this remote period the fish of the little bay [. . .] explored their forests of tangle in pursuit of their prey; the ammonite raised its sails to the wind as it passed from sea to sea [. . .] and the oyster ripened its pearl in the ooze below.

The lapse of time is conveyed by a concatenation of brief clauses:

> generation succeeded generation; the living inhabited the place of the dead; [. . .] vapours arose; rains descended; winds darkened the sea; waves whitened the shore; [. . .] the broken fragments of every earlier generation were buried beneath the remains of the generation which flourished after; [. . .] hills became less lofty, and the depths of the sea less profound [. . . .][18]

Finally, the landscape erupts before our eyes, at last fulfilling the promise of the Keats epigraph at the beginning ("Drowned wast thou, till an earthquake made thee steep").[19] Denying that such a scene can be pictured as a single image, he describes it as a chaos of separate and "unimaginable" elements. The result strongly resembles Sharon Turner's description of the Deluge. It is a turbulent composite image in the style of Martin, unified only by a "mingled darkness" which is all the more visible for its contrast with the sunlit, panoramic view with which the chapter opened. Once again we are given a visionary diptych:

> But what imagination can conceive of all the circumstances of grandeur [. . .] when the lower rocks, glowing like the moon as she rises amid a sea of vapour, raised their immense heads above the crust that for ages had ac-

16. Miller 1835, 46–7.
17. Miller 1835, 51–61.
18. Miller 1835, 61–2.
19. Miller 1835, 44, quoting Keats, "To Ailsa Rock", line 13 (Keats 1926, 325).

cumulated over them; and the granitic band [. . .] was snapped asunder like the magic girdle of Florimel, or the withs of Delilah. We fail in recalling, amid the brightness and tranquillity of a scene so lovely as that which now lies spread before us, the heave of earthquakes, the bursting of flames, the roar of waves, the rush of cataracts, lands rising, seas receding, the yawning of valleys, the upheaving of hills, and thickening over the whole, the canopy of a mingled darkness of smoke, ashes, and vapour.[20]

The "brightness and tranquillity" of the present-day view serves as a foil for the description of the past catastrophe, rather as Martin had used Byron's description of a sunlit Mount Ararat in his *Deluge* guidebook, or as Lyell had described the peaceful Auvergne landscape before the Tertiary volcanic eruptions.[21] On top of this diptych Miller deposits further strata of incongruously magical imagery, and his Turnerian technique of piling on the participial clauses heightens the sense of cataclysmic simultaneity. Our mind's eyes are sorely taxed: we may "fail in recalling" such a scene, but, as Miller has already informed us, it is in such imaginative straining that geology may be dignified with the status of poetry.

The profusion of techniques by which Miller manipulates his reader's visual imagination, centring on a single, all-embracing, but shifting view, strongly suggests that panoramas and dioramas were hovering at the back of his mind. In the 1820s, when living near Edinburgh, he recorded his visit to the panorama building on the Mound (Fig. 10.2) which jostled alongside other attractions "from the smoking baboon to the giant of seven feet and a half". There he saw a dioramic series of scenes representing the Battle of Trafalgar which, for all his dislike of the theatre, made quite an impression on him.[22]

Miller's theatrical metaphors soon became more overt, presenting earth history as a discontinuous succession of scenes rather than as a gradually evolving narrative.[23] In *The Old Red Sandstone,* for instance, he enlisted Shakespeare's Perdita in an assault on transmutationist theories. Perdita appears as a baby in Act III of *A Winter's Tale,* and as a fifteen-year-old shepherdess in Act IV, but Miller asserts that no such "genealogical link" could be shown to connect one race of fossil fish to another: "The scene shifts as we pass from formation to formation; we are introduced in each to a new *dramatis personæ*".[24] The book concludes with a three-chapter-long narrative "history". This is no ordinary story: it unfolds

20. Miller 1835, 63–4. Compare S. Turner 1832, 520.
21. See above, pp. 167–8 and 303.
22. Bayne 1871, I, 147–50; Miller 1995, 209–10. This is contrary to what I have suggested elsewhere (R. O'Connor 2003a, 246).
23. Merrill 1989, 248; R. O'Connor 2003a.
24. Miller 1841, 43.

the leaves still unwithered; and there floats a tuft of fern. Land, from the mast-head! land! land!—a low shore thickly covered with vegetation. Huge trees of wonderful form stand out far into the water [. . . .] A river of vast volume comes rolling from the interior, darkening the water for leagues with its slime and mud [. . . .] We near the coast, and now enter the opening of the stream. A scarce penetrable phalanx of reeds [. . .] is ranged on either hand. The bright and glossy stems seem rodded like Gothic columns; [. . . .] Can that be a club-moss that raises its slender height for more than fifty feet from the soil? [. . . .] Have we arrived at some such country as the continent visited by Gulliver [. . .]?

It is an oppressive, inhuman realm: "there is silence all around, uninterrupted save by the sudden splash of some reptile fish [. . .] or when a sudden breeze stirs the hot air, and shakes the fronds of the giant ferns".[31] Yet even here human analogies make an appearance: Gothic architecture, *Gulliver's Travels,* Columbus, and more military imagery ("phalanx"). This technique is characteristic of Miller's associative method, charging the outward forms of wild nature with associations drawn from human life, artefacts, and history, to infuse extinct worlds with human meaning. As one reviewer put it, "By spreading what Coleridge calls 'the mist of obscure feeling' over the cold forms with which he has to deal, our sympathy is secured even for the fortunes of a fossil fish."[32] Finally Miller halts this "voyage of discovery" and declares, "We pursue our history no further". The time-travel episode is revealed as part of a larger "history", or "series" of "scenes". The overseeing Creator is dimly glimpsed in the closing evocation of the "onward and upward march" of nature, which rejoins the drama of sacred history by culminating in Christ Himself, "a new heaven and a new earth", and "the deep echoes of eternity".[33]

## Panoramas of Creation

Over the next decade and a half Miller found plenty of opportunities to practise his new-found dramaturgical skills. His sequences of scenes became steadily more unified and self-contained as literary performances, often occupying the final pages of an essay, lecture, or book. One of his earliest large-scale visions of earth history was printed in a collaborative volume on the Bass Rock, edited by Thomas McCrie and published in 1848.[34] McCrie himself wrote an essay on the Rock's history, but Miller,

31. Miller 1841, 269–72.
32. Anon. 1841–2, 211.
33. Miller 1841, 274–5.
34. Miller 1848. This essay had initially been published piecemeal in the *Witness* in 1846–7, and was reprinted in 1864: see Miller 1869, 223–313.

whose brief was to cover geology, characteristically allowed himself ample space for snippets of local history too.

Miller's essay concludes with a seventeen-page geological "history" of the Rock imagined as the scenes in a diorama, from the period of the Primary rocks up to recent times. As in *Scenes and Legends,* he prepares the ground for his visions by assuming his chatty autobiographical mode, telling us about the sunset he had enjoyed when visiting the Rock. This gannet-haunted island, rearing like a mountain out of the sea near the mouth of the Firth of Forth, was an ideal spot for a panoramic view. As usual, Miller takes every opportunity to charge its individual elements with poetic significance. The island's shadow "stretched in darkness towards the east, like the shadow of the mysterious pillar of cloud of old along the sands of the desert"; while in the dim distance the "white Pharos" on the Isle of May "seemed a sheeted spectre,—the solitary inhabitant of some island of Cloudland." In short, "There was a magnificent combination of fairy wildness and beauty in the scene. And yet it was all a reality, though a transitory one." This observation leads Miller to consider the passing of time, and the scene before his eyes becomes a "canvass" which periodically "exhibit[s], within the old outlines, another and fresher succession of colours." As the geologist knows, the "outlines" themselves shift no less dramatically, if infinitely more slowly, than the colours that fill them.[35]

Standing beside the "little pyramid on the summit of the Rock", Miller now makes his magic wish:

> I could not help wishing that, under the influence of some such vision as fell upon Mirza in the "long hollow valley of Bagdad," I could see scene succeed scene in the surrounding area, from the early dawn of being [. . .] down to those historic periods during which, doing or suffering, man enacted his part upon the stage.[36]

Miller then invokes Moses himself by quoting twenty-eight lines of a poem by Charles Caleb Colton, asking the patriarch to tell the story of the rocks:

> "[. . . .] Oh, deign to tell,
> Seer of the pillared flame and granite well!
> Who taught old Mother Earth to hide
> The lava's age-repeated tide; [. . . .]
> These are earth's secrets,—but to gain

---

35. Miller 1848, [120]–[121].

36. Miller 1848, [121]–[122], alluding to Addison's oriental fantasy *The Vision of Mirza* (1711).

> Those of the Deep *thou* rent in twain,
> 'Twere worth a dull eternity
> Of common life,—to question thee."

> The curtain rises, and there spreads out a wide sea, limited, however, in its area by a dark fog that broods along the horizon [. . . .] It is the ocean of our Scotch Grauwacke that rolls beneath and around us [. . . .][37]

Framed as an incantation to raise the dead, this piece of verse works as a curtain-raiser, in the manner of the quotations scattered throughout the guidebook accompanying the Cyclorama's Lisbon Earthquake show.[38] With bracing present-tense immediacy, we are thrust into the deep past with our guide: poetry quotation again mediates the gulf between present and past. Indeed, the entire sequence is framed by poetry, for Coleridge and Milton are enlisted at the end to carry us forward into the present day. Coleridge's *Rime of the Ancient Mariner* gives a Gothic colouring to the gloomy, iceberg-ridden epoch of the boulder-clay, and, as the modern period dawns, lines from *Paradise Lost* bring the vision (and the whole essay) to a close:

> My history speeds on to its conclusion. We dimly descry, amid fog and darkness, yet one scene more. [. . . .] But this last scene in the series I find drawn to my hand, though for another purpose, by the poet who produced the "Ancient Mariner:"—

> > "Anon there come both mist and snow,
> >    And it grows wondrous cold;
> > And ice mast-high comes floating by,
> >    As green as emerald; [. . . .]

> > The ice is here, the ice is there,
> >    The ice is all around;
> > It cracks and growls, and roars and howls,
> >    Like noises in a swound."

> But the day breaks, and the storm ceases, and the submerged land lifts up its head over the sea; and the Bass, in the fair morn of the existing creation, looms tall and high to the new-risen sun,—then, as now,

> > "An island salt and bare,
> > The haunt of seals and orcs, and sea-mews' clang."[39]

37. Miller 1848, [122]–[123].

38. This show, discussed above (pp. 301–2), opened in the London Colosseum shortly after Miller's trip to the Bass.

39. Miller 1848, [138]–[139], quoting from Part I of Coleridge's revised *Rime of the Ancient Mariner* (Coleridge 1985, 48) and *Paradise Lost* XI.834–5 (Milton 1971, 605).

Miller has certainly appropriated Coleridge's vision to his own "purpose", altering all the original past-tense verbal forms (*came, grew*) to the present tense (*comes, grows*). The final quotation represents a still bolder appropriation. It comes from a passage in which Milton speculates on what happened to Eden during the Flood, suggesting that it might have become detached from the land and floated off into the ocean, where it survives to this day as a barren skerry. By wresting this passage out of its original context and pressing it into service here, Miller maps the imagery of sacred legend directly onto that of earth history. This seems appropriate enough in the light of his initial request for the vision, addressed to the author of Genesis; but it also has the curious and haunting effect of displaying the Bass Rock as a lost fragment of Paradise, fit representative of "the fair morn of the existing creation"—and a suitable retreat for the saintly Baldred the Culdee, with whom the human history of the Bass Rock begins in McCrie's essay.[40]

In the extended sequence which these poetry quotations frame, each scene is separated from its predecessor by the formation and dissipation of mist or darkness:

> The light brightens over the wide expanse, and the fog rises; myriads of ages have passed by; the countless strata of the Grauwacke are already deposited; and we have entered on the eras of the Old Red Sandstone.[41]

The theatrical agency of mist, functioning like curtains separating scenes, imitates the standard practice of dioramas. Scene-changing served as a useful metaphor in its own right in *First Impressions of England,* written around the same time as the Bass essay, in a discussion of the puzzling boundary between the Palaeozoic and Mesozoic periods (today interpreted as the greatest mass-extinction in history). Miller does not commit himself to any explanation of the sudden end of "the long drama of the Palaeozoic period, with all its distinct acts", but he uses the scene-changing analogy to convey the ignorance of the scientific spectator:

> A strange shifting of scenes took place on that rough stratum at our feet; but it would seem as if the theatre had been darkened when the alterative process was going on. The lamps burnt low, and concealed the machinery of the stage.[42]

40. See Miller 1848, [52].
41. Miller 1848, [123].
42. Miller 1847b, 219.

Miller continued to develop and recycle these techniques and frameworks in his geological dioramas of the late 1840s and 1850s.[43] Scattered across his *oeuvre,* these appear as varying aspects of a single visionary pageant which was becoming increasingly present to Miller's mind's eye. The *Sketch-Book of Popular Geology,* based on six lectures given to the Edinburgh Philosophical Institution in the early 1850s, forms a magnificent summation of all these techniques. Miller had meant to publish these lectures piecemeal in his projected *magnum opus, The Geology of Scotland,* in order to "lighten the drier details";[44] in the event, they were edited by his widow Lydia after his untimely death. In this form they concentrate a gallery of restored landscapes into six chapters, structured (like Mantell's) as a voyage back in time or down the strata, but punctuated (more than Mantell's) by scenes and sequences in chronological order. Nowhere else in nineteenth-century literature do we find such a profusion of prehistoric pageantry. When the book appeared in 1859, unadorned by a single picture, its "masterpieces of geological landscape" won high praise for having been achieved purely by verbal means.[45]

In these lectures Miller made overt use of the diorama and panorama (these terms were by now interchangeable in common parlance). At the end of one lecture, a lengthy description of Cretaceous Scotland is introduced as follows:

> The geologic diorama abounds in strange contrasts. When the curtain last rose upon our country, we looked abroad over the amber-producing forests of the Tertiary period, with their sunlit glades and brown and bosky recesses, and we saw, far distant on the skirts of the densely wooded land, a fire-belching volcano, over-canopied by its cloud of smoke and ashes. And now, when the curtain again rises, we see the same tract occupied, far as the eye can reach, by a broad ocean, traversed by a pale milky line, that wends its dimpling way through the blue expanse, like a river through a meadow.[46]

At the close of another lecture Miller develops his tour-guide persona in a powerful piece of time travel into the Age of Reptiles. He first apologizes for giving the audience nothing but a "dry list" of Oolitic "productions", then offers an alternative:

> Could we but [. . .] travel backwards into the vanished past, as we can descend into the strata that contain their remains, and walk out into the

43. For example, Miller 1849, 201–3; Miller 1869, 80-2.
44. Miller 1859, xv.
45. Anon. 1859, 446.
46. Miller 1859, 117.

404
~
405

woods, or along the sea-shores of old Oolitic Scotland,—we should be greeted by a succession of marvels strange beyond even the conceptions of the poet, or at least only equalled by the creations of him who, in his adventurous song, sent forth the Lady Una to wander over a fairy land of dreary wolds and trackless forests, whose caverns were haunts of dragons and satyrs, and its hills the abodes

> "Of dreadful beasts, that, when they drew to hande,
> Half-flying and half-floating, in their haste,
> Did with their largeness measure o'er much lande,
> And made wide shadow under bulksome waist,
> As mountain doth the valley overcaste;
> And trailing scaly tails did rear afore
> Bodies all monstrous, horribill, and vaste." [47]

Miller here adapts canto I, stanza 8 of the first book of Edmund Spenser's Elizabethan epic *The Faerie Queene,* and these lines about Satanic dragons perform the same lens-like function as the Milton quotation used by Buckland to depict his pterodactyle. [48] Through this curtained portal we embark on a "short walk into the wilds of the Oolite".

The nightmarish light projected by these lines is enhanced by Miller's substantial alterations to Spenser's original. Spenser himself had described only one "Beast", as follows:

> By this the dreadfull Beast drew nigh to hand,
> Halfe flying, and halfe footing in his hast,
> That with his largenesse measured much land,
> And made wide shadow vnder his huge wast;
> As mountaine doth the valley ouercast.
> Approaching nigh, he reared high afore
> His body monstrous, horrible, and vast [. . . .] [49]

All the other monsters have been invented by Miller, as have their "scaly tails" and semi-aquatic character. He has also added orthographic archaisms, some reminiscent of mediaeval Scots (*horribill*). The result is more than mere pastiche: Spenser becomes a geological poet, while Oolitic Sutherland becomes a realm of dark romance. Gillian Beer has described Darwin's literary project as the "recast[ing] of "inherited mythologies" and "narrative orders", based on subtle transformations of symbols such

47. Miller 1859, 148–9.

48. That passage in Milton's *Paradise Lost* may itself have been inspired by the very same stanza in Spenser's *Faerie Queene* (Milton 1971, 134 n.).

49. Spenser 1977, 144.

> "Here awhile the Muse,
> High hovering o'er the broad cerulean scene,
> Sees Caledonia in romantic view;
> Her airy mountains, from the waving main,
> Invested with a keen diffusive sky,
> Breathing the soul acute; her forests huge,
> Incult, robust, and tall, by Nature's hand
> Planted of old; her azure lakes between, [. . . .]"

After eight more lines, Miller invites his audience "in like manner" to call up "the features of our country in one continuous landscape, as they appeared at the commencement of the glacial period".[59] For the next five pages we stand before a shifting series of magnificent scenes, with Miller standing beside us and pointing to important or picturesque details. After restoring us to the present day, Miller challenges his audience:

> It is said that modern science is adverse to the exercise and development of the imaginative faculty. But is it really so? Are visions such as those in which we have been indulging less richly charged with that poetic pabulum on which fancy feeds and grows strong, than those ancient tales of enchantment and *faery* which beguiled of old, in solitary homesteads, the long winter nights. Because science flourishes, must poesy decline?[60]

The answer is, of course, "no", and Miller concludes his lecture by recalling Wordsworth's predictions about the future relations of poetry and science. One day, says Miller, "whenever a truly great poet arises,—one that will add a profound intellect to a powerful imagination," that poet will find

> more of the suggestive and the sublime in a few broken scaurs of clay, a few fragmentary shells, and a few green reaches of the old coast line, than versifiers of the ordinary calibre in their once fresh gems and flowers,—in sublime ocean, the broad earth, or the blue firmament and all its stars.[61]

The campaign continues in the introductory portion of lecture 3, which directly combats Moir's claim that "geological exposition" was one of many "staggering blows which science has inflicted" on poetry.[62] Between the publication of his book and Miller's lecture, Moir had soft-

---

59. Miller 1859, 74–5, quoting from *Autumn*, lines 878–93 (Thomson 1987, 112–13).

60. Miller 1859, 79–80.

61. Miller 1859, 80. Miller's choice of final word is no coincidence: each of the three books of Dante's *Divine Comedy* ends with the word *stelle* ("stars"; Alighieri 1981, 254, 512, 768). For the comparison with Wordsworth see Paradis 1996, 123–4.

62. Moir 1851, 114. Compare the opinion mentioned by Francis 1839, 152–6.

ened his views in a letter to Miller, and had subsequently died; but the views were still out there in print, and called for a firm answer. Miller ingeniously combined this with a heartfelt homage to Moir, whose reflective poetry he admired. The Thomsonian "panorama" of lecture 2 itself contains an illustrative quotation from "Mary's Mount", a prospect poem by Moir in which the speaker stands on a hill and imagines himself carried back several centuries to witness a historic battle. The geohistorical sequence in lecture 3 is broadly modelled on the same poem.[63]

This time the leap into the past is effected by a poetic "spot of time": a memory within a memory, itself nested within an autobiographical reminiscence, like the "apparition" recorded in *Scenes and Legends*. Miller recollects attending Moir's funeral at Inveresk, then recalls how, on that melancholy occasion, "From the sadness of the present my thoughts let themselves out upon the past." He remembers having reflected on the site's Roman origins, and on a previous visit to the same spot a year earlier. On that previous visit he had hoped to trace "the remains of that stern old people whose thirst of conquest and dominion had led them so far." The act of unearthing these buried strata of memory prompts a two-page voyage back in time:

And lo! like a dream remembered in a dream, as the crowd broke up and retired, the visions of that quiet day were again conjured up before me [. . .]

On that rising ground, so rich in historic associations, [. . . .] the vision of a forest-covered country rose before me,—a vision of the ancient aboriginal woods rising dusky and brown in one vast thicket [. . . .] The grim legionaries of the Proconsul of Augustus were opening with busy axes a shady roadway through the midst; and the incessant strokes of the axe and the crash of falling trees echoed in the silence throughout the valley. And then there arose another and earlier vision, when [. . .] the site of the town itself [existed] as a sandy bay, swum over by the sea-wolf and the seal; [. . . .] and the Esk, fed by the glaciers of the interior, whose blue gleam I could mark on the distant Lammermoors [. . .] rolled downwards, a vast stream [. . . .] And then there arose yet other and remoter scenes. From a foreground of weltering sea I could mark a scattered archipelago of waste uninhabited islands, picturesquely roughened by wood and rock; and near where the Scottish capital now stands, a submarine volcano sent forth its slim column of mingled smoke and vapour into the sky. And then there rose in quick succession scenes of the old Carboniferous forests: long withdrawing lakes, fringed with dense thickets of the green Calamite, tall and straight as the masts of pinnaces, and inhabited by enormous fishes,

63. Miller 1859, 78.

that glittered through the transparent depths in their enamelled armour of proof; [. . .] or yet again, there rose a scene of coral bowers and encrinal thickets, that glimmered amid the deep green of the ancient ocean, and in which, as in the groves sung by Ovid, the plants were sentient, and the shrinking flowers bled when injured. And, last of all, on the further limits of organic life a thick fog came down upon the sea, and my excursions into the remote past terminated, like the voyage of an old fabulous navigator, in thick darkness.[64]

Inhuman scenes are again overlaid with human associations: the fishes' mediaeval "armour of proof", the ship-like trees of the Coal Measures, and a glimpse of the mythical Golden Age among the Carboniferous corals. Miller has become a "fabulous navigator" in more than one sense, using the theatrical form of the dream-vision to navigate the high seas of geological restoration.

Having awakened his audience's imaginations, Miller now invites them to reflect on what they have just heard. Whereas it might be easy to tell, looking back, "where the historic ended and the geologic began", still "no corresponding line" indicated "where the poetry ended and the prose began." Moir, in his dismissal of geology, had been thinking only of "stiff diagrams and hard names", whereas Miller "was luxuriating among the strange wild narratives and richly poetic descriptions of which its pregnant records consist."[65] Finally, before turning to the official subject of the lecture (the Tertiary and Cretaceous periods), he asks, "What is it [. . .] that imparts to Nature its poetry?" The answer includes both Author and reader:

> Nature is a vast tablet, inscribed with signs, each of which has its own significancy, and becomes poetry in the mind when read; and geology is simply the key by which myriads of these signs, hitherto undecipherable, can be unlocked and perused, and thus a new province added to the poetical domain.[66]

This sentence crystallizes the relation of theatricality to geological popularization: in order to demonstrate how geological "signs" can become "poetry" in their audience's "mind" (in Miller's terms, to show why fossils were fascinating), its popularizers unfolded the interpretative visions in their own minds. Since the imagination was often conceived in theatrical terms, these visions assumed theatrical forms.

64. Miller 1859, 83–5.
65. Miller 1859, 85–6.
66. Miller 1859, 86–7.

Yet their significance ran deeper. Miller compares nature's "signs" with the mysterious inscriptions in the Sinai desert, then thought to date back to the Israelites' exodus from Egypt. To the geologist, every rock bears similar "hieroglyphic characters",

> that tell of the Creator's journeyings of old, of the laws which He gave, [. . .] and the marvels which He wrought,—of mute prophecies wrapped up in type and symbol,—of earth gulfs that opened, and of reptiles that flew,—of fiery plagues that devastated on the dry land, and of hosts more numerous than that of Pharaoh, that "sank like lead in the mighty waters;" and [. . .] we [. . .] refer, in asserting the poetry of our science, to the sublime revelations with which they are charged, and the vivid imagery which they conjure up.[67]

That term "revelations" encapsulates the conceptual proximity of Scripture and the "book of nature" when the latter was viewed geologically and theatrically. Sacred history and geohistory, two stately dramas marking "a progress Godwards" and directed by Providence,[68] are here superimposed, their "vivid imagery" mingling in a *martinien* riot of colour. The poetry of geology is like the poetry of Scripture: sublime, prophetic, and infallible. They become allegories of each other, both pointing to the same infinite yet earthly truth: Christ.

## Upstaging Milton

Miller developed this allegorical unity in his most daring piece of geohistorical spectacle, which occupied the final pages of "The Mosaic Vision of Creation". In this lecture from the 1850s, printed in *The Testimony of the Rocks* (1857), he built on recent German and Scottish exegesis to advance a new scheme for harmonizing geology with the text of Genesis 1. He took his cue from "the two great episodes" of Milton's *Paradise Lost*. One was the angel Raphael's report of Creation, as told to Adam; the other was Adam's own vision of future antediluvian history, culminating in the Flood. In the former episode, observes Miller (quoting Addison), the "scenes [. . .] rise up" at Raphael's narration so that "the reader seems present at this wonderful work, and to assist among the choirs of angels who are spectators of it"; whereas the latter is "a series of magnificent pictures, that form and then dissolve before the spectator".[69]

---

67. Miller 1859, 87.

68. Miller 1857, 156.

69. Miller 1857, 157–8. For a modern reprint see Miller 2001. For the "episodes" themselves, see *Paradise Lost* VII and XI.376–901 (Milton 1971, 355–94, 581–608).

Moses's Creation-narrative in Genesis must have been based, says Miller, on one of these two kinds of testimony, narrative report or visual representation, since Moses could not have been present at the event itself. Miller takes Genesis 1 as an example of retrospective prophecy, suggesting that "the scenes of the chapter are prophetic tableaux, each containing a leading phase of the drama of creation". Genesis 1 is thus read as "an exhibition of the actual phenomena of creation presented to the mental eye of the prophet under the ordinary laws of perspective, and truthfully described by him in the simple language of his time."[70] Any apparent conflicts between the content of Genesis and the evidence of geology are to be explained by Moses's earthbound viewpoint: the patriarch described simply what he had seen. Here Miller's useful technique of placing the more uncertain elements of his restorations in the distance, or behind fog, is extended to the patriarch's own imagined experience.[71]

Genesis 1 thus becomes an account of what the phases of Creation might look like if they were viewed by someone who was on close terms with God but was ignorant of modern science. Moses is first shown the primitive earth, but the all-enveloping cloud and steam prevent him from seeing anything but thick darkness. The next scene (the first "day") displays the end of the Primary period, where the fog has lightened enough for Moses to see an expanse of steaming water. He next sees a Silurian scene, by which time atmospheric conditions allow him to distinguish between sea and a cloudy sky. The Carboniferous forests are then revealed, but the sky is still thick with cloud. Only when the New Red Sandstone period dawns do the clouds disperse and the sun and stars become visible: the result is that Genesis records the creation of the heavenly bodies on the fourth day. Then comes the "Age of Reptiles" (*taninim*), followed at last by a Tertiary scene peopled by larger mammals, and by man.[72]

This vision was, we are told, conveyed to Moses in "the successive scenes of a great air-drawn panorama" (which Miller also calls a "drama" and a "diorama"). Each scene was representative of a whole epoch, and over these "shifting pictures the curtain rose and fell six times in succession", replicating the darkened auditorium of the diorama and giving Moses the impression of days and nights.[73] Miller concludes his lecture by taking his audience back in time to show them Moses watching "the

70. Miller 1857, 160–1. On the exegetical context see Haber 1959, 236–40.

71. This means of explaining the appearance of light before the sun had been used by several literalist writers: see, for example, Mary Roberts (1837, 3–10).

72. Miller 1857, 175–84.

73. Miller 1857, 175, 184.

chant teeth, barbed sting, and sharp spine, and enveloped in glittering armour of plate and scale![81]

The narrative ends, like *The Old Red Sandstone,* with Christ's Incarnation. Lucifer realizes that he has been a pawn in a greater narrative, and that the "destiny of all creation" had been "fore-ordained", namely, that its "eternal lord and monarch" should be "God [. . .] in the form of primæval man".[82] In uniting the "two books" in one theatrical story, Miller's assumption of Milton's role seems clear.

Miller was not the only writer wishing to soar beyond Milton. Indeed, the necessary foil for his own Lucifer-epic was provided by one of the poems for which the label "hyper-Miltonic" had been coined, Robert Montgomery's cosmic epic *Satan* (1830).[83] Despite Miller's slighting remarks, Montgomery had his scientific admirers: Henry Christmas, James Rennie, and Thomas Hawkins all quoted him with evident admiration.[84] More recently Montgomery has been called "one of the most spectacularly successful bad poets ever";[85] but if his works can be dismissed as inflations of sacred history, so—in a sense—can those of many popularizers of geology.

Hawkins rewrote Genesis in a particularly hyper-Miltonic manner. The exegetical first chapter of his *Book of the Great Sea-Dragons* concludes with a visionary geohistorical sequence, grafted onto Genesis's post-Adamite episodes of antediluvian warfare and the Deluge. The act of grafting is encapsulated by the first three words:

"In the Beginning," chaotic, void, nascent, shadowy, nerveless, negative: such in the Primary, non-fossiliferous, and purely mechanical Granite, is the interpretation of Nature, and Science, and Reason, and History, and Revelation.

In the instant Coal-Measures, behold, a Vision of the torpid Deep, and the New-born Earth unconscious with Floral Robes.

Then a Vision of Abysmal Waters, swarming with all wondrous creatures of Life, and gelid Swamps with amphibious things, and Dragon Pterodactyles flitting in the hot air with Vampire Wing.

Then Mammoth, and Sivatheria, and a great Company of docile Giants,

81. Miller 1857, 259–61.

82. Miller 1857, 264.

83. Miller 1857, 258. For the term "hyper-Miltonic" see Anon. 1832b, 227, discussed in R. O'Connor 2003b, 227–8. Robert Montgomery is not to be confused with his older, better-known, and more restrained namesake James, though his publishers made capital out of this similarity.

84. Christmas 1850, 3–4, 287; [Rennie] 1828, xi; Hawkins 1834a, 51.

85. N. Parsons 1988, 123.

sublime panorama of creation".[74] The opening gives some flavour of the whole, which included sound effects:

A "great darkness" first falls upon the prophet [. . . .] Unreckoned ages, condensed in the vision into a few brief moments, pass away; the creative voice is again heard, "Let there be light," and straightway a gray diffused light springs up in the east, and, casting its sickly gleam over a cloud-limited expanse of steaming vaporous sea, journeys through the heavens towards the west. One heavy, sunless day is made the representative of myriads; the faint light waxes fainter,—it sinks beneath the dim undefined horizon; the first scene of the drama closes upon the seer; and he sits awhile on his hill-top in darkness, solitary but not sad, in what seems to be a calm and starless night.

The light again brightens,—it is day; and over an expanse of ocean without visible bound the horizon has become wider and sharper of outline than before. There is life in that great sea,—invertebrate, mayhap also ichthyic life; but, from the comparative distance of the point of view occupied by the prophet, only the slow roll of its waves can be discerned [. . . .][75]

The content of each scene replicates that of Miller's other sequences, even at the level of phrasing ("an ocean without visible shore"); but this is a vision within a vision. We see through Moses's eyes as well as through Miller's.

In effect, Miller has followed Milton's example, updating Genesis 1 for his own times. Having argued that Moses's account is a retrospective prophecy, he presents the infallible key: modern geology. In the hands of some other popularizers, geological spectacle had the (often intended) effect of throwing Genesis 1 into the shade. But Miller, like Milton, used science to enrich, rather than to eclipse, the received narrative of sacred history, and thus "justify the ways of God to man".[76] Miltonic undercurrents flow beneath much of *The Testimony of the Rocks.* This particular lecture is strewn with quotations from *Paradise Lost*: not only does Miller use it as a conceptual framework for his project, but it literally frames the lecture, appearing at the beginning and end. He presents his concluding vision as a hint of what might be achieved by a modern Milton: "Such a description of the creative vision of Moses [. . .] would be a task rather for the scientific poet than for the mere practical geologist or sober theologian."[77] Yet many readers considered Miller himself to be the

74. Miller 1857, 191.

75. Miller 1857, 187.

76. Miller 1857, 259, adapting *Paradise Lost* I.26 ("men") (Milton 1971, 44).

77. Miller 1857, 186–7.

foremost "poet of geology", and his "Mosaic Vision" to be "an epic in prose with which few epics in verse can be compared."[78] That Miller presents his stories of Creation in the form of prose summaries, as mere materials for poems, need not suggest a lack of Miltonic ambition on his part.[79]

Elsewhere in the same book Miller proposes another possible updating of *Paradise Lost,* this time a projected poem. In this poem Lucifer, defeated in the primal "war in heaven", would be shown falling from heaven onto the primitive earth, and then witnessing the birth and development of organic life over countless aeons. Such a poem, Miller says, might be written by "a poet of the larger calibre, who to the divine faculty and vision added such a knowledge of geologic science as that which Virgil possessed of the natural history of his time, or as that which Milton possessed of the general learning of *his*".[80] Like his "Mosaic Vision", Miller's six-page resumé of "a *possible* poem" presents us with a vivid present-tense narrative, poetically impressive in its own right. It demonstrates how the narrative of sacred history may "assert Eternal Providence" all the more powerfully when extended geologically. It does so by counterpointing the providential drama of creation with the bafflement, stratagems, apparent victory, and final confusion of Lucifer:

> There has been war among the intelligences of God's spiritual creation. Lucifer, son of the morning, has fallen like fire from heaven; and our present earth, existing as a half-extinguished hell, has received him and his angels. [. . . .] [A]nimal life, to even the profound apprehension of the fallen angel, is an inconceivable idea. Meanwhile, as the scarce reckoned centuries roll by, vacantly and dull, [. . .] the miserable prisoners of our planet become aware that there is a slow change taking place in the condition of their prison-house. Where a low, dark archipelago of islands raise their flat backs over the thermal waters, the heat glows less intensely than of old; [. . . .] The hitherto verdureless land bears the green flush of vegetation; [. . . .] the comprehensive intellect of the great fallen spirit [. . .] would have found ample employment in attempting to fathom the vast mystery, and in vainly asking what these strange things might mean.
>
> [. . . .] With what wild thoughts must that restless and unhappy spirit have wandered amid the tangled mazes of the old carboniferous forests! With what bitter mockeries must he have watched the fierce wars which raged in their sluggish waters, among ravenous creatures horrid with tren-

---

78. Anon. 1841–2, 211; 1857a, 600.
79. Compare Beer 1983, 29–36.
80. Miller 1857, 258.

roaming the Universal Land, crunching the succulent limb of the Forest Tree, or banqueting upon his juicy root: a Vision of peace and rest, pachydermata, and other huge herbivorous animals having with Man undivided dominion over the Euthanasian Earth.

Then a Vision of brute Savages haunting Eldritch Caves: of gaunt Lords of wassail, war, blood, and perdition: Blasted Continents, and withering pines, and briars and thorns: Rebellion, Violence, horrors manifold: Prometheus chained, the Vulture, the Liver: The World at the brink of Death.

Apollo transfixing Python, The booming Flood, driving, rolling, roaring, wrenching, wrecking, whelming the accursed Titans in endless destruction.[86]

Like Montgomery and similar poets, Hawkins tortured his words into distended Miltonic formations, and used words either newly-coined or no longer in use.[87] Like them, he followed Milton in juxtaposing lengthy classical compounds with curt Anglo-Saxonisms (*Euthanasian Earth*) and employing shorter Latinate terms whose sound is as important as their meaning (*gelid*). Nouns frequently function as adjectives (*Vampire Wing*); the syntax, with its consistent lack of main verbs and its Latinate noun-adjective constructions (*horrors manifold*), is knotted and double-knotted; and capital letters run rampant. Besides these Miltonisms, the "hyper-Miltonics"—and especially their later offshoot known as the "Spasmodic School"—overlaid a more personal, rapt strain drawn from Byron and Ossian. Hawkins enthusiastically followed suit. Great visionary lists of images and synonyms tumble down on the reader, with alliterating strings of verbs or adjectives whose linkages seem not so much logical as musical; and inanimate objects are unexpectedly gendered (*his juicy root*). What initially set Hawkins apart from his poetic contemporaries, apart from his palaeontological theme, was that he used these common techniques in prose rather than verse (perhaps imitating Carlyle's *French Revolution*).

It is hard to think of a prose style more distant from Miller's Augustan elegance; yet the underlying concerns show some convergence. Like Miller, Hawkins framed the fossil record as "the Panorama of Ages",[88] and he used similar travel-analogies in viewing it: "we cross the Silent Seas of Time, and explore the Solitary Countries of the Past."[89] But, like the literalists, Hawkins had come to believe by 1840 that death—and

---

86. Hawkins 1840a, 5.
87. For comments on hyper-Miltonic neologisms see Anon. 1830a, 1; Anon. 1832b, 233; Birley 1962, 203–5.
88. Hawkins 1840a, 25.
89. Hawkins 1840a, 20.

hence carnivores—could not have existed before Adam's sin. Unwilling to assent to a literal six-day creation, Hawkins found an alternative explanation. The toothy denizens of the "Age of Reptiles" must have been created by fallen angels, and that era must have been a kind of geological hell: "Emulous, Satan seems to have been thrice seized, generating Horrors commanding, and realizing a teeming Spawn fitted for the lowest Abysm of Chaos."[90]

In later years Hawkins would assert in Manichaean vein that the whole of creation, including humans, was the work of demons, signified by the Hebrew plural אלהים (elōhīm, literally "Gods", as opposed to the true "God");[91] but in 1840 only the saurians came under his *odium theologicum*. God had granted these "ugly Aborigines" an age in which to ravage the primordial oceans, and then he had annihilated them, leaving their fossils as an allegorical sign so that a latter-day prophet of geology (i.e. Hawkins) would interpret them and foretell Satan's final overthrow.[92] Martin's frontispiece for *Sea-Dragons* (Fig. 10.3) thus adds another epoch to that artist's list of divinely-ordained cataclysms in sacred history, drawing on his depiction of the floating Satan in *Paradise Lost* (Fig. 10.4): the saurians are being punished by mutual torment. Hawkins's lurid prose repeatedly depicts these "Hordes of the Spiritual Attila which ruled in the dark Ages before Adam" engaged in melodramatic warfare, "enacting Perdition".[93] The latter phrase captures the role which Hawkins's monsters perform on the "crimson Stage" of deep time.[94] Five years earlier he had presented saurians as minor characters in that part of the "gorgeous antediluvian drama" left untold by Moses and Milton;[95] he now saw them as "carnivorous automata", simulacra of Satan;[96] and their terrestrial stage was, to use Burnet's phrase, "a lively representation of *Hell* it self".[97]

Hawkins's geo-demonology may seem bizarre, but in view of Miller's proposed rewriting of *Paradise Lost,* staging the ancient earth as Lucifer's domain, it is hardly unexpected. Of course, Miller's "poem" was not meant to narrate literal truth, whereas Hawkins seems to have really believed in his tales of ancient devilry. Yet the boundary between truth and fable was not so simple, least of all when Milton was hovering in the background. Miller speculated in his Bass Rock essay that the angels'

90. Hawkins 1840a, 22; see R. O'Connor 2003b, 235–8.

91. Hawkins 1887, 71–6.

92. Hawkins 1840a, 26.

93. Hawkins 1840a, 25, 27.

94. Hawkins 1840a, 18.

95. Hawkins 1834b, viii. This edition of *Memoirs of Ichthyosauri* was printed in 1835, despite the date on the title-page.

96. Hawkins 1840a, 25.

97. Burnet 1965, 305.

*Figure 10.3.* "The Sea-Dragons as They Lived", John Martin's mezzotint engraving for Thomas Hawkins's *Book of the Great Sea-Dragons.* Hawkins 1840a, frontispiece.

*Figure 10.4.* Satan on the burning lake of hell. This mezzotint, engraved by Martin in 1825 for the illustrated *Paradise Lost,* represents Milton's description in Book I (Milton 1971, 55–7) comparing Satan to the huge and treacherous sea-monster Leviathan. Note here the little monsters swimming on the right. Milton 1853, facing p. 7.

revolt might have "exerted a malign influence on the pre-Adamite ages of suffering, violence, and death", hinting that saurian ferocity might indeed have demonic origins. His cosmic speculations were not limited to the earth: he also suggested that Christ's Crucifixion might have somehow saved the rational inhabitants of other inhabited worlds, or alternatively that other planets, populated only by plants and animals, might still be awaiting their destined lords in the form of blest humans.[98]

Forward movement, a "progress Godwards", underpins these narrative speculations. In his projected Lucifer-epic, Miller's primordial earth is described as a "half-extinguished hell"—but a hell progressing towards heaven.[99] In his final *Sketch-Book* lecture he describes the lifeless Primary landscapes in recognizably Miltonic terms:

> I dare not speak of the scenery of the period. We may imagine, however, a dark atmosphere of steam and vapour, which for age after age conceals the face of the sun [. . .]; oceans of thermal water heated in a thousand centres to the boiling point; low half-molten islands, dim through the fog, and scarce more fixed than the waves themselves [. . .]; roaring geysers [. . .]; and, in the dim outskirts of the scene, the red gleam of fire, shot forth from yawning cracks and deep chasms, and that bears aloft fragments of molten rock and clouds of ashes.[100]

He then calls our infant planet "a solitary hell, without suffering or sin"—words which might seem to dissociate the scene from theology, using the term "hell" to denote a comet or other inhospitable planetary body.[101] Yet the Miltonic imagery rekindles Miltonic narrative patterns, as do the twelve powerful lines with which the lecture ends:

> But should we continue to linger amid a scene so featureless and wild, or venture adown some yawning opening into the abyss beneath, where all is fiery and yet dark, [. . .]—we would do well to commit ourselves to the guidance of a living poet of true faculty,—Thomas Aird,—and see with his eyes, and describe in his verse:—
>
> > "The awful walls of shadows round might dusky mountains seem,
> > But never holy light hath touched an outline with its gleam;
> > 'Tis but the eye's bewildered sense that fain would rest on form,
> > And make night's thick blind presence to created shapes conform.

98. Miller 1848, [101]; 1847b, 337–8.
99. Compare "Christian Philosopher" 1824, 42–4.
100. Miller 1859, 238–9.
101. OED s.v. *hell* 4b.

No stone is moved on mountain here by creeping creature cross'd,
No lonely harper comes to harp upon this fiery coast;
Here all is solemn idleness; no music here, no jars,
Where silence guards the coast ere thrill her everlasting bars;
No sun here shines on wanton isles; but o'er the burning sheet
A rim of restless halo shakes, which marks the internal heat;
As in the days of beauteous earth we see, with dazzled sight,
The red and setting sun o'erflow with rings of welling light." [102]

These sonorous heptameters, originally penned to describe the hell of Protestant theology, come from Thomas Aird's "colossal fragment" *The Devil's Dream on Mount Aksbeck* (1839), compared by some contemporaries with *Paradise Lost* and Dante's *Inferno*.[103] In creative hands the concept of a physical hell could hardly avoid erupting into theological life. When, in Byron's dream-vision "Darkness" (1816) and John Edmund Reade's suspiciously similar "Darkness" (1829), the earth "wander[s] darkling" from the sun and becomes a kind of comet, its desperate inhabitants enact scenes of damnation, grimacing at each other in mutual torment like Martin's sea-dragons.[104] In the geological imagination, hellish associations were reinforced by the prevailing metaphor of the underworld quest, cast by Miller in Dantean terms as we "commit ourselves to the guidance" of the "poet";[105] and many Christians followed Dante in believing hell to be literally at the centre of the earth.[106] The so-called nineteenth-century decline of hell may have drained the underworld of much of its doctrinal significance, but it retained its atavistic potency in visions of the deep past.

In *A Tour through the Northern Counties of Scotland,* Miller's speculations came still closer to the Miltonic dramaturgy of his projected Lucifer-epic. Here he concludes a "vision" of an Oolitic landscape in which

the gigantic iguanodon stretched his long length of eighty feet in the sand. But who shall reveal the higher history of the time? The reign of war and death had commenced long before; and who shall assert that moral evil had not long before cast its blighting shadows over the universe,—that there had not been that war in heaven in which the uncreated angel had overthrown the dragon,—or that unhappy intelligences did not wander,

102. Miller 1859, 239, quoting from Aird, *The Devil's Dream on Mount Aksbeck* (1839), printed in Aird 1878, 40-52 (p. 48).

103. Aird 1878, xxv. On Aird see *ODNB*.

104. Byron 1986, 272–3; [Reade] 1829, 171–8. Compare *Manfred* I.i.43–6, 110-31 (Byron 1986, 276–8). On Byronic images of the end of the world, see F. Stafford 1994, 160-288; Paley 1999, 193–219. See also Neswald 2006.

105. On this topos see Sommer 2003.

106. For example, "Christian Philosopher" 1824, 4, 11.

"seeking rest, but finding none," in an earth of "waste places," whose future sovereign still lay hid in the deep purposes of Eternity?[107]

The Satan-epithet "dragon" reflects back onto Miller's Miltonic image of a recumbent, Leviathan-like Iguanodon. He nowhere suggests that the saurians were of Satanic origin, but their world is discreetly demonized. This procedure differed from Hawkins's in degree rather than kind. For all geological writers, extinct animals and landscapes represented, in their very remoteness, sites at which the material and spirit worlds overlapped. In Hawkins's *Sea-Dragons* these overlap entirely; in Miller's speculations a subtler alchemy takes place, and these vanished scenes occupy a poetic realm between history and theology.

In *First Impressions of England,* Miller drew attention to his habit of "letting the mind loose to expatiate on those historic periods to which the [fossil] record so graphically refers". The writer describing "extinct states" necessarily

> delight[s] in the metaphor and the simile,—in pictures of the past and dreams of the future,—in short, in whatever introduces amid one set of figures palpable to the senses, another visible but to the imagination, and thus blends the ideal with the actual, like some fanciful allegorist, sculptor, or painter, who mixes up with his groupes of real personages, qualities and dispositions embodied in human form,—angelic virtues with wings growing out of their shoulders, and brutal vices furnished with tails and claws.[108]

It sounds like a recipe for creating monsters after the manner of Hieronymus Bosch, and Hawkins's saurians certainly qualify as "brutal vices" of this kind. In fact, as James Paradis has shown, Miller was simply defending his freedom to employ poetic language when describing past scenery, and to superimpose further layers of meaning on his descriptions.[109] Yet, as we have seen, he often exercised this freedom in loose allegory, concerned as he was to yoke all histories—sacred, secular, geological—to a single Christian narrative of progress. These histories were made to run in a single course, causing a typological interference of imagery and meaning.

Such interference was inevitable wherever geological writers allowed both Providence and poetry a role in their popularizations. Even when sacred history was ostensibly ignored, the commanding narrative form

107. Miller 1859, 250.
108. Miller 1847b, ix–x.
109. Paradis 1996, 123–5.

fighting", or else by implacable hate.[119] Literalists like Andrew Ure called these creatures "the victim[s] of sin", observing that they suffered extinction for the sins of Cain's "brood";[120] old-earth geologists' occasional verses, by contrast, suggested that they were prototypes of Cain himself in their "fratricidal" cannibalism, and that they brought destruction upon themselves.[121] But however much their depravity was disapproved of, it was usually constructed as amoral rather than immoral. Miller suggested that God had reduced the reptiles to their present size in order to allow for the introduction of large mammals, with whom any surviving dinosaurs would have engaged in "horrid, exterminating war", making the earth "a place of torment".[122] Pictures like Martin's presented saurians as embodiments of primordial depravity, prefiguring the destructions of Martin's more culpable human and superhuman subjects.

These images conflicted radically with the prevailing doctrine of natural theology to which most of these writers adhered, according to which these apparently monstrous creatures had in fact been perfectly designed by God.[123] Far from being freaks of nature doomed to a life of violence, they were claimed to have enjoyed the happiest of possible worlds in "the great drama of universal life", since a quick death by sharp teeth was the least unpleasant solution to overcrowding.[124] If contradiction seemed to loom here, most of its adherents sidestepped difficulty by employing images of warfare and monstrosity only metaphorically. The tension, however, remained, and Hawkins and Miller were unusual in grabbing the bull by both horns. Hawkins opted for a kind of "natural demonology", while Miller developed a complex doctrine of "degradation" by which humanity's fallen state (and, in some cases, future damnation) was typologically prefigured in the geological evidence for the gradual decline and "deformity" of each class of life-form after its initial creation in a perfect, "monarchical" form.[125] More commonly, though, the ambiguity was left unresolved. Even for writers who did not engage philosophically with the concept of monstrosity, the tension between positive and negative images of the ancient earth remained potent. Buckland's *Geology and Mineralogy* made the strata into a vast cathedral of natural

---

119. Mill [*c.* 1855], 52; Hawkins 1840a.

120. Ure 1829, 349; compare J. Brown 1838, 41.

121. Daubeny 1869, 72–3, 94, 119–20; R. O'Connor 2006a. On the myth of Cain as the progenitor of monsters see D. Williams 1982, 19–39.

122. Miller 1849, 296.

123. W. Buckland 1836, I, 214, 221–2, 233; see also Bell 1834, 35.

124. W. Buckland 1836, I, 129–34. For discussion see Robson 1990, 104–5.

125. Miller 1849, 297–301. On the aesthetic and palaeontological dimensions of this doctrine see Brooke 1996; Henry 1996; Janvier 2003.

theology, but within some of its niches, gargoyle-like, lurked Miltonic "fiends" and Spenserian "dragons".[126]

Curiosity-value besides, this monstrosity was typically constructed not only as a quality in itself, but equally as a quality which strikes *us* as we gaze in awe upon an intractably primitive world, a "turbulent" and "infant" planet not yet prepared for its future "lords". If transported there, we should be appalled at the monstrosity of its tyrants, and we should either die in the noxious atmosphere which sustained them or be instantly eaten.[127] This awareness sharpens our sense of the distance between them and us. Their monstrous qualities are a function of their radical pastness, reinforced by persistent analogies with *anciens régimes* and crumbling "dynasties". The implication is that civilized humans have legitimately inherited the land once tyrannized by primeval monsters.[128] This underlying mythic pattern runs right through Western culture, from mediaeval legends and heroic sagas to Victorian fantasies of imperial exploration; the admixture of evolutionary theory later in the century would only reinforce the sense of rightful inheritance by providing genealogical links for the pageant's separate scenes.

This onward and upward march rested on the dramatic forms of (sacred-) historical narrative.[129] But such narrative could serve many purposes depending on the identity of the goal. For Miller, the goal was the kingdom of heaven, and hell itself could be imagined as an early stage in the spiritual evolution of the physical world. In similar vein, the late-Victorian woman of letters Frances Power Cobbe would paint a demonic picture of our bestial ancestors, in direct opposition to Milton:

> our first human parents, far from resembling Milton's glorious couple, were hideous beings covered with hair, with pointed and movable ears, beards, tusks, and tails,—the very Devils of mediæval fancy.[130]

When combined with the view that some present-day peoples were primitive relics of a past order, this progressionist narrative from hell to heaven could play a more aggressive role, legitimizing the forcible dethronement and displacement of Australian aborigines and native Americans by civilized white men. In mid-nineteenth-century America, for instance, both the pageant of geohistory and panoramic paintings of

---

126. W. Buckland 1836, I, 74–5, 224–5.
127. W. Buckland 1836, I, 74–5; Broderip 1848, 326; Francis 1839, 158.
128. Compare the Classical accounts cited in Mayor 2000.
129. Miller 1849, 282–3.
130. Cobbe 1872, 3–4.

phasizing the moral distance between the ancient earth and the present world of which man is the rightful "lord".[140] This is the gulf that Tennyson shockingly elides. In similar vein, and around the same time, Miller remarked that unregenerate man often appeared barely more civilized than the saurians, and no less liable to "oppress and devour his fellowmen".[141] But all these critiques, Conrad's included, took for granted the savagery of the ancient earth and the duty of humans (whatever their ancestry) to transcend their baser instincts. Humans belonged not with the monsters, but with the angels. In its basic form, the Victorian panorama of earth history celebrated the exaltation of Man over an indefinitely extended and visible natural world prepared for him alone.

Looking was not a neutral act. The panorama of the past did more than just celebrate human dominion over nature; it *enacted* this dominion by laying out all nature, past and present, beneath the human gaze. Unearthing and restoring fossils was perceived as a heroic act of conquest and colonization—shackling the monster—and by the 1830s this topos was familiar enough to be mocked (Fig. 10.5).[142] Man's angelic destiny was inscribed into the pageant of geohistory, by its panoramic breadth on the one hand and by the hellish otherness of the worlds viewed on the other. Monsters, as the word's etymology suggests, were made chiefly for looking at. The pageant of earth history was directly comparable to the celebration of the British Empire's remoter acquisitions in urban show-culture, with its commanding views separating empowered colonial viewer from exotic subject.[143]

By 1848 Robert Hunt could write that "Geology teaches us to regard our position upon the earth as one far in advance of all former creations. It bids us look back through the enormous vista of time".[144] Hunt then develops the theatrical significance of the words "regard" and "look", urging us to assert our "strength" in visual terms. He invokes an older version of the theatre metaphor even as he promotes the new:

When man places himself in contrast with the Intelligences beyond him, he feels his weakness; [. . . .] But looking on the past, surveying the progress of matter through the inorganic forms up to the higher organizations, until at length man stands revealed as the chief figure in the foreground of the picture, [. . .] the absolute ruler of all things around him, he rises like a giant in the conscious strength of his far-reaching mind. That so great, so

140. Anon. 1834e, 358; Broderip 1848, 379. On the slime motif see Shatto 1976, 148–9.
141. Miller 1858, 88–9; see also 26–7.
142. On the conquest motif see J. Secord 1982; Rudwick 1992, 244; Sommer 2003, 193–4.
143. T. Mitchell 1992; Leask 1998a; 2002, 226–7; Aguirre 2002.
144. R. Hunt 1854, 352.

*Figure 10.5.* "Dr M. in extasies at the approach of his pet Saurian", a sketch by Henry De la Beche celebrating Mantell's acquisition in 1834 of the "Maidstone Iguanodon", the best-preserved specimen of the time. The laurel-crowned geologist has tamed the monster, which now wags its tail like a dog. Ink on paper, 229 × 184 mm, Alexander Turnbull Library, Wellington, New Zealand, E-295-q-003.

noble a being, should suffer himself to be degraded by the sensualities of life to a level with the creeping things [. . .] is a lamentable spectacle, over which angels must weep.[145]

If man does not behave like an angel, he will become the theatrical object of the pitying angelic gaze, reduced to the "sensualit[y]" of the "creeping things" far below and far back, on which he is meant to assert his own "ruling" gaze. Stationed at this panoramic viewpoint, far from dissolving in the wastes of time, man anticipates his angelic destiny by "deciphering [. . .] the *ideas* of God".[146] In Hawkins's words, "we navigate the pre-Adamite Seas to their very Margin, and look over the Edge of Matter into Chaos, lifting the Veil of Isis, metamorphosed as gods."[147]

Here, as in Davy's vision of the tentacled philosophers of Jupiter, the work of science merged imperceptibly with speculations regarding a future state and the possibilities of extraterrestrial intelligence. The

145. R. Hunt 1854, 353.
146. Miller 1854, 508; 1993, 506–7.
147. Hawkins 1840a, 9.

evangelical image of science as visible and comprehensible took on eschatological force in the title-page epigraph of Stephen Fullom's *Marvels of Science*: "THERE IS NOTHING HIDDEN THAT SHALL NOT BE KNOWN."[148] Perfect knowledge would be attained only in heaven, where (as Thomas Dick suggested) Christ and his angels might offer lecture-courses and comets might be used as vehicles for celestial guided tours;[149] but geology's marvellous capacity for resurrection made it a good starting-point in this world. As Fullom saw it, geology proved "that man has been ordained, from the beginning, to be the spectator and expounder of the works of God" and that "ultimately all things will be placed under his feet."[150]

To the same effect, and by analogy with fossil evidence, the Presbyterian geologist Edward Hitchcock developed a theory of the "Telegraphic System of the Universe", an epistemological extrapolation of Newtonian mechanics according to which all phenomena in the universe—past, present, and future—were the effects of definite causes, indissolubly connected in "a tremulous mass of jelly". Exalted and supernaturally sensitive beings, such as humans in a future blessed state, would be able to perceive, by every minute movement in this web, everything that has ever happened. Such beings would witness a "panorama of the indefinite past", and the whole universe would become one "vast picture gallery".[151] Hitchcock transformed Mantell's fictive device of an extraterrestrial observer into potential reality, suggesting that "there may be in the universe created beings, with powers of vision acute enough" to view the ancient earth simply by taking up a viewpoint distant enough for the speed of light to take millions of years to reach them: "Suppose such a being at this moment upon a star of the twelfth magnitude, with an eye turned toward the earth. He might see the deluge of Noah, just sweeping over the surface." Dick, for his part, proposed that nebulae might serve as colossal magnifying lenses for these angelic astronomers.[152]

In Miller's writing the act of looking on the past is seen to impel the gaze towards the future, to the imminent end of this world, and to the individual's urgent need for repentance. In *First Impressions of England*, Miller presents the geologist viewing a metaphorical sea-scene, his eye

---

148. Fullom 1854, title-page.

149. Astore 2001, 56–7, 77. Compare Davy 1851, 52–64; Hitchcock 1851, 324–9.

150. Fullom 1854, 63, 85–6. This image of geology reflects contemporary natural theologies of vision identified by Lightman (2000, 656): "The universe was transparent to humans, and God intended it to be that way". Compare also Linnaeus's use of the "earth as theatre" metaphor (Brooke 1991, 197).

151. Hitchcock 1851, 337, 353.

152. Hitchcock 1851, 337. For Dick's suggestion see Astore 2001, 89. The astronomical popularizer Richard Proctor (1870, 295–324) later developed similar astronomical speculations in order to render intelligible the concept of God's omniscience. Compare Whewell 1852, 14–15.

"purged and strengthened by the euphrasy of science"[153] as he sees "the many vast regions of other creations" stretching out before him. Each succeeding "region" is "roughened with graves", prompting the question: "What mean the peculiar place and standing of our species in the great geologic week?" The answer, with which Miller's chapter concludes, reveals that to gaze upon the past is to enact our fundamental distinction from the creatures that preceded us. So forceful is the narrative direction of the pageant viewed, and so visible its happy ending, that we can but watch and wait:

> all these unreasoning creatures of the bygone periods [. . . .] perished ignorant of the past, and unanticipative of the future [. . . .] But not such the character of the last born of God's creatures,—the babe that came into being late on the Saturday evening, and that now whines and murmurs away its time of extreme infancy during the sober hours of preparation for the morrow. Already have the quick eyes of the child looked abroad upon all the past, and already has it noted why the passing time should be a time of sedulous diligence and expectancy. The work-day week draws fast to its close, and to-morrow is the Sabbath![154]

Spectacle takes on a significance beyond entertainment: the life of this world ends with the geologically brief appearance of its quick-eyed infant lord, whose act of gazing upon the past both embodies its higher destiny and ensures that it will achieve that destiny. De Quincey's drowning girl comes to mind, the drama of her life ending (or so it seems) by suddenly making itself panoramically visible as a "mighty theatre [. . .] within her brain";[155] but Miller's theatre was not just a diverting show. Views of the distant past, rich in pictorial immediacy, were a mandate for present and immediate action.

On one level, then, the monstrous colouring of the ancient earth in these writings was designed to attract attention and excite interest from an audience attuned to weird and wonderful spectacles. But on another level this characterization was a logical consequence of the kind of story geologists were trying to tell—a story of upward progress, and a story which was pre-eminently visible. In both religious and secular versions (and the many versions in between), the underlying thrust of sacred history not only determined the direction of the story, but also informed its visual immediacy and helped to shape the conventional imagery of

---

153. This alludes to *Paradise Lost* XI.414 (Milton 1971, 585), where Michael "purges" Adam's eyes with "euphrasy and rue" in order to open his eyes to the panorama of future history.

154. Miller 1847b, 341–2.

155. De Quincey 1985, 145.

its scenery and characters, even when it was being subverted. By embodying the conquest of brute nature, exotic nations, and ancient worlds, spectacle offered pleasures beyond mere aesthetic gratification. Looked at another way, aesthetic gratification showed itself to have deeper roots than the mere appreciation of formal qualities.

432
~
433

This self-perpetuating fantasy of total visibility, with which earth-historians awakened and exercised the Victorian mind's eye, lives on today. It continues to be played out in restorations of the prehistoric world, on our screens and in our museums, which seek to persuade us that we really are looking at that world, and which claim to turn fantasy into reality. New palaeontological knowledge and new imaging technologies may have made extinct animals more visible for today's general public; but they have not made them any less distant from us. Until someone invents a real time-machine, poetry and fiction will continue to shape the depths of the past.

# Epilogue:

## *New Mythologies of the Ancient Earth*

### The Queen of the Sciences?

Our story ends in suicide and dementia. In his last months, his health under massive strain from lung disease and overwork, Miller had been plagued by vivid nightmares and a terrible fear that his mind was giving way. On the night before Christmas Eve 1856 he suddenly awoke, penned a hasty note to his wife and children, took up the loaded revolver which lay by his bedside, and shot himself in the chest. Miller's extraordinary literary output, which had done more than any other to conjure up phantasmal visions of the monsters of a former world, was cut short by a monstrous dream. His funeral in Edinburgh was one of the largest that city had ever known.[1]

Buckland had died earlier that year in circumstances hardly less unfortunate, following several years of progressive dementia in which he had

1. On Miller's death see Porter 1996; Sutherland and McKenzie Johnston 2002, 104–11; and especially Taylor 2007.

taken no interest in science and barely recognized those around him.[2] Mantell had died by his own hand back in 1852, taking an overdose of opiates in a last attempt to still the agonizing neuralgia caused by severe damage to his spine many years earlier.[3] George Richardson, driven to despair by bankruptcy, had cut his own throat in 1848. Hawkins was becoming increasingly violent and unstable, obsessed with the dishonour he felt his fossil dragons had suffered at the hands of the British Museum, and now produced thunderous epic poems instead of fossil books. Those placid Scotsmen Lyell and Chambers, by contrast, continued to revise their respective masterpieces for new editions, and Lyell was shortly to turn his hand to a rather different field of inquiry in his last book, *Geological Evidences of the Antiquity of Man* (1863).

New visions of the old reptile tyrants had stopped flowing from these prolific pens. Yet their books continued to be printed and reprinted, copied and excerpted, read and reread—in Miller's case, reprinted for more than half a century after his death. And of course new books and articles poured from the pens of a younger generation of writers: the literary career of British earth history did not end in 1856. What is remarkable, however, is the extent to which later representations of the ancient earth reproduced earlier narrative forms and styles, even after the new harvest of North American dinosaurs greatly expanded the repertoire of possible monsters in the late nineteenth century, and after new evolutionary theories revolutionized the perceived relationship between humans and extinct beasts.

Take Henry Hutchinson's *Extinct Monsters* (1892), one of the first British books to popularize the new American fossils. Its introduction begins by arguing that "Truth is stranger than fiction", and that "this panorama of scenes that have for ever passed away" is "a veritable fairy-land". Hutchinson proposes to present his extinct monsters to our imaginations "as they really were when alive", using "the testimony of the rocks beneath our feet" to "conjure up scenes of the past, and paint them as on a moving diorama."[4] These are familiar patterns, and indeed the book is littered with quotations and paraphrases from Buckland, Miller, Mantell, and Hawkins, even to the extent of reproducing their poetry quotations as chapter-epigraphs. Hutchinson's self-promoting announcement of novelty ("Few popular writers have attempted to depict, as on a canvas,

2. On Buckland's dementia see F. Buckland 1858, who suggested that it resulted from complications following a fall (see also Boylan 1984, 265–6). Cadbury (2000, 284–5) tells a different story, according to which Buckland went mad because he could no longer bridge the "unbridgeable gap" between his religion and "the relentless onslaught of new evidence" from geology. There is no evidence to support this myth, which has also been perpetuated with regard to Miller's death (e.g. Barber 1980, 238; Macmillan 2000, 212).

3. Mantell's illness is discussed in Dean 1999 and Cadbury 2000.

4. Hutchinson 1892, 1–5. On Hutchinson see *DNBS*.

the great earth-drama that has, from age to age, been enacted on the terrestrial stage")[5] must not be taken literally. It obscures not only his many early-Victorian sources but also the intervening work of writers such as Arabella Buckley, whose account of earth history in *Winners in Life's Race* (1883) begins by inviting readers to "allow the shifting scene to pass like a panorama before us", and manages scene-changes with a familiar device: "When we look again, 'a change has come o'er the spirit of the dream'".[6]

These late-Victorian replications of an earlier, non-evolutionary narrative of earth history tend to be missed when such writers are seen purely in the context of their own times, and especially when Charles Darwin is taken as the main narrative point of reference. Just as Bernard Lightman seems to have taken Hutchinson's claim of originality at face value, Barbara Gates, in her ground-breaking study of women science-writers, has called Buckley's story of nature a "brilliantly original" way of handling "narrative problems that had also faced Darwin".[7] Although both Hutchinson and Buckley were indeed highly original in other ways, the forms of their geohistorical narratives were lifted bodily from the early Victorians, whose dynamic reworkings of still older story-patterns had set the terms for subsequent retellings of earth history.

Since the Victorian period, pictures have increasingly taken over the explanatory burden from words in restorations of the ancient earth, but textual restorations have continued to employ the same structures. Virtual panoramas, time-travel tours, sublime views, and a dramatic or pageant-like schema hinting at larger questions of human destiny may be found, not only in the seminal prehistory books of Josef Augusta and Zdeněk Burian, but also in present-day palaeontological writing, from Simon Conway Morris's *Crucible of Creation* to the children's book *Dinosaurs of Eden* by the young-earth creationist Ken Ham.[8] The replacement of the dream-vision by the time machine may introduce a new vein of science-fiction realism (and Conway Morris's nifty submersible is a very different vehicle to Ham's "Bible Time-Gate"), but the narrative patterns remain remarkably constant. Museums and television programmes maintain these patterns to a surprising extent: for all its technological innovation, the pterosaur episode of Tim Haines's series *Walking with Dinosaurs* (BBC, 1999) ends with a tragic tableau which all but replicates Martin's apocalyptic engraving in Hawkins's *Book of the Great Sea-Dragons*.[9]

5. Hutchinson 1892, x.
6. Buckley 1927, 334, 211.
7. Lightman 1999, 1–3; Gates 1998, 58–9 (also in Gates 1993, 299–301; 1997, 165–7).
8. Conway Morris 1998; Ham 2001.
9. R. O'Connor 2003b, 239. A variant version of the pterosaur's tale is told in Haines 1999, 158–97.

This continuity may seem surprising given the huge theoretical gulfs between early-Victorian progressionism, late-Victorian evolutionism, present-day Darwinism, and present-day young-earth creationism. Can the same church really serve different faiths? When, late in the sixth century A.D., Pope Gregory the Great sent missionaries to Anglo-Saxon England, he advised them not to destroy the old pagan shrines, but to consecrate and adapt them for Christian use, turning familiar sites to new ends.[10] Likewise, Mantell, Miller, and Buckland (among others) built a sturdy edifice whose stonework was redecorated rather than re-designed by later popularizers, and which served the purposes of many scientific (and religious) creeds.[11] The content shifted, the moral of the story changed, but the narrative techniques remained largely fixed. This was not because later writers were any less able than their predecessors, but because there was little need for them to seek new solutions to an old narrative problem which had already been solved so effectively. Some later popularizers even went back to the eighteenth-century models on which the early Victorians had themselves built. Henry Knipe's lavishly illustrated evolutionary epic *Nebula to Man* was cast entirely in heroic couplets, in imitation of Erasmus Darwin:

> But dinosaurs abound that peaceful roam,
> With vegetarian tastes. Well armed are some,
> Though meek disposed. See Polacanthus here,
> With dorsal shield, and rows of spiky gear.[12]

And this while James Joyce was writing *Dubliners*. Knipe's poem—and scientific literature generally—radically undercuts the old-fashioned view that literary history moves in a single, stately pageant from one stylistic "period" to another.

After the 1850s the focus for literary innovation in popular-science writing turned elsewhere, such as to the relatively new problem of representing the mechanisms and purposes of evolution, a challenge which spawned some striking literary innovations by Charles Darwin, Thomas Huxley, Charles Kingsley, and Buckley herself.[13] For new directions in the representation of earth history, we must turn to fiction—to the new

---

10. Niles 1991, 129–30.

11. Later popularizers using old narrative frameworks included Thomas Huxley, W. S. Ormerod, Archibald Geikie, Agnes Giberne, J. William Dawson, J. E. Taylor, C. G. Bonney, and E. Ray Lankester. Compare Landau's observations (1991, 16) on the folktale structure of evolutionary theory.

12. Knipe 1905, 88.

13. Beer 1983; Lightman 1997b, 197–9; Gates 1998, 50–64.

genre of the scientific romance, as pioneered by Jules Verne, H. G. Wells, and Arthur Conan Doyle, to its disturbing transformations by Conrad, and to the unexpected intrusions of extinct monsters into the work of Walt Whitman, Thomas Hardy, and Virginia Woolf.[14] Yet even here the novelty lay in the way earth history was used to reflect on human concerns, rather than in the representation of earth history itself, for which novelists leaned heavily on the popular-science writing of their day.[15] The old narratives were enclosed within new structures, where they played roles their early-Victorian creators would not have dreamt of. Or would they? Some of the narrative tropes associated with literary Modernism find striking analogues in early-Victorian geological literature: the visionary juxtaposition of civilized present with primeval past, the reliving of the past in fragmentary moments of heightened consciousness, the collapsing of time and space, the extinct monster as emblem of technological modernity.[16] Science writers may have played a more central role in canonical literary movements than the standard histories, bent on excluding most non-fiction, would have us believe.

However, the historical significance of early-Victorian geological writing goes beyond questions of literary influence or of how the ancient earth was imagined. Geology became so popular among the British middle classes that it came to be seen as the paradigmatic science, the yardstick by which other disciplines measured their scientific rigour and imaginative power.[17] It is well known that Darwin's theory of evolution by natural selection was presented as a biological application of geological principles (specifically Lyell's);[18] but further-flung disciplines were also in thrall to geology's siren song. Astronomy, traditionally the most exalted of the sciences, was promoted in geological terms by its foremost late-Victorian popularizer, Richard Proctor. On opening Proctor's best-known work, *Other Worlds than Ours* (1870), we might think we have stumbled upon a geology book:

> Geology teaches us of days when this earth was peopled with strange creatures such as now are not found upon the surface [. . . .] Strange forms of vegetation clothe the scene which the mind's eye dwells upon. The air is heavily laden with moisture [. . .]; hideous reptiles crawl over their slimy

14. On the scientific romance see Stableford 2003.

15. On some of these debts see Beer 1983 (on Hardy); Beer 1992 (on Woolf); Ian Duncan 1995, ix (on Doyle); Breyer and Butcher 2003 (on Verne).

16. On this last trope see W. Mitchell 1998.

17. On the concept of a hierarchy of truth among Victorian sciences, see Cannon 1978, 272–4.

18. J. Secord 1997, xxxvi; Herbert 2005.

domain, battling with each other [. . .] weird monsters pursue their prey amid the ocean depths [. . . .]

Proctor draws the curtain on this vision, ushering in his own subject with the words "Astronomy has a kindred charm."[19]

Three years later, W. K. Sullivan wrote in the opening paragraph of his introduction to ancient Irish history and philology:

The great stone book, in which are written the annals of the globe, [. . .] has been carefully examined; and from those annals science has enabled us to construct the geography of our globe at the dawn of its time [. . .] and people it with birds and beasts. We may, as it were, walk by the shores of its ancient sea, follow the tracks of the marine animals that crawled upon its sands [. . . .]

The second paragraph introduces his chosen subject in geological terms:

Adopting the spirit and method of physical science, the sounds of dead languages [. . .] have been reawakened; we can rebuild the fallen cities, people them anew with their ancient inhabitants, [. . .] and partake in imagination of their daily life.[20]

Ethnographers and folklorists were staking claims to the scientific status of their discipline in a similar manner,[21] and Frances Power Cobbe described the comparative mythologist and anthropologist James Fergusson to the readers of *Fraser's Magazine* in 1869 as "the Murchison of a new Siluria" for having "traced out and described a buried world".[22] Add to this the fact that the early history of humanity was now being pushed back into the geologists' territory, that Carlyle conceived of the historian's task as a form of palaeontology, that Leslie Stephen wanted to make literary criticism as scientific as "ticketing a fossil in a museum", and that psychological models were increasingly taking on stratigraphic forms—and it becomes possible to call geology the "queen of the sciences" in late-Victorian Britain.[23]

In the late eighteenth and early nineteenth centuries, geology had

19. Proctor 1870, 1–2. On Proctor's work see Lightman 1997b, 199–200; 2000, 661–71; *DNBS*.

20. Sullivan 1873, i–ii.

21. G. Bennett 1994.

22. Cobbe 1872, 179.

23. On Palaeolithic research, see A. O'Connor 2007 in press; on Carlyle, see Ulrich 2006; on Stephen, see Beer 1990, 785. The influence of positivist science on late-Victorian historical research is discussed in Heyck 1982, 120–54.

been very much the poor relation, developing its disciplinary integrity by borrowing the tools and conceptual models of history and philology, the mystique of antiquarianism, and the prestige of astronomy. Only a few decades on, the tables were turned: these same disciplines (or their newly "scientific" descendants), and many others, were now borrowing the "spirit and method" of geology. Fossils were now far more than petrified organic remains: they were the very language of the earth, talismans enabling its crust to be ordered and exploited and its lost worlds recovered. For many onlookers in the second half of the century, beset by a sense that time itself was speeding up and increasingly uncertain about the future of the human race, fossils represented the supreme symbols of science's power to control brute nature and reveal the vanished past: to know and to show.[24]

How did geology rise to this position of authority? On one level, simply because it worked. By mid-century several national geological surveys had mapped the structure of large tracts of the earth's crust, an achievement which compelled wonder in its own right. But the science's power to make the past spectacularly visible had been constructed by its popularizers, above all in its literature. During the first half of the nineteenth century, writers associated with the Geological Society's redefinition of geology fashioned new ways of narrating earth history from the ruins of exploded cosmologies. To begin with, these stories were told at the peripheries of the science, in privately-circulated comic verses; but, as the conflict between rival theories became more apparent, writers of both literalist and non-literalist persuasions increasingly courted the burgeoning middle-class reading public. Using the sites and genres that this public was already familiar with, they developed a rhetoric of spectacular display which dovetailed the theatrical tropes of eighteenth-century theories of the earth with the new vogue for virtual tourism and apocalyptic drama, mediated by the culturally authoritative voice of the poet. Geology's truth-claims were built on an uneasy yet highly effective patchwork of image and text, prose and poetry, history and prophecy, fact and fiction. The reader's imagination became a stage, equipped with all the latest technological innovations, on which the invisible was made visible. Binding archaic mythic patterns with the sinews of sacred history, these writers offered their readers a God's-eye view of the process of Creation itself, staged in the mind's eye. This was no mere science, but a new form of consciousness. For some, it was a foretaste of a higher existence.

On the other hand, these fantasies of total visibility often went hand in hand with a humbling awareness that, sublime as these visions were,

24. On speed and uncertainty, see Kern 1983.

*Figure E.1.* "Extinct Monsters of the Ancient Earth", John Templeton's engraving for Thomas Hawkins's *Memoirs of Ichthyosauri and Plesiosauri*. Hawkins 1834a, frontispiece.

they offered only tiny glimpses of the whole. In this sense fossils made the past seem all the more unattainable, marooned in the ocean of time. This combination of vividness and evanescence underpinned geological writing just as it underpinned the poetry of John Keats and the novels of Marcel Proust. Each fossil relic offered a concentrated dose of eternity, which then dissolved and awakened an addictive craving for more. As Hawkins expressed it:

> Over these vestiges of Ichthyos and Plesion-sauri [. . .] we love to dwell. Such countless hosts of associations are connected with these gone-by things [. . .]—they are sensations—operations—that concentrate infinity and identifies it, a something that the human understanding can grasp bodily and be satisfied therewith, like the opium-eater, and his drug, for awhile.[25]

This passage, with which *Memoirs of Ichthyosauri* concludes, seems to encapsulate the mood of the frontispiece engraving at the very beginning of the book (Fig. E.1). In John Templeton's picture, an alienating sense of

25. Hawkins 1834a, 51.

otherness is curiously enriched by a feeling of nostalgia, conveyed perhaps by the familiar texture of the cloudscape above and the stillness of that ancient sea. Sound is inscribed into the scene, but it is irrecoverable: the flapping of "leathern wings" breaks the silence above, while the fish-lizard in the sea has its jaws slightly open, as if uttering some unearthly cry.[26]

Even Miller, who more than any other writer celebrated the visibility of the ancient earth, admitted that its imaginative force lay not so much in its certainties as in its obscurities, "which bid fair to be quite dark and uncertain enough for all the purposes of poesy for centuries to come." [27] This recalls one of Charles Darwin's jottings:

> I a geologist have illdefined notion of land covered with ocean, former animals, slow force cracking surface &c truly poetical. (V. Wordsworth about science being sufficiently habitual to become poetical) [28]

For Edward Fitzgerald, contemplating "a Past of which we can just fathom so much as to know that it is unfathomable [. . .] fills, and immeasurably over-fills, the human Soul, with Wonder and Awe and Sadness!" [29] Vast obscurity was integral to the concept of the sublime, and the success of an artist like Martin rested on his ability to convey to the imagination not only the precise outward forms of another time, but also its mystery, suggesting infinite vistas receding beyond the foregrounded scene. When Mantell described "the *Granite*" as "that shroud which conceals for ever from human ken the earliest scenes of the earth's physical drama", he was speaking not simply as a frustrated investigator but also as a poet revelling in truths beyond the visible world.[30] There was, as Sedgwick put it, "an intense and poetic interest in the very uncertainty and boundlessness of our speculations." [31] More irrecoverable than the lost landscapes of pre-industrial Britain or the vanished societies of Scott's Waverley novels, the ancient earth had gone forever. In striving to restore these scenes, geology fulfilled its "poetical" promise.[32]

This plangent awareness that the past can never really be seen, only glimpsed in the imagination, underpins even the most celebratory of Miller's visionary moments. In this light we may revisit the passage quoted at the beginning of this book in which the unbounded ocean

26. Hawkins 1834a, 5. On the picture see also Rudwick 1992, 65–7.
27. Miller 1859, 83.
28. Notebook M [1838], 40 (Barrett *et al.* 1987, 529). See Herbert 2005, 136.
29. Terhune and Terhune 1980, I, 569.
30. Mantell 1844, 873.
31. Sedgwick 1826–33a, 212.
32. On the uses of nostalgia, see Lowenthal 1985.

serves as a Byronic emblem of the geologist's attempt to scan the vastness of deep time and glimpse the world beyond:

> a bit of fractured slate, embedded among a mass of rounded pebbles,
> proves voluble with idea of a kind almost too large for the mind of man
> to grasp. The eternity that hath passed is an ocean without a further shore
> [. . . .] But from the beach, strewed with wrecks, on which we stand to
> contemplate it, we see far out towards the cloudy horizon many a dim
> islet and many a pinnacled rock, the sepulchres of successive eras,—the
> monuments of consecutive creations: the entire prospect is studded over
> with these landmarks of a hoar antiquity [. . .]; nor can the eye reach to the
> open shoreless infinitude beyond, in which only God existed: and—as in
> a sea-scene in nature, in which headland stretches dim and blue beyond
> headland, and islet beyond islet, the distance seems not lessened, but in-
> creased, by the crowded objects—we borrow a larger, not a smaller idea of
> the distant eternity, from the vastness of the measured periods that occur
> between.
>
> Over the lower bed of conglomerate [. . .] we find a bed of gray stratified
> clay, containing a few calcareo-argillaceous nodules.[33]

Looked at another way, the rock had become a window upon eternity,
a "magic casement", "opening on the foam / Of perilous seas, in faery
lands forlorn"; but now, "Fled is that music": the casement has closed
onto bare rock, and we are left like Keats to wonder if that was "a vision, or a waking dream".[34] Some such yearning seems to have underlain
speculations that heaven will allow us to see everything, including the
drama of earth history (this time in the director's cut). Even in the late
twentieth century Stephen Jay Gould, a vigorous opponent of the biblically inspired "pageant" tradition, could conclude his reflections on the
inadequacy of palaeontological restoration with an allusion to St Paul:
"I love this book of life, but it continues to bring us the past as through
a glass darkly. Someday, perhaps, we shall meet our ancestors face to
face."[35]

What these writers seem to have strained against, to have found impossible to accept, was the idea that this magnificent pageant of former
worlds could have been called into being without rational eyes to see it.
One chapter in Miller's *Old Red Sandstone* protests that earth history "exhibited" God's wisdom before humans existed, but

33. Miller 1858, 7.
34. *Ode to a Nightingale* (Keats 1926, 232–3).
35. Gould 1993b, 21.

exhibited surely not in vain. May we not say with Milton,—

> Think not though men were none,
> That heaven could want spectators, God want praise;
> Millions of spiritual creatures walked the earth,
> And these with ceaseless praise his works beheld.[36]

The angelic spectators of Burnet's frontispiece (see Fig. 9.6) are retained to ensure that the panorama was meaningful from the beginning of time. Miller's semi-fictional speculations about Lucifer and his legions literally walking the ancient earth perhaps answered the same need. Any rational spectator, however depraved, seems to have been better than the prospect of worlds born to blush unseen.

This need for a close connection with the lost past may have helped sustain speculations on the survival of the saurians into human times, such as the "Great Sea Serpent" craze of the 1840s (and its present-day analogues) and theories about biblical monster-names.[37] Literalists used the imaginative links between dragon-legends and fossil saurians to suggest that the latter had lived alongside humans for several generations. According to most adherents of the new geology, humans could not possibly have seen a living Iguanodon—but this only made the link with dragon-legends all the more impressive. As Robert Hunt put it, "The process of communion between man of the present, and the creations of a former world, we know not; it is mysterious, and for ever lost to us." Hunt invoked a more mystical, proto-Jungian connection between the inexhaustibly bizarre realities of the past and the creations of human fiction:

> to a great extent, fiction is dependent upon truth for its creations; [. . . .]
> There are floating in the minds of men certain ideas which are not the result of any associations drawn from things around; we reckon them amongst the mysteries of our being. May they not be the truths of a former world, of which we receive the dim outshadowing in the present, like the faint lights of a distant Pharos, seen through the mists of the wide ocean?[38]

Yet the relations of truth and fiction were more complicated than Hunt suggests. As we have seen in this book, the truths of geology—its wondrous visions of battling dragons and shifting scenery—were dependent

---

36. Miller 1841, 102, adapting *Paradise Lost* IV.675–9 but omitting line 678 (Milton 1971, 234). Compare the more cautious use of this passage in [C. Lyell] 1826b, 538.

37. Rupke 1983b, 219–22. On sea serpents see Lyons 2006.

38. R. Hunt 1854, 354–5.

on fiction for their embodiment, using the "floating" forms of poetry to make a convincing mythology of the ancient earth. It was, as John Mill recognized, "almost impossible" to describe these scenes "without passing from the real to the ideal, and casting an air of romance" over the earth's life-history.[39] The educational imperative to channel readers' immature enjoyment of fairy-tales into a mature delight in nature continued to underpin the work of popularizers, and became increasingly overt in books like Buckley's *Fairyland of Science*. The prevalence of dragon imagery to assert geology's superiority to "the fictions of romance"[40] was still strong enough at the turn of the twentieth century to irritate the young J. R. R. Tolkien, later to become an unusually perceptive writer on fantasy and its charged boundary with the real. As a boy, Tolkien resented being told that dinosaurs were "dragons":

> I did not want to be quibbled into Science and cheated out of Faërie by people who seemed to assume that by some kind of original sin I should prefer fairy-tales, but according to some kind of new religion I ought to be induced to like science.

Nevertheless, Tolkien "liked the 'prehistoric' animals best"—better, in fact, than "most fairy-stories". Their allure was in their temporal distance: "they had at least lived long ago, and hypothesis (based on somewhat slender evidence) cannot avoid a gleam of fantasy."[41] In later years he enthused about pterodactyls as products of "the new and fascinating semi-scientific mythology of the 'Prehistoric'".[42] Like his own stratified mythology of Middle-earth, that of "the 'Prehistoric'" combined novelty of conception with antiquity of content, and was built from the ruins of older myths. The image of the broken sword forged anew serves for both: this was the cutting edge of science.

Geology's "semi-scientific mythology" was developed most fully in its literary manifestations. "Facts", already richly freighted with conjecture, were brought into the arena of fancy, clustered about with poetic quotation, allusion, simile, and metaphor, entwined with myth, history, and folklore, and infused with the growing legendary stature of their discoverers and the quest images surrounding the science.[43] The landscapes of prehuman Britain were mapped onto the Old North, the American West, the exotic East, and outer space; they were depicted with the help

39. Mill [*c*. 1855], 53.
40. Mantell 1827, 78, 83–4.
41. Tolkien 1983, 158–9. The biologist E. Ray Lankester (1905, 59) also hated the concept of "fairy tales of science", but more because it seemed to detract from science's truth-content.
42. Carpenter 1995, 282; see also *ODNB*.
43. On the making of the "Heroic Age" of geology see Knell 2000, 305–36.

of Norse myths, Egyptian hieroglyphs, Native American legends, Hindu cosmologies, and Arabian romance, not to mention the invisible realms of heaven and hell, and the world of dream. As Wordsworth had predicted, these writers brought poetry, with all its incongruities and tensions, "into the midst of the objects of the science itself."[44] The poetic qualities of the resulting literature, according to the *Presbyterian Review*, propelled geology's meteoric rise to celebrity: "A science like this [. . .] needed a patron to introduce it. Literature has been its patron."[45]

## Literary Criticism and the History of Science

If literature was central to the business of science popularization in the nineteenth century—and it seems unlikely that geology was alone here—then one would expect the tools of literary criticism and a clear sense of literary history to be of central importance to historians of science. In the latter discipline, such interests have tended to take two overlapping forms. First, since 1980, historians have been increasingly concerned to apply rhetorical analysis to scientific texts, showing how (usually) men of science have manipulated the conventions of scientific discourse to stake claims of truth, significance, and objectivity.[46] This approach has contributed to a widespread "demystification" of scientific texts, and in some quarters to a full-scale rejection of science's truth-claims. Second, a more recent emphasis on the role of figures other than prestigious men of science in producing knowledge has led several historians to draw on book history as a means of studying popularization in action, from the educational and commercial priorities of scientific authors and publishers to the variety of meanings given to texts by readers. Exemplifying this approach are James Secord's monumental *Victorian Sensation*, Paul White's penetrating studies of the late-Victorian professionalization of science and literature, and the collaborative volumes associated with Leeds University's "Science in the Nineteenth-Century Periodical" project.[47] These and related studies have enhanced our sense of public science as a set of active transformations of (and claims about) natural knowledge, rather than a simple diffusion of facts from on high.

Yet, although both these kinds of history have illuminated the texts and books on which they focus, literary criticism and literary history as

---

44. W. Wordsworth 1944–72, II, 396.

45. Anon. 1841–2, 209.

46. Examples include Myers 1990; Harré 1990; and the essays in Shinn and Whitley 1985 and Dear 1991.

47. J. Secord 2000a; P. White 2002; 2005; Cantor *et al.* 2004a; Henson *et al.* 2004; Cantor and Shuttleworth 2004. See also Topham 1998; Frasca-Spada and Jardine 2000; A. Secord 2002; Fyfe 2004a.

traditionally understood have slipped through the historical net. Despite a widespread feeling among social and cultural historians that the textual strategies of authors have been awarded a disproportionate amount of attention,[48] we still have only a vestigial sense of where scientific texts fit into literary history as a whole. Individual scientific texts have often been treated as *sui generis,* without indicating their relation to other texts of the same literary genre. The bibliographic objects themselves are often placed within a constellation of publication formats, particularly those which literary critics traditionally neglect (encyclopaedia, miscellany, textbook, newspaper), but, with a few exceptions, historians have remained oddly wary of engaging with the better-known spectrum of literary genres such as novels, dramas, and epic poems. We have an increasingly clear sense, especially for the late-Victorian period, of how scientific theories in popular works took different forms to those presented in more specialized writings, but—with the notable exception of the dialogue genre, which has been well served[49]—we have only a limited sense of the literary forms which effected these transformations. Narrative devices are often alluded to, but rarely examined in depth. The lack of attention paid to the aesthetic component of scientific culture (except as a sign of its middle-class "debasement") seems once again to reflect our highly artificial convention of distinguishing between "literary" and "scientific" texts, allotting the former to literary scholars and the latter to historians. Such a carve-up does no justice to the common literary culture of the nineteenth-century reading public.

On the other side of the disciplinary gulf, literary scholars have enthusiastically examined science's aesthetic possibilities, again yielding valuable insights into science's wider cultural embedment. Yet the framing of most "science and literature" studies falls short of challenging the view that scientific writing was never "literature" in any meaningful sense. Typically, a canonical representative of Literature with a capital L—Wordsworth, Percy or Mary Shelley, Byron—is presented alongside a set of scientific ideas, which he or she is then shown to have transformed into something more interesting. Titles or subtitles often take the form "poet *x* and science *y*": "Tennyson and Geology", "Wordsworth and the Geologists", "Shelley and Huttonian Earth Science", "Byron and the Geology of Cuvier".[50] These *ands* conceal a colonial order in which Literature represents the centre and Science one of several possible peripheries, supplying the raw material for poetic production: geology and chemistry

48. Cooter and Pumfrey 1994, 244; Topham 2000, 570.
49. See the works of Myers, Gates, and Shteir.
50. Dean 1985; Wyatt 1995; Leask 1998b; R. O'Connor 1999.

bear fruit in *Prometheus Unbound,* galvanism in *Frankenstein,* but the scientific texts involved are usually gutted of their content and cast aside. More general literary explorations of "science and literature" tend likewise to focus on the relations between scientific ideas and literary forms, often from a philosophical perspective; but the literary forms in question rarely include science writing itself, still less popular-science writing.

Again, there are notable exceptions, such as Noah Heringman's work on early geology and the Romantic poets, and above all Gillian Beer's pioneering study of Darwin and the language of evolution, *Darwin's Plots.*[51] With its close attention to Darwin's personal development as a writer, this book set a new agenda for "science and literature" studies: Darwin already had a secure place in the Victorian canon thanks to the perceived pre-eminence of his ideas, but few scholars had dug so deeply into his use of language. Yet the book's centre of gravity is still Literature with a capital L: the first part's analysis of Darwin's prose becomes a rich resource for illuminating the narrative devices of later novelists like George Eliot, who are seen transforming Darwin's ideas in unexpected and imaginative ways. To this extent, though revolutionary in its procedure, *Darwin's Plots* does not challenge the old hierarchy. Heringman's *Romantic Rocks, Aesthetic Geology* has taken the unusual step of drawing on relatively unknown geological texts to illuminate the grounding of geology and landscape poetry in a common aesthetic discourse; but here, too, poetry is the central focus. Heringman's subtle analyses of the rhetorical tropes used by early geologists result in a deepened understanding of the poetic strategies of Wordsworth, Blake, Shelley, and Erasmus Darwin, whose work is analysed in detail and quoted at length; but the geological prose still functions as a "source", and one comes away with no clear sense of its authors' larger literary or imaginative aims. Overall, the recent drive to burst the limits of canons has tended to restrict itself to fictional texts, and relatively few critics have taken up Beer's challenge and applied close literary analysis to scientific texts. This job has, it seems, been left largely to the historians of science.

So, despite many superb studies of the interrelations between literature and science, and despite the increasingly sophisticated literary-critical and cultural-historical techniques brought to bear on the material, much basic but rewarding groundwork still needs to be done. How rewarding this can be is demonstrated by Marjorie Hope Nicolson's classic studies of seventeenth- and eighteenth-century scientific sensibilities: her analytical frameworks may have been superseded, but the inclusivity

---

51. Heringman 2004; Beer 1983.

and sensitivity of her literary analysis remain unsurpassed.[52] More recent "science and literature" studies have often focused on the old debate about whether these two categories represent two cultures or one, a debate which is all too often premised on a prior distinction between "scientific" texts and "literary" texts. I hope to have shown in this book that, for at least the first two-thirds of the nineteenth century, this distinction cannot be assumed. Even today, to judge by the tone (and the mere presence) of reviews of science books in literary periodicals such as the *Times Literary Supplement,* many popular-science books are still firmly part of literary culture, capable of being appreciated for their aesthetic qualities as well as for the truth-value of their content. This is as one would expect, given the literary craftsmanship evident in reflective treatises like E. O. Wilson's *Consilience* and Conway Morris's *Life's Solution.*[53] A scientific text published for a general readership was, and is, self-evidently an intervention both in the field of literature and in the field of (public) science, and should not be dismissed out of court as second-rate literature or second-rate science.

Applying numerals to the slippery word "culture" does not help to pin anything down, and keeping in mind the basic distinction between form and content will save us from entangling ourselves in truisms. This applies to one-culture views no less than to those who emphasize the chasm between science and literature. Here is not the place to enter into the so-called "science wars", but it is worth pointing out that to call science "just another form of literature" (or cultural artefact), as the extreme-constructivist denial of science's truth-value is often paraphrased, misses the mark on two fronts. From a moderate-realist perspective, such a statement brushes aside science's power to manipulate the physical world:[54] to put it crudely, *Prometheus Unbound* never cured anyone of mumps. But, with its weasel word "just", this statement also implicitly devalues literature: the use of such a formulation to deny truth-value rests on an assumption that literature and other cultural artefacts are necessarily devoid of truth-value. And this assumption, as we have seen, is a cultural construct deriving from the redefinition of "literature" by a subgroup of late-Victorian literary professionals.

In this book, drawing on the different branches of scholarship mentioned above, I have tried to show how geology was moulded into certain shapes when it took literary form, and that these shapes were informed

52. Nicolson 1946; 1960; 1963.

53. E. Wilson 1998; Conway Morris 2003. On Wilson see Myers 1990, 193–246; on the wider genre, see Eger 1997.

54. On the implications of realism and constructivism for the history of science and literature see Levine 1987b.

by political, social, commercial, educational, aesthetic, and religious factors, as well as by the often-disregarded factors of literary tradition and the momentum of what Beer has called "inherited mythologies".[55] As Misia Landau has shown how folktale patterns underpin modern theories of human evolution,[56] I have traced the ways in which the old "drama of history" template was put to new uses by progressionist and literalist geological writers, culminating in a fully-fledged pageant of earth history. These writers were all "popularizers" in the broad sense I have been advocating. They all made geology known outside specialist circles, and the variety of their motives for doing so (didactic zeal, religious commitment, pleasure, quests for scientific or literary status, financial necessity) only makes the common features of their palaeontological word-painting all the more striking.

Following Nicolson, I have tried to ground this analysis firmly in literary history—examining the texts in their generic frameworks, attending to the development and innovation of specific narrative devices, taking account of authors' literary ambitions, and opening up new ways in which the culture of spectacle and its textual accompaniments can be used to inform such analysis. Popular scientific writing needs to be taken seriously as literature, because it was as literature that it reached its public (which included canonical writers like Dickens). This means attending not only to the bibliographic object or the theories it promoted, but equally to textual features such as style, rhythm, voice, and structure, and to a text's place in the larger history of its genre. All these features helped constitute the science for its public, and—at least in part—for its practitioners as well.

Identifying the ways in which literary culture has shaped a science's textual manifestations does not necessarily compromise its truth-value, at least not in terms of the provisional truth aimed at by modern scientists. Science depends on language for its expression, and individual theories have always been open to attack on this front. Particularly since the late seventeenth century, rhetorical ornament has been seen as the bane of clear-sighted scientific thought, and the detached, objective style of modern professional scientific discourse is one product of this long-standing suspicion. As Heringman has put it, "Specialist discourses have relied, from the beginning, on the literary as a category against which the scientific may be set off"[57]—although I should be inclined to replace the word "literary" with "fictional". The pejorative use of words

55. Beer 1983, 5.
56. Landau 1991.
57. Heringman 2004, 271.

like "poetry", "fiction", "romance", "dream", and "imagination" to deni-
grate the truth-value of rival theories was a standard part of the rhetorical armoury of Enlightenment science. As we have seen, however, such
concepts were also paradoxically used to raise geology's public profile
and claims to cultural authority. This was a tricky tightrope to walk,
especially for a science whose most public claims rested not on experimental demonstration, but on reconstructing the past from fragmentary
evidence and developing plausible scenarios. As Claudine Cohen has
shown in her study of interpretations of the mammoth, the "play of the
imagination" has always lain at the heart of palaeontological reconstruction.[58] Geologists have therefore found it particularly necessary to keep
an eagle eye on the deceptive power of storytelling, even as they have
found it necessary to use stories to think about and communicate their
ideas.

The resulting tension between the repudiation and the cultivation
of the poetic imagination can be traced right through geology's literary history, as writers have sought to engage enthusiasm whilst warn-
ing against the deceptions of rival theories. Thus, in the late seventeenth
century Thomas Burnet dismissed an earlier theory as emanating from a
mere "Oratour", then proceeded to display his own oratory in narrating
the earth's formation and destruction as a grand dramatic spectacle.[59]
Burnet's *Sacred Theory* and its kin were dismissed by John Playfair in 1811
as the products of "imagination" and "mental derangement"; but Playfair
also advertised geology's pleasurable power to unhinge the mind, recalling with pleasure how "the mind seemed to grow giddy by looking so far
into the abyss of time".[60] For the literalist John Murray (no relation of the
publisher), Playfair's own "visionary" theories seemed too smoothly narrated to be true. Heringman has called the Playfair-Murray debate a "late
stage in the transition from 'letters' to 'science' as the cultural rubric for
study of the natural world",[61] but this dialectic would seem rather to be a
cultural constant in the historical sciences. Lyell went on to portray literalists and catastrophists as deluded visionaries while himself indulging in
spectacular Dantean visions;[62] and modern evolutionary biologists and
palaeontologists likewise point out the deceptively poetic nature of rival
theories while making positive use of fictive and poetic techniques to
convey their own ideas.[63] This double rhetoric is a necessary and healthy
part of public science, where the need to communicate and display goes

58. C. Cohen 2002, 250. See also Sommer 2008 forthcoming.
59. Burnet 1965, 90–1, discussed in Heringman 2004, 271.
60. [Playfair] 1811–12, 207; Playfair 1805, 73.
61. Heringman 2004, 270–1.
62. See above, pp. 165–74.
63. Dawkins 1998, 209 and 61; Conway Morris 2003, 64–5 and 197.

hand in hand with the need to demonstrate a commitment to truth and objectivity. To dismiss such practice as hypocrisy would be to evade the issues it raises, especially the continuing complexity of the relation between truth and aesthetics.

Ever since the earth's past became the subject of study, visions of earth history have sprung up and flowered within the fertile territory where science and the imagination overlap. Exploring the uses of narrative devices and literary forms in the production of scientific knowledge offers historians a powerful conceptual tool for understanding the rise of public science, and it offers scientists the means to cultivate a higher level of self-awareness when constructing scenarios and hypotheses. Finally, it offers literary scholars the chance to remove the blinkers imposed by late-Victorian pundits on the limits of their field of inquiry. The realization that popular scientific writing was and is "literature" opens up a vast range of texts for the attention of the literary critic or literary historian. Even for scholars whose interests remain focused on canonical figures, studies of the relations between science and poetry may be taken beyond the "poet $x$ and science $y$" framework, since a given poet's response to a given science was, in large part, a response to the literary forms which that science took.

Surveying such responses has been beyond the scope of this book. I have concentrated on making a case for the literary nature of popular-science writing, and this case has had to be constructed from the ground up. In the light of our exploration of the literature of early geological popularization, allusions to fossils in early-nineteenth-century poetry can only take on a new richness. In this connection the last words may be left to the poet Thomas Lovell Beddoes, whose musical gifts and morbid fascination with death combined to produce some of the most haunting lyrics in early-Victorian literature. In the voice of his female protagonist Thanatos, Beddoes encapsulated that nameless nostalgia for the vanished past which seems to have suffused so much geological writing then, and which continues to do so today.[64]

"But the sight of a palm with its lofty stem and tuft of long grassy leaves, high in the blue air, or even such a branch as this" (breaking off a large fern leaf) "awake in me a feeling, a sort of nostalgy and longing for ages long past. When my ancient sire used to sit with me under the old dragon tree or Dracaena, I was as happy as the ephemeral fly balanced on his wing in the sun, whose setting will be his death-warrant. But why do I speak to you so? You cannot understand me."—And then she would sing whisperingly to herself:

64. "Thanatos to Kenelm", *c.* 1837 (Beddoes 1935, 142). On Beddoes see Potter 1923; *ODNB*.

The mighty thoughts of an old world
Fan, like a dragon's wing unfurled,
   The surface of my yearnings deep;
And solemn shadows then awake,
Like the fish-lizard in the lake,
   Troubling a planet's morning sleep.

My waking is a Titan's dream,
Where a strange sun, long set, doth beam
   Through Montezuma's cypress bough:
Through the fern wilderness forlorn
Glisten the giant harts' great horn
   And serpents vast with helmed brow.

The measureless from caverns rise
With steps of earthquake, thunderous cries,
   And graze upon the lofty wood;
The palmy grove, through which doth gleam
Such antediluvian ocean's stream,
   Haunts shadowy my domestic mood.

# Appendix: Currencies, and Sizes of Books

**British Currency before 1971**

£1 or 1*l* (pound) = 20*s* (shillings) − 240*d* (pence)
1*s* (shilling) = 12*d* (pence)
1 guinea = 21*s* = £1 1*s*

**Book Sizes**

folio (fol.)
quarto (4to)
octavo (8vo)
duodecimo (12mo)
octodecimo (18mo)

These terms refer to the number of times the printer folds each sheet of paper before printing and stitching them. For a folio, each sheet of paper is folded once, yielding two pages (i.e. four printed sides); for a quarto, it is folded twice, yielding four pages (hence *quarto*); for an octavo, eight pages; and so on. Folios are therefore generally bigger than quartos. But these terms only give a relative idea of book size, since they depend entirely on the original paper size. The following names were given to paper sizes (in inches):

| | |
|---|---|
| Imperial | 30" × 22" |
| Elephant | 28" × 23" |
| Royal | 25" × 20" |

————, 1978. *The Shows of London: A Panoramic History of Exhibitions, 1600–1862,* Cambridge, Massachusetts: Belknap

Anderson, George, and Peter Anderson, 1834. *Guide to the Highlands and Islands of Scotland,* London: John Murray

Anderson, Patricia J., 1994. *The Printed Image and the Transformation of Popular Culture 1790–1860,* 2nd ed., Oxford: Clarendon Press

Anon., 1782. "A View of the Eidophusikon", *European Magazine,* 1, 180–2

Anon., 1790. "Of the Enormous Bones Found in America", *American Museum,* 8, 284–5

Anon., 1819. *Description of the View of Venice,* London: n.p.

Anon., 1821a. *Description of a View of Bern, and the High Alps,* London: n.p.

Anon., 1821b. "Macculloch *on Rocks*", *Eclectic Review,* 2nd series, 15, 430–41

Anon., 1822a. *Another Cain. A Poem,* London: Hatchard

Anon., 1822b. "On Reading 'Cain, A Mystery' ", *Manchester Iris,* 1, 44

Anon., 1822c. "Antient Cave", *Gentleman's Magazine,* 92, 161

Anon., 1822d. "Antediluvian Cave", *Gentleman's Magazine,* 92, 352–3

Anon., 1822e. *Uriel: A Poetical Address to the Right Honourable Lord Byron,* London: the author

Anon., 1823. "Moore's Loves of the Angels", *Blackwood's Edinburgh Magazine,* 13, 63–71

Anon., 1827. *Description of a View of the City and Lake of Geneva,* London: n.p.

Anon., 1829a. *Description of an Attempt to Illustrate Milton's Pandemonium,* London: n.p.

Anon., 1829b. "Niebuhr's *History of Rome*", *Eclectic Review,* 3rd series, 1, 189–98

Anon., 1829c. "The March of Intellect: Learned Cats", *Literary Gazette,* 1829, 132

Anon., 1829d. "*Anon.*: Attributed to J. Rennie, [. . .] Conversations on Geology", *Magazine of Natural History,* 1, 280

Anon., 1830a. "Satan", *Athenaeum,* 1830, 1–2

Anon., 1830b. "Principles of Geology", *Athenaeum,* 1830, 595–7

Anon., 1831. "Principles of Geology", *Eclectic Review,* 3rd series, 6, 75–81

Anon., 1832a. "Lyell's Principles of Geology", *Spectator,* 14 January 1832, 39–40

Anon., 1832b. "Sacred Poetry", *Eclectic Review,* 3rd series, 8, 226–39

Anon., 1832c. "Revelation Consistent with Science", *Eclectic Review,* 3rd series, 8, 14–33

Anon., 1833a. *Description of a View of the Falls of Niagara,* London: n.p.

Anon., 1833b. "Wolves", *Penny Magazine,* 2, 396

Anon., 1833–58. *The Penny Cyclopædia of the Society for the Diffusion of Useful Knowledge,* 27 vols, London: Charles Knight

Anon., 1834a. "Principles of Geology", *Metropolitan Magazine,* 10, 4–5

Anon., 1834b. "Martin's *Illustrations of the Bible*", *Westminster Review,* 20, 452–65

Anon., 1834c. "Montgomery's Lectures on Poetry", *Presbyterian Review,* 5, 114–24

Anon., 1834d. "Memoirs of Ichthyosauri and Plesiosauri", *Metropolitan Magazine,* 11, second run of pages, 1–2

Anon., 1834e. "Popular Geology Subversive of Divine Revelation", *Literary Gazette,* 1834, 358–9

Anon., 1835. *Description of a View of Isola Bella, the Lago Maggiore, and the Surrounding Country,* London: n.p.

Anon., 1836a. "A Popular Course of Geology", *Magazine of Popular Science,* 2, 1–13, 223–35, 267–78, 437–47

Anon., 1836b. "British Association", *Literary Gazette*, 1836, 634–5

Anon., 1836–7. "Buckland's Bridgewater Treatise", *Presbyterian Review*, 9, 222–46

Anon., 1837a. *Description of a View of Mont Blanc, the Valley of Chamounix, and the Surrounding Mountains*, London: n.p.

Anon., 1837b. "Geology and Natural Theology", *Eclectic Review*, 4th series, 1, 23–37

Anon., 1838. "The Wonders of Geology", *Literary Gazette*, 1838, 376

Anon., 1841a. "Pterodactyle", in Anon. 1833–58, XIX, 96–9

Anon., 1841b. "The Fossil Reptiles of England", *Literary Gazette*, 1841, 513–19

Anon., 1841–2. "Miller's *Old Red Sandstone*", *Presbyterian Review*, 14, 208–17

Anon., 1841–5. *London Interiors*, 2 vols, London: the proprietors

Anon., 1844a. "Transactions of the Association of American Geologists and Naturalists", *Athenaeum*, 1844, 727

Anon., 1844b. "Geology, Introductory, Descriptive, and Practical", *Athenaeum*, 1844, 1136–7

Anon., 1845. *A Description of the Colosseum as Re-Opened in M.DCCC.XLV*, London: n.p.

Anon., 1847. *Sacred Geology; or, The Scriptural Account of the World's Creation Maintained*, London: William Painter

Anon., 1848. *Description of the Royal Cyclorama, or Music Hall*, London: n.p.

Anon., ed., 1852. *Lectures on the Results of the Great Exhibition of 1851*, London: David Bogue

Anon., 1853. "Princess's Theatre", *The Times*, 14 June 1853, 7

Anon., 1857a. "Hugh Miller and Geology", *Dublin University Magazine*, 50, 596–610

Anon., 1857b. "The Book and the Rocks", *Blackwood's Edinburgh Magazine*, 82, 312–29

Anon., 1859. "Sketch-Book of Popular Geology", *Athenaeum*, 1859, 445–7

Ansted, D. T., 1847. *The Ancient World; or, Picturesque Sketches of Creation*, London: John Van Voorst

[Apjohn, James], 1833. "Principles of Geology", *Athenaeum*, 1833, 409–11

Ashe, Thomas, 1806. *Memoirs of Mammoth, and Various Other Extraordinary and Stupendous Bones, of Incognita, or Non-Descript Animals*, Liverpool: n.p. ["Introduction" paginated separately from rest of book]

———, 1815. *Memoirs and Confessions of Captain Ashe*, 3 vols, London: Henry Colburn

Ashfield, Andrew, and Peter de Bolla, eds, 1996. *The Sublime: A Reader in British Eighteenth-Century Aesthetic Theory*, Cambridge: Cambridge University Press

Astore, William J., 2001. *Observing God: Thomas Dick, Evangelicalism, and Popular Science in Victorian Britain and America*, Aldershot: Ashgate

Augusta, Josef, 1961. *Prehistoric Reptiles and Birds*, trans. Margaret Schierl, London: Hamlyn

"The Author of 'A Portrait of Geology'", 1839. *Christian Observer*, 39, 400–1

Bainbridge, Simon, 1995. *Napoleon and English Romanticism*, Cambridge: Cambridge University Press

Bakewell, Robert, 1813. *An Introduction to Geology, Illustrative of the General Structure of the Earth*, London: J. Harding

———, 1828. *An Introduction to Geology: Comprising the Elements of the Science in Its Present Advanced State*, 3rd ed., London: Longman *et al.*

———, 1833. *An Introduction to Geology [. . .] Greatly Enlarged*, 4th ed., London: Longman *et al.*

Brock, M. G., and M. C. Curthoys, eds, 1997. *The History of the University of Oxford Volume VI: Nineteenth-Century Oxford, Part I*, Oxford: Clarendon Press

Brock, W. H., 1980. "The Development of Commercial Science Journals in Victorian Britain", in Meadows 1980, 95–122

Broderip, W. J., 1848. *Zoological Recreations*, 2nd ed., London: Henry Colburn

Brook, Peter, 1990. *The Empty Space*, new ed., London: Penguin

Brooke, John Hedley, 1979. "The Natural Theology of the Geologists: Some Theological Strata", in Jordanova and Porter 1979, 39–64

———, 1991. *Science and Religion: Some Historical Perspectives*, Cambridge: Cambridge University Press

———, 1996. "Like Minds: The God of Hugh Miller", in Shortland 1996a, 171–86

———, 1999. "The History of Science and Religion: Some Evangelical Dimensions", in Livingstone *et al.* 1999, 17–40

Brooke, John, and Geoffrey Cantor, 1998. *Reconstructing Nature: The Engagement of Science and Religion*, Edinburgh: Clark

Brooks, Chris, 1998. "Introduction: Historicism and the Nineteenth Century", in Brand 1998, 1–19

Brown, Eluned, ed., 1966. *The London Theatre 1811–1866: Selections from the Diary of Henry Crabb Robinson*, London: Society for Theatre Research

Brown, J. Mellor, 1838. *Reflections on Geology: Suggested by the Perusal of Dr Buckland's Bridgewater Treatise*, Edinburgh: James Nisbet

Browne, Janet, 1992. "Squibs and Snobs: Science in Humorous British Undergraduate Magazines around 1830", *HS*, 30, 165–97

Buckland, Francis T., 1858. "Memoir of the Very Rev. William Buckland", in W. Buckland 1858, I, xvii–lxxxiii

———, 1903. *Curiosities of Natural History: First Series*, 2nd ed., London: Macmillan

Buckland, William, 1820. *Vindiciæ Geologicæ; or the Connexion of Geology with Religion Explained*, Oxford: Oxford University Press

———, 1822. "Account of an Assemblage of Fossil Teeth and Bones of Elephant, Rhinoceros, Hippopotamus, Bear, Tiger, and Hyæna, and Sixteen Other Animals; Discovered in a Cave at Kirkdale, Yorkshire, in the Year 1821", *Philosophical Transactions of the Royal Society of London*, 112, 171–236

———, 1823. *Reliquiæ Diluvianæ; or, Observations on the Organic Remains Contained in Caves, Fissures, and Diluvial Gravel, and on Other Geological Phenomena, Attesting the Action of an Universal Deluge*, London: John Murray

———, 1835. "On the Discovery of a New Species of Pterodactyle in the Lias at Lyme Regis" [1829], *Transactions of the Geological Society of London*, 2nd series, 3 (1), 217–22

———, 1836. *Geology and Mineralogy Considered with Reference to Natural Theology*, 2 vols, London: William Pickering

———, 1858. *Geology and Mineralogy Considered with Reference to Natural Theology* [. . .] *with Additions*, 3rd ed., ed. Francis T. Buckland, 2 vols, London: George Routledge

Buckley, Arabella B., 1927. *Winners in Life's Race or The Great Backboned Family*, new ed., London: Macmillan

Buffetaut, Eric, 1987. *A Short History of Vertebrate Palaeontology*, London: Croom Helm

Buffon, Georges Leclerc, comte de, 1988. *Des Époques de la nature*, ed. Jacques Roger, 2nd ed., Paris: Éditions du Muséum

[Bugg, George], 1826–7. *Scriptural Geology; or, Geological Phenomena Consistent Only with the Literal Interpretation of the Sacred Scriptures,* 2 vols, London: Hatchard and Son

Bullock, William, 1812. *A Companion to Mr. Bullock's London Museum and Pantherion,* 12th ed., London: William Bullock

Bulwer-Lytton, Edward, 1970. *England and the English,* ed. Standish Meacham, Chicago, Illinois: University of Chicago Press

Burek, Cynthia V., 2001. "Where are the Women in Geology?", *Geology Today,* 17, 110–14

Burke, Edmund, 1878. *Select Works: Four Letters on the Proposals for Peace with the Regicide Directory of France,* ed. E. J. Payne, Oxford: Clarendon Press

Burkhardt, Frederick, and Sydney Smith, eds, 1987. *The Correspondence of Charles Darwin: Volume 3 1844–1846,* Cambridge: Cambridge University Press

Burnet, Thomas, 1722. *The Sacred Theory of the Earth,* 5th ed., 2 vols, London: J. Hooke

———, 1965. *The Sacred Theory of the Earth,* ed. Basil Willey, London: Centaur Press

Burnett, John, 1969. *A History of the Cost of Living,* London: Penguin

Butler, Marilyn, 1993. "Culture's Medium: The Role of the Review", in Curran 1993, 120–47

Buzard, James, 1993. *The Beaten Track: European Tourism, Literature, and the Ways to Culture, 1800–1918,* Oxford: Clarendon Press

Byron, George Gordon, Lord, 1821. *Sardanapalus, A Tragedy. The Two Foscari, A Tragedy. Cain, A Mystery,* London: John Murray

———, 1980–93. *The Complete Poetical Works,* ed. Jerome J. McGann and Barry Weller, 7 vols, Oxford: Clarendon Press

———, 1986. *The Major Works,* ed. Jerome J. McGann, Oxford: Oxford University Press

Cadbury, Deborah, 2000. *The Dinosaur Hunters: A Story of Scientific Rivalry and the Discovery of the Prehistoric World,* London: Fourth Estate

Callaway, Jack M., and Elizabeth L. Nicholls, eds, 1997. *Ancient Marine Reptiles,* San Diego, California: Academic Press

Campbell, Mary B., 1988. *The Witness and the Other World: Exotic European Travel Writing, 400–1600,* Ithaca, New York: Cornell University Press

Campbell, Thomas, 1904. *Poems,* ed. Lewis Campbell, London: Macmillan

Cannon, Susan Faye, 1978. *Science in Culture: The Early Victorian Period,* New York: Science History

Cantor, Geoffrey, 2005. *Quakers, Jews, and Science: Religious Responses to Modernity and the Sciences in Britain, 1650–1900,* Oxford: Oxford University Press

Cantor, Geoffrey, and Sally Shuttleworth, eds, 2004. *Science Serialized: Representations of the Sciences in Nineteenth-Century Periodicals,* Cambridge, Massachusetts: MIT Press

Cantor, Geoffrey, *et al.,* 2004a. *Science in the Nineteenth-Century Periodical: Reading the Magazine of Nature,* Cambridge: Cambridge University Press

———, 2004b. "Introduction", in Henson *et al.* 2004, xvii–xxv

Carlile, Richard, 1822a. "*Queen Mab; Cain, a Mystery*; and a Royal Reviewer", *Republican,* 5, 192

———, ed., 1822b. *Cain; A Mystery, by Lord Byron,* London: R. Carlile

———, 1823. Reply to William Fitton, *Republican,* 7, 396–411

Carlson, Julie A., 1994. *In the Theatre of Romanticism: Coleridge, Nationalism, Women,* Cambridge: Cambridge University Press

Reflections on the History of Science Popularization and Science in Popular Culture", *HS*, 32, 237–67

Copeman, W. S. C., 1951. "Andrew Ure, M.D., F.R.S. (1778–1857)", *Proceedings of the Royal Society of Medicine*, 44, 655–62

Corsi, Pietro, 1988. *Science and Religion: Baden Powell and the Anglican Debate, 1800–1860*, Cambridge: Cambridge University Press

Crary, Jonathan, 1990. *Techniques of the Observer: On Vision and Modernity in the Nineteenth Century*, Cambridge, Massachusetts: MIT Press

Crawford, Robert, 2000. *Devolving English Literature*, 2nd ed., Edinburgh: Edinburgh University Press

[Croly, George], 1837. "The World We Live In: No. XIII", *Blackwood's Edinburgh Magazine*, 42, 673–92

[Crosse, J.], 1845. "The Vestiges, etc.", *Westminster Review*, 44, 152–203

Crowe, Michael J., 1986. *The Extraterrestrial Life Debate 1750–1900: The Idea of a Plurality of Worlds from Kant to Lowell*, Cambridge: Cambridge University Press

Cunningham, Andrew, and Nicholas Jardine, eds, 1990. *Romanticism and the Sciences*, Cambridge: Cambridge University Press

Curran, Stuart, ed., 1993. *The Cambridge Companion to British Romanticism*, Cambridge: Cambridge University Press

Curtius, Ernst Robert, 1953. *European Literature and the Latin Middle Ages*, trans. Willard R. Trask, London: Routledge and Kegan Paul

Curwen, E. Cecil, ed., 1940. *The Journal of Gideon Mantell, Surgeon and Geologist: Covering the Years 1818–1852*, London: Oxford University Press

Cuvier, Georges, 1812. *Recherches sur les ossemens fossiles de quadrupèdes, où l'on rétablit les caractères de plusieurs espèces d'animaux que les révolutions du globe paroissent avoir détruites*, 4 vols, Paris: Deterville

———, 1813. *Essay on the Theory of the Earth*, trans. Robert Kerr, ed. Robert Jameson, Edinburgh: Blackwood *et al.*

———, 1815. *Essay on the Theory of the Earth*, trans. Robert Kerr, ed. Robert Jameson, 2nd ed., Edinburgh: Blackwood *et al.*

———, 1817. *Essay on the Theory of the Earth*, trans. Robert Kerr, ed. Robert Jameson, 3rd ed., Edinburgh: Blackwood *et al.*

———, 1821–4. *Recherches sur les ossemens fossiles, où l'on rétablit les caractères de plusieurs animaux dont les révolutions du globe ont détruit les espèces*, 2nd ed., 5 vols in 7, Paris: G. Dufour and E. D'Ocagne

———, 1822. *Essay on the Theory of the Earth*, trans. Robert Kerr, ed. Robert Jameson, 4th ed., Edinburgh: Blackwood and Cadell

———, 1827. *Essay on the Theory of the Earth*, trans. Robert Kerr, ed. Robert Jameson, 5th ed., Edinburgh: Blackwood and Cadell

Cuvier, Georges, *et al.*, 1827–35. *The Animal Kingdom Arranged in Conformity with Its Organization*, ed. Edward Griffith, 16 vols, London: Whittaker, Treacher, & Co.

Czerkas, Sylvia J., and Stephen A. Czerkas, 1990. *Dinosaurs: A Global View*, Limpsfield: Dragon's World

Czerkas, Sylvia Massey, and Donald F. Glut, 1982. *Dinosaurs, Mammoths, and Cavemen: The Art of Charles R. Knight*, New York: Dutton

Dahl, Curtis, 1953. "Bulwer-Lytton and the School of Catastrophe", *Philological Quarterly*, 32, 428–42

Darwin, Charles, 1839. *Journal of Researches into the Geology and Natural History of the Various Countries Visited by H.M.S. Beagle*, London: Henry Colburn.

————, 1958. *The Autobiography of Charles Darwin, 1809–1882,* ed. Nora Barlow, London: Collins

Darwin, Erasmus, 1803. *The Temple of Nature; or, The Origin of Society,* London: J. Johnson

Daston, Lorraine, 1991–2. "Marvelous Facts and Miraculous Evidence in Early Modern Europe", *Critical Inquiry,* 18, 93–124

————, 1995. "Curiosity in Early Modern Science", *WI,* 11, 391–404

————, 2001. "Fear and Loathing of the Imagination in Science", in Galison *et al.* 2001, 73–95

Daston, Lorraine, and Peter Galison, 1992. "The Image of Objectivity", *Representations,* 40, 81–128

Daubeny, C. G. B., ed., 1869. *Fugitive Poems Connected with Natural History and Physical Science,* Oxford: James Parker

Davidson, Peter, 2005. *The Idea of North,* London: Reaktion

Davis, R. W., and R. J. Helmstadter, eds, 1992. *Religion and Irreligion in Victorian Society: Essays in Honor of R. K. Webb,* London: Routledge

Davy, Humphry, 1813. *Elements of Agricultural Chemistry,* London: Longman *et al.*

————, 1827. *Six Discourses Delivered before the Royal Society,* London: John Murray

————, 1851. *Consolations in Travel, or, The Last Days of a Philosopher,* ed. John Davy, 5th ed., London: John Murray

Dawkins, Richard, 1998. *Unweaving the Rainbow: Science, Delusion and the Appetite for Wonder,* London: Penguin

Dawson, Gowan, *et al.,* 2004. "Introduction", in Cantor *et al.* 2004a, 1–34

Dawson, J. W., 1882. *The Story of the Earth and Man,* 7th ed., London: Hodder and Stoughton

Dean, Dennis R., 1968. "Geology and English Literature: Crosscurrents, 1770–1830", unpublished doctoral thesis, University of Wisconsin

————, 1979. "The Word 'Geology'", *AS,* 36, 35–43

————, 1981. "'Through Science to Despair': Geology and the Victorians", in Paradis and Postlewait 1981, 111–36

————, 1985. *Tennyson and Geology,* Lincoln: Tennyson Society

————, 1992. *James Hutton and the History of Geology,* Ithaca, New York: Cornell University Press

————, 1999. *Gideon Mantell and the Discovery of Dinosaurs,* Cambridge: Cambridge University Press

Dear, Peter, ed., 1991. *The Literary Structure of Scientific Argument: Historical Studies,* Philadelphia: University of Pennsylvania Press

De Johnsone, Fowler, 1838. *A Vindication of the Book of Genesis,* London: Groombridge

De la Beche, Henry T., 1834. *Researches in Theoretical Geology,* London: Charles Knight

Delair, Justin B., and William A. S. Sarjeant, 1975. "The Earliest Discoveries of Dinosaurs", *Isis,* 66, 5–25

Demaray, John G., 1980. *Milton's Theatrical Epic: The Invention and Design of Paradise Lost,* Cambridge, Massachusetts: Harvard University Press

De Quincey, Thomas, 1985. *Confessions of an English Opium-Eater and Other Writings,* ed. Grevel Lindop, Oxford: Oxford University Press

Desmond, Adrian, 1989. *The Politics of Evolution: Morphology, Medicine, and Reform in Radical London,* Chicago, Illinois: University of Chicago Press

Dibdin, C., Jr., 1807. *Mirth and Metre: Consisting of Poems, Serious, Humorous, and*

*Satirical; Songs, Sonnets, Ballads, & Bagatelles,* London: Vernor, Hood and Sharpe

Dibdin, Thomas Frognall, 1836. *Reminiscences of a Literary Life,* 2 vols, London: John Major

[Dickens, Charles], 1850. "Some Account of an Extraordinary Traveller", *Household Words,* 1, 73–7

[———], 1851. "The Wind and the Rain", *Household Words,* 3, 217–22

———, 1998. *Pictures from Italy,* ed. Kate Flint, London: Penguin

Dirks, Nicholas B., ed., 1992. *Colonialism and Culture,* Ann Arbor: University of Michigan Press

Disraeli, Benjamin, 1826–7. *Vivian Grey,* 5 vols, London: Henry Colburn

Dowden, Wilfred S. ed., 1964. *The Letters of Thomas Moore,* 2 vols, Oxford: Clarendon Press

Doyle, Arthur Conan, 1995. *The Lost World,* ed. Ian Duncan, Oxford: Oxford University Press

Doyle, P., and E. Robinson, 1993. "The Victorian 'Geological Illustrations' of Crystal Palace Park", *PGA,* 104, 181–94

Drachman, Julian M., 1930. *Studies in the Literature of Natural Science,* New York: Macmillan

Duff, David, 2001. "Antididacticism as a Contested Principle in Romantic Aesthetics", *Eighteenth Century Life,* 25, 252–70

Duncan, Ian, 1995. "Introduction", in Doyle 1995, vii–xxi

[Duncan, Isabella], 1860. *Pre-Adamite Man; or, The Story of Our Old Planet & Its Inhabitants, Told by Scripture & Science,* London: Saunders and Otley

Eastmead, William, 1824. *Historia Rievallensis: Containing the History of Kirkby Moorside, and an Account of the Most Important Places in Its Vicinity,* London: Baldwin

Edmonds, J. M., 1979–80. "The Founding of the Oxford Readership in Geology, 1818", *NRRSL,* 34, 33–51

———, 1991. "*Vindiciae Geologicae,* Published 1820; The Inaugural Lecture of William Buckland", *ANH,* 18, 255–68

Edmonds, J. M., and J. A. Douglas, 1975–6. "William Buckland, F.R.S. (1784–1856) and an Oxford Geological Lecture, 1823", *NRRSL,* 30, 141–67

Eger, Martin, 1997. "Hermeneutics and the New Epic of Science", in McRae 1997, 186–209

Eliot, George (= Marian Evans), 1980a. *Adam Bede,* ed. Stephen Gill, London: Penguin

———, 1980b. *The Mill on the Floss,* ed. Gordon S. Haight, Oxford: Clarendon Press

———, 1994. *Middlemarch,* ed. Rosemary Ashton, London: Penguin

Eliot, Simon, 1995. "Some Trends in British Book Production, 1800–1919", in Jordan and Patten 1995, 19–43

Esslin, Martin, 1994. "Romantic Cosmic Drama", in Gillespie 1994, 413–27

Eyles, Victor A., ed., 1970. *James Hutton's System of the Earth, 1785; Theory of the Earth, 1788; Observations on Granite, 1794; together with Playfair's Biography of Hutton* [facsimile reprints], Darien, Connecticut: Hafner

Fairholme, George, 1833. *A General View of the Geology of Scripture,* London: James Ridgway

[———], "A Layman", 1834a. "On the Infidel Tendency of Certain Scientific Speculations", *Christian Observer,* 34, 199–207

[———], "A Layman", 1834b. "On Scriptural Geology: With Observations Thereon", *Christian Observer,* 34, 479–96

Farlow, James O., and M. K. Brett-Surman, 1999. *The Complete Dinosaur,* Bloomington: Indiana University Press

"F. E——s", 1816. "On the Cosmogony of Moses", *Philosophical Magazine and Journal,* 47, 9–12

Feaver, William, 1975. *The Art of John Martin,* Oxford: Clarendon Press

"Fides", 1839. "Scriptural Geology", *Christian Observer,* 39, 403–5

Figuier, Louis, 1863. *La Terre avant le déluge,* 2nd ed., Paris: Hachette

———, 1865. *The World before the Deluge,* 4th ed., trans. W. S. Ormerod, London: Chapman and Hall

Finnegan, Diarmid A., 2004. "The Work of Ice: Glacial Theory and Scientific Culture in Early Victorian Edinburgh", *BJHS,* 37, 29–52

Fleming, John, 1825–6. "The Geological Deluge, as Interpreted by Baron Cuvier and Professor Buckland, Inconsistent with the Testimony of Moses and the Phenomena of Nature", *Edinburgh Philosophical Journal,* 14, 205–39

Flint, Kate, 2000. *The Victorians and the Visual Imagination,* Cambridge: Cambridge University Press

Foote, George, 1952. "Sir Humphry Davy and His Audience at the Royal Institution", *Isis,* 43, 6–12

[Forbes, Edward], 1851. "A Discourse on the Studies of the University of Cambridge", *Literary Gazette,* 1851, 5–7

[———], 1853. "Geology, Popular and Artistic", *Dublin University Magazine,* 42, 338–49

Forbes, James D., 1843. *Travels through the Alps of Savoy,* Edinburgh: Adam and Charles Black

Forgan, Sophie, 1994. "The Architecture of Display: Museums, Universities and Objects in Nineteenth-Century Britain", *HS,* 32, 139–62

———, 1999. "Bricks and Bones: Architecture and Science in Victorian Britain", in Galison and Thompson 1999, 181–208

Fortey, Richard, 1997. *Life: An Unauthorised Biography: A Natural History of the First Four Thousand Million Years on Earth,* London: HarperCollins

Fowles, John, 1984. *Lyme Regis Museum Curator's Report 1983,* Lyme Regis: n.p.

Francis, Frederick John, 1839. *A Brief Survey of Physical and Fossil Geology,* London: J. Hatchard and Son

Franklin, Caroline, 2000. *Byron: A Literary Life,* Basingstoke: Macmillan

Frasca-Spada, Marina, and Nick Jardine, ed. 2000. *Books and the Sciences in History,* Cambridge: Cambridge University Press

Freeman, Michael, 2004. *Victorians and the Prehistoric: Tracks to a Lost World,* New Haven, Connecticut: Yale University Press

Frei, Hans W., 1974. *The Eclipse of Biblical Narrative: A Study in Eighteenth and Nineteenth Century Hermeneutics,* New Haven, Connecticut: Yale University Press

Fulford, Tim, and Peter J. Kitson, eds, 1998. *Romanticism and Colonialism: Writing and Empire, 1780–1830,* Cambridge: Cambridge University Press

Fuller, J. G. C. M., 2001. "Before the Hills in Order Stood: The Beginning of the Geology of Time in England", in Lewis and Knell 2001, 15–23

———, 2005. "A Date to Remember: 4004 BC", *ESH,* 24, 5–14

Fullom, S. W., 1854. *The Marvels of Science, and Their Testimony to Holy Writ,* 8th ed., London: Longman *et al.*

Fyfe, Aileen, 2000. "Young Readers and the Sciences", in Frasca-Spada and Jardine 2000, 276–90

———, 2004a. *Science and Salvation: Evangelical Popular Science Publishing in Victorian Britain,* Chicago, Illinois: University of Chicago Press

————, 2004b. "Introduction", in Marcet 2004, xxi–xxvii

————, 2004c. "Periodicals and Book Series: Complementary Aspects of a Publisher's Mission", in Henson *et al.* 2004, 71–82

Fyfe, Aileen, and Paul Smith, 2003. "Telling Stories", *BJHS*, 36, 471–6

Galison, Peter, and Emily Thompson, eds, 1999. *The Architecture of Science*, Cambridge, Massachusetts: MIT Press

Galison, Peter, *et al.*, eds, 2001. *Science in Culture*, New Brunswick, New Jersey: Transaction

Gallaway, W. F., Jr., 1940. "The Conservative Attitude toward Fiction, 1770–1830", *Publications of the Modern Language Association of America*, 55, 1041–59

Galperin, William H., 1993. *The Return of the Visible in British Romanticism*, Baltimore, Maryland: Johns Hopkins University Press

Garfinkle, Norton, 1955. "Science and Religion in England, 1790–1800: The Critical Response to the Work of Erasmus Darwin", *Journal of the History of Ideas*, 16, 376–88

Gates, Barbara T., 1993. "Retelling the Story of Science", *Victorian Literature and Culture*, 21, 289–306

————, 1997. "Revisioning Darwin with Sympathy: Arabella Buckley", in Gates and Shteir 1997a, 164–76

————, 1998. *Kindred Nature: Victorian and Edwardian Women Embrace the Living World*, Chicago, Illinois: University of Chicago Press

Gates, Barbara T., and Ann B. Shteir, eds, 1997a. *Natural Eloquence: Women Reinscribe Science*, Madison: University of Wisconsin Press

————, 1997b. "Introduction", in Gates and Shteir 1997a, 3–24

Gauthier, Jacques A., 1999. *China's Feathered Dinosaurs*, New Haven, Connecticut: Peabody Museum of Natural History

Geikie, Archibald, 1895. *Memoir of Sir Andrew Crombie Ramsay*, London: Macmillan

Gibson, J., 1822. "Organic Remains, Kirkdale, near Kirbymoorside", *Yorkshire Gazette*, 9 March 1822

Gillespie, Gerald, ed., 1994. *Romantic Drama*, Amsterdam: John Benjamins

Gillispie, Charles Coulston, 1951. *Genesis and Geology: A Study in the Relations of Scientific Thought, Natural Theology, and Social Opinion in Great Britain, 1790–1850*, Cambridge, Massachusetts: Harvard University Press

Gisborne, Thomas, 1818. *The Testimony of Natural Theology to Christianity*, 2nd ed., London: T. Cadell and W. Davies

Gliserman, Susan, 1974–5. "Early Victorian Science Writers and Tennyson's *In Memoriam*: a Study in Cultural Exchange", *Victorian Studies*, 18, 277–308, 437–59

Glut, Donald F., 1980. *The Dinosaur Scrapbook*, Secaucus, New Jersey: Citadel

Godden, Malcolm, and Michael Lapidge, eds, 1991. *The Cambridge Companion to Old English Literature*, Cambridge: Cambridge University Press

Goldstein, Stephen L., 1975. "Byron's *Cain* and the Painites", *SIR*, 14, 391–410

Golinski, Jan, 1992. *Science as Public Culture: Chemistry and Enlightenment in Britain, 1760–1820*, Cambridge: Cambridge University Press

Goodman, Lawrence P., 1969. "More Light on the Limelight", *Theatre Survey*, 10, 114–20

Goodrich, Samuel, 1849. *The Wonders of Geology*, new ed., Boston: Rand and Mann

Gordon, Mrs. [Elizabeth Oke], 1894. *The Life and Correspondence of William Buckland*, London: John Murray

Gornall, Thomas, *et al.*, eds, 1961–77. *The Letters and Diaries of John Henry Newman*, 31 vols, Oxford: Clarendon Press

Gould, Stephen Jay, 1987. *Time's Arrow, Time's Cycle: Myth and Metaphor in the Discovery of Geological Time,* Cambridge, Massachusetts: Harvard University Press

———, ed., 1993a. *The Book of Life,* London: Ebury Hutchinson

———, 1993b. "Reconstructing (and Deconstructing) the Past", in Gould 1993a, 6–21

———, 1997. "The Invisible Woman", in Gates and Shteir 1997a, 27–39

Grafton, Anthony, 1991. *Defenders of the Text: The Traditions of Scholarship in an Age of Science, 1450–1800,* Cambridge, Massachusetts: Harvard University Press

Gray, Thomas, 1786. *Poems,* new ed., London: John Murray

———, 1973. *Poems* [facsimile reprint of 1st ed.], ed. Arthur Sherbo, Menston: Scolar Press

Green, Robert J., 1968. "Some Notes on the St. George Play", *Theatre Survey,* 9, 21–35

Greene, John C., 1959. *The Death of Adam: Evolution and Its Impact on Western Thought,* Ames: Iowa State University Press

Guðrún P. Helgadóttir, ed., 1987. *See under* Helgadóttir.

Guralnick, Stanley M., 1972. "Geology and Religion before Darwin: The Case of Edward Hitchcock, Theologian and Geologist (1793–1864)", *Isis,* 63, 529–43

"H.", 1829. "Misstatements in Ure's New System of Geology", *Magazine of Natural History,* 2, 465–6

Haber, Francis C., 1959. *The Age of the World: Moses to Darwin,* Baltimore, Maryland: Johns Hopkins University Press

Hack, Maria, 1835. *Geological Sketches, and Glimpses of the Ancient Earth,* 2nd ed., London: Darton and Harvey

Hagstrum, Jean, 1958. *The Sister Arts: The Tradition of Literary Pictorialism and English Poetry from Dryden to Gray,* Chicago, Illinois: University of Chicago Press

Haines, Tim, 1999. *Walking with Dinosaurs: A Natural History,* London: BBC Worldwide

Hallam, A., 1989. *Great Geological Controversies,* 2nd ed., Oxford: Oxford University Press

Ham, Ken, 2001. *Dinosaurs of Eden: A Biblical Journey through Time,* Green Forest, Arkansas: Master Books

Hamblyn, Richard, 1996, "Private Cabinets and Popular Geology: The British Audiences for Volcanoes in the Eighteenth Century", in Chard and Langdon 1996, 179–205

Hamilton, James, 2001. *Fields of Influence: Conjunctions of Artists and Scientists, 1815–1860,* Birmingham: University of Birmingham Press

Hanham, Harry, and Michael Shortland, 1995. "Introduction", in Miller 1995, 1–86

Hanna, William, 1849. *Memoirs of the Life and Writings of Thomas Chalmers,* 4 vols, Edinburgh: Sutherland and Knox

Harré, Rom, 1990. "Some Narrative Conventions of Scientific Discourse", in Nash 1990, 81–101

Harrison, J. F. C., 1979. *The Second Coming: Popular Millenarianism, 1780–1850,* New Brunswick, New Jersey: Rutgers University Press

Harvie, Christopher, 2003. "Hugh Miller and the Scottish Crisis", in Borley 2003, 34–47

Haste, Helen, 1993. "Dinosaur as Metaphor", *Modern Geology,* 18, 349–70

Hawkins, Thomas, 1834a. *Memoirs of Ichthyosauri and Plesiosauri, Extinct Monsters of the Ancient Earth,* London: Relfe and Fletcher

———, 1834b. *Memoirs of Ichthyosauri and Plesiosauri*, 2nd ed., London: Relfe and Fletcher

———, 1840a. *The Book of the Great Sea-Dragons, Ichthyosauri and Plesiosauri*, גרלים תנינם *Gedolim Taninim, of Moses. Extinct Monsters of the Ancient Earth*, London: William Pickering

———, 1840b. *The Lost Angel, and the History of the Old Adamites, Found Written on the Pillars of Seth*, London: William Pickering

———, 1841. *One Centenary of Sonnets Dedicated to Her Most Gracious Majesty Queen Victoria*, London: William Pickering

———, 1844. *The Wars of Jehovah, in Heaven, Earth, and Hell*, London: Francis Baisler

———, 1853. *The Christiad*, London: n.p.

———, 1887. *My Life and Works. (Block-Plan.); Prometheus (Second Edition), Volume I*, London: n.p.

[Hawthorne, Nathaniel], "Peter Parley", 1860. *Universal History, on the Basis of Geography*, 7th ed., London: William Tegg

Hays, J. N., 1983. "The London Lecturing Empire, 1800–50", in Inkster and Morrell 1983, 91–119

Hebblethwaite, Kate, 2006. "Hunting, Shooting, Straightening and the American Way", *Viewpoint*, no. 81, 9

Helgadóttir, Guðrún P., ed., 1987. *Hrafns saga Sveinbjarnarsonar*, Oxford: Clarendon Press

Helmstadter, Richard J., and Bernard Lightman, eds, 1990. *Victorian Faith in Crisis: Essays on Continuity and Change in Nineteenth-Century Religious Belief*, Basingstoke: Macmillan

Henry, John, 1996. "Palaeontology and Theodicy: Religion, Politics and the *Asterolepis* of Stromness", in Shortland 1996a, 151–70

Henson, Louise, *et al.*, eds, 2004. *Culture and Science in the Nineteenth-Century Media*, Aldershot: Ashgate

Herbermann, Charles G., *et al.*, eds, 1907–22. *The Catholic Encyclopaedia*, 17 vols, London: Caxton

Herbert, Sandra, 1992. "Between Genesis and Geology: Darwin and Some Contemporaries in the 1820s and 1830s", in Davis and Helmstadter 1992, 68–84

———, 2005. *Charles Darwin, Geologist*, Ithaca, New York: Cornell University Press

Heringman, Noah, 2003a. *Romantic Science: The Literary Forms of Natural History*, Albany, New York: SUNY Press

———, 2003b. "Introduction: The Commerce of Literature and Natural History", in Heringman 2003a, 1–19

———, 2004. *Romantic Rocks, Aesthetic Geology*, Ithaca, New York: Cornell University Press

Herschel, John, 1830. *A Preliminary Discourse on the Study of Natural Philosophy*, London: Longman *et al.*

Heyck, T. W., 1982. *The Transformation of Intellectual Life in Victorian England*, London: Croom Helm

Higgins, W. M., 1842. *The Book of Geology*, London: R. Tyas

Hilton, Boyd, 1988. *The Age of Atonement: The Influence of Evangelicalism on Social and Economic Thought, 1785–1865*, Oxford: Clarendon Press

Hitchcock, Edward, 1851. *The Religion of Geology and Its Connected Sciences*, Glasgow: Collins

Hole, Robert, 1989. *Pulpits, Politics and Public Order in England 1760–1832*, Cambridge: Cambridge University Press

Hollis, Patricia, 1970. *The Pauper Press: A Study in Working-Class Radicalism of the 1830s*, Oxford: Oxford University Press

Hontheim, Joseph, 1907–22. "Hell", in Herbermann 1907–22, VII, 207–11

Hopkins, Justine, 2001. "Phenomena of Art and Science: The Paintings and Projects of John Martin", in Hamilton 2001, 51–92

Howell, Margaret J., 1982. *Byron Tonight: A Poet's Plays on the Nineteenth Century Stage*, Windlesham: Springwood

[Hunt, Leigh], 1834. "On a Stone", *Leigh Hunt's London Journal*, 1, 9–10

Hunt, Robert, 1848. *The Poetry of Science, or Studies of the Physical Phenomena of Nature*, London: Reeve

———, 1854. *The Poetry of Science*, 3rd ed., London: Bohn

Hunter, William, 1768. "Observations on the Bones, Commonly Supposed to be Elephants Bones, Which Have Been Found near the River *Ohio* in *America*", *Philosophical Transactions of the Royal Society of London*, 58, 34–45

Hutchinson, H. N., 1891. *The Autobiography of the Earth: A Popular Account of Geological History*, new ed., New York: D. Appleton

———, 1892. *Extinct Monsters: A Popular Account of Some of the Larger Forms of Ancient Animal Life*, London: Chapman and Hall

Hutton, James, 1788. "Theory of the Earth; or an Investigation of the Laws Observable in the Composition, Dissolution, and Restoration of Land upon the Globe", *Transactions of the Royal Society of Edinburgh*, 1, Part 2, 209–304

———, 1795. *Theory of the Earth, with Proofs and Illustrations*, 2 vols, Edinburgh: William Creech

Hyde, Ralph, 1977. "Thomas Hornor: Pictural Land Surveyor", *Imago Mundi*, 29, 23–34

———, 1982. *The Regent's Park Colosseum: or, "without hyperbole, the wonder of the world": Being an Account of a Forgotten Pleasure Dome and Its Creators*, London: Ackermann

———, 1988. *Panoramania! The Art and Entertainment of the "All-Embracing" View*, London: Trefoil

Inkster, Ian, 1976. "The Social Context of an Educational Movement: a Revisionist Approach to the English Mechanics' Institutes, 1820–1850", *Oxford Review of Education*, 2, 277–307

———, 1979. "London Science and the Seditious Meetings Act of 1817", *BJHS*, 12, 192–6

———, 1980. "The Public Lecture as an Instrument of Science Education for Adults—The Case of Great Britain, *c.* 1750–1850", *Paedagogica Historica*, 20, 80–107

———, ed., 1985. *The Steam Intellect Societies—Essays on Culture, Education and Industry circa 1820–1914*, Nottingham: Department of Adult Education, University of Nottingham

Inkster, Ian, and Jack Morrell, eds, 1983. *Metropolis and Province: Science in British Culture, 1780–1850*, London: Hutchinson

Irving, Washington, 1996. *The Sketch-Book of Geoffrey Crayon, Gent.*, ed. Susan Manning, Oxford: Oxford University Press

James, Edward, and Farah Mendlesohn, eds, 2003. *The Cambridge Companion to Science Fiction*, Cambridge: Cambridge University Press

Janvier, Philippe, 2003. "Armoured Fish from Deep Time: From Hugh Miller's Insights to Current Questions of Early Vertebrate Evolution", in Borley 2003, 177–96

Jauss, Hans Robert, 1982. *Toward an Aesthetic of Reception,* trans. Timothy Bahti, Brighton: Harvester Press

Jones, Frederick L., ed., 1964. *The Letters of Percy Bysshe Shelley,* 2 vols, Oxford: Clarendon Press

Jones, William Powell, 1966. *The Rhetoric of Science: A Study of Scientific Ideas and Imagery in Eighteenth-Century English Poetry,* London: Routledge and Kegan Paul

Jordan, John O., and Robert L. Patten, eds, 1995. *Literature in the Marketplace: Nineteenth-Century British Publishing and Reading Practices,* Cambridge: Cambridge University Press

Jordanova, L. J. ed., 1986. *Languages of Nature: Critical Essays on Science and Literature,* London: Free Association

Jordanova, L. J., and Roy S. Porter, eds, 1979. *Images of the Earth: Essays in the History of the Environmental Sciences,* Chalfont St Giles: British Society for the History of Science

Keats, John, 1926. *The Poetical Works,* ed. H. Buxton Forman, London: Oxford University Press

Keene, Melanie, 2007 forthcoming. "'An object in every walk': Gideon Mantell and the Art of Seeing Pebbles", in Macdonald and Reid 2007 forthcoming

Kepler, Johannes, 1965. *Kepler's Dream,* trans. Patricia Kirkwood, Berkeley: University of California Press

Kern, Stephen, 1983. *The Culture of Time and Space 1880–1918,* Cambridge, Massachusetts: Harvard University Press

Kingsley, Charles, 1855. *Glaucus; or, The Wonders of the Shore,* 2nd ed, Cambridge: Macmillan

——, 1967. *Alton Locke: Tailor and Poet,* ed. Herbert van Thal, London: Cassell

Kirby, William, 1835. *On the Power Wisdom and Goodness of God as Manifested in the Creation of Animals,* 2 vols, London: William Pickering

"Kirkdaliensis", 1822. "The Wonders of the Antediluvian Cave!", *Gentleman's Magazine,* 92, 491–4

Kitis, Eliza, 1997. "Ads—*Part of Our Lives:* Linguistic Awareness of Powerful Advertising", *WI,* 13, 304–13

Kitteringham, Guy, 1982. "Science in Provincial Society: The Case of Liverpool in the Early Nineteenth Century", *AS,* 39, 329–48

Klancher, Jon P., 1987. *The Making of English Reading Audiences, 1790–1832,* Madison: University of Wisconsin Press

——, 1994. "Romanticism and its Publics: A Forum: Introduction", *SIR,* 33, 523–5

Klaver, J. M. I., 1997. *Geology and Religious Sentiment: The Effect of Geological Discoveries on English Society and Literature between 1829 and 1859,* Leiden: Brill

Klonk, Charlotte, 1996. *Science and the Perception of Nature: British Landscape Art in the Late Eighteenth and Early Nineteenth Centuries,* New Haven, Connecticut: Yale University Press

Knell, Simon J., 2000. *The Culture of English Geology, 1815–1851: A Science Revealed through Its Collecting,* Aldershot: Ashgate

Knell, Simon J., and Michael A. Taylor, 2006. "Hugh Miller: Fossils, Landscape and Literary Geology", *PGA,* 117, 85–98

Knight, David M., 1967. "The Scientist as Sage", *SIR,* 6, 65–88

——, 1992. *Humphry Davy: Science and Power,* Oxford: Blackwell

Knipe, Henry R., 1905. *Nebula to Man,* London: Dent

Kölbl-Ebert, Martina, 1997. "Mary Buckland (née Morland) 1897–1857", *ESH,* 16, 33–8

Laissus, Yves, ed., 1998. *Il y a 200 ans, les savants en Égypte,* Paris: Nathan

Landau, Misia, 1991. *Narratives of Human Evolution,* New Haven, Connecticut: Yale University Press

Lankester, E. Ray, 1905. *Extinct Animals,* London: Constable

Larrington, Carolyne, tr., 1996. *The Poetic Edda,* Oxford: Oxford University Press

Larrissy, Edward, 1999. "The Celtic Bard of Romanticism: Blindness and Second Sight", *Romanticism,* 5, 43–57

Laurance, John, 1835. *Geology in 1835; A Popular Sketch of the Progress, Leading Features, and Latest Discoveries of this Rising Science,* London: Simpkin & Marshall

Lean, E. Tangye, 1970. *The Napoleonists: A Study in Political Disaffection 1760–1960,* London: Oxford University Press

Lear, John, 1965. *Kepler's Dream: with the Full Text and Notes of* Somnium, sive astronomia lunaris Joannis Kepleri, trans. Patricia Frueh Kirkwood, Berkeley: University of California Press

Leask, Nigel, 1998a. "'Wandering through Eblis'; Absorption and Containment in Romantic Exoticism", in Fulford and Kitson 1998, 165–88

———, 1998b. "Mont Blanc's Mysterious Voice: Shelley and Huttonian Earth Science", in Shaffer 1998, 182–203

———, 2002. *Curiosity and the Aesthetics of Travel Writing, 1770–1840: "From an Antique Land",* Oxford: Oxford University Press

Levere, Trevor H., 1981. *Poetry Realized in Nature: Samuel Taylor Coleridge and Early Nineteenth-Century Science,* Cambridge: Cambridge University Press

Levine, George, 1987a. *One Culture: Essays in Science and Literature,* Madison: University of Wisconsin Press

———, 1987b. "One Culture: Science and Literature", in Levine 1987a, 3–32

[Lewes, George Henry], 1847. "The Condition of Authors in England, Germany and France", *Fraser's Magazine,* 35, 285–95

Lewis, C. L. E., and S. J. Knell, eds, 2001. *The Age of the Earth: From 4004 BC to AD 2002,* London: Geological Society

Lightman, Bernard, ed., 1997a. *Victorian Science in Context,* Chicago, Illinois: University of Chicago Press

———, 1997b. "'The Voices of Nature': Popularizing Victorian Science", in Lightman 1997a, 187–211

———, 1999. "The Story of Nature: Victorian Popularizers and Scientific Narrative", *Victorian Review,* 25/2, 1–29

———, 2000. "The Visual Theology of Victorian Popularizers of Science: From Reverent Eye to Chemical Retina", *Isis,* 91, 651–80

———, ed., 2004. *The Dictionary of Nineteenth-Century British Scientists,* 4 vols, Bristol: Thoemmes

Lindberg, David C., and Ronald L. Numbers, eds, 1986. *God and Nature: Historical Essays on the Encounter between Christianity and Science,* Berkeley: University of California Press

[Lindley, John], 1833. "A General View of the Geology of Scripture", *Athenaeum,* 1833, 228–9

[———], 1854. "The Crystal Palace Garden", *Athenaeum,* 1854, 780

Lindqvist, Svante, 1992. "The Spectacle of Science: An Experiment in 1744 Concerning the Aurora Borealis", *Configurations,* 1, 57–94

Lindsay, Gillian, 1996. "Mary Roberts: A Neglected Naturalist", *Antiquarian Book Monthly,* 23, 20–2

Livingstone, David N., *et al.,* eds, 1999. *Evangelicals and Science in Historical Perspective,* New York: Oxford University Press

Lonsdale, Roger, ed., 1969. *The Poems of Thomas Gray, William Collins, Oliver Gold-smith,* London: Longmans, Green and Co.

Lowenthal, David, 1985. *The Past is a Foreign Country,* Cambridge: Cambridge University Press

———, 2003. "Caring for Nature: The Transatlantic Canvas of the Nineteenth Century", in Borley 2003, 14–33

Lowry, Delvalle, 1822. *Conversations on Mineralogy,* 2 vols, London: Longman *et al.*

Luckhurst, Roger, and Josephine McDonagh, eds, 2002. *Transactions and Encounters: Science and Culture in the Nineteenth Century,* Manchester: Manchester University Press

[Lyell, Charles], 1826a. "Scientific Institutions", *Quarterly Review,* 34, 153–79

[———], 1826b. "Transactions of the Geological Society", *Quarterly Review,* 34, 507–40

[———], 1827. "Scrope's *Geology of Central France*", *Quarterly Review,* 36, 437–83

———, 1830–3. *Principles of Geology, Being an Attempt to Explain the Former Changes of the Earth's Surface, by Reference to Causes Now in Operation,* 3 vols, London: John Murray

———, 1835. *Principles of Geology,* 3rd ed., 4 vols, London: John Murray

———, 1841. *Elements of Geology,* 2nd ed., 2 vols, London: John Murray

———, 1845. *Travels in North America; with Geological Observations on the United States, Canada, and Nova Scotia,* 2 vols, London: John Murray

———, 1990–1. *Principles of Geology,* 3 vols, ed. Martin Rudwick [facsimile reprint of 1st ed.], Chicago, Illinois: University of Chicago Press

———, 1997. *Principles of Geology,* ed. James A. Secord [abridged ed.], London: Penguin

Lyell, Katherine M., ed., 1881. *Life, Letters and Journals of Sir Charles Lyell, Bart.,* 2 vols, London: John Murray

Lynch, John M., ed., 2002a. *Creationism and Scriptural Geology, 1817–1857,* 7 vols, Bristol: Thoemmes

———, 2002b. "Introduction", in Lynch 2002a, I, ix–xxiv

———, 2006. "'Scriptural Geology', *Vestiges of the Natural History of Creation* and Contested Authority in Nineteenth-Century British Science", in Clifford *et al.* 2006, 131–41

Lyons, Sherrie, 2006. "Swimming at the Edges of Scientific Respectability: Sea Serpents in the Victorian Era", in Clifford *et al.* 2006, 31–44

McCalla, Arthur, 2006. *The Creationist Debate: The Encounter between the Bible and the Historical Mind,* London: T&T Clark International

McCalman, I. D., 1992. "Popular Irreligion in Early Victorian England: Infidel Preachers and Radical Theatricality in 1830s London", in Davis and Helm-stadter 1992, 51–67

McCartney, Paul J., 1977. *Henry De la Beche: Observations on an Observer,* Cardiff: National Museum of Wales

McCrie, Thomas, ed., 1848. *The Bass Rock: Its Civil and Ecclesiastic History, Geology, Martyrology, Zoology and Botany,* Edinburgh: W. P. Kennedy *et al.*

McDonagh, Josephine, 1987. "Writings on the Mind: Thomas De Quincey and the Importance of the Palimpsest in Nineteenth Century Thought", *Prose Studies,* 10, 207–24

Mac Donald, George, 1881. *Mary Marston,* 3 vols, London: Sampson Low *et al.*

Macdonald, Helen, and Francis Reid, eds, 2007 forthcoming. *The Cabinet of Natural History Garden Party Talks IV: Objects,* Cambridge: Department of History and Philosophy of Science, University of Cambridge

Macfarlane, Robert, 2003. *Mountains of the Mind*, New York: Pantheon

McGowan, Christopher, 2001. *The Dragon Seekers: How an Extraordinary Circle of Fossilists Discovered the Dinosaurs and Paved the Way for Darwin*, Cambridge, Massachusetts: Perseus

Mackenzie, W. M., 1905. *Hugh Miller: A Critical Study*, London: Hodder and Stoughton

McKillop, Alan D., 1925. "A Victorian Faust", *Publications of the Modern Language Association of America*, 40, 743–68

McKinsey, Elizabeth, 1985. *Niagara Falls: Icon of the American Sublime*, Cambridge: Cambridge University Press

Macmillan, Duncan, 2000. *Scottish Art 1460–2000*, Edinburgh: Mainstream

McNeil, Maureen, 1986. "The Scientific Muse: The Poetry of Erasmus Darwin", in Jordanova 1986, 159–203

McPhee, John, 1980. *Basin and Range*, New York: Farrar, Strauss, Giroux

McRae, Murdo William, ed., 1997. *The Literature of Science: Perspectives on Popular Scientific Writing*, Athens: University of Georgia Press

Manning, Peter J., 1995. "Wordsworth in the *Keepsake, 1829*", in Jordan and Patten 1995, 44–73

Mannoni, Laurent, 2000. *The Great Art of Light and Shadow: Archaeology of the Cinema*, Exeter: University of Exeter Press

Mantell, Gideon Algernon, 1822. *The Fossils of the South Downs; or Illustrations of the Geology of Sussex*, London: Lupton Relfe

———, 1827. *Illustrations of the Geology of Sussex: Containing a General View of the Geological Relations of the South-Eastern Part of England*, London: Lupton Relfe

———, 1831. "The Geological Age of Reptiles", *Edinburgh New Philosophical Journal*, 11, 181–5

———, 1833. *The Geology of the South-East of England*, London: Longman *et al.*

[———], 1834a. "More Thoughts 'On a Stone'", *Leigh Hunt's London Journal*, 1, 110

———, 1834b. *A Descriptive Catalogue of the Collection Illustrative of Geology and Fossil Comparative Anatomy, in the Museum, of Gideon Mantell*, London: Relfe and Fletcher

———, 1836a. *Thoughts on a Pebble; or, a First Lesson in Geology*, London: Relfe and Fletcher

———, 1836b. *A Descriptive Catalogue of the Objects of Geology, Natural History, and Antiquity, (Chiefly Discovered in Sussex,) in the Museum, Attached to the Sussex Scientific and Literary Institution, at Brighton*, 6th ed., London: Relfe and Fletcher

———, 1838. *The Wonders of Geology*, 2 vols continuously paginated, London: Relfe and Fletcher

———, 1839. *The Wonders of Geology*, 3rd ed., 2 vols continuously paginated, London: Relfe and Fletcher

———, 1842. *Thoughts on a Pebble*, 6th ed., London: Relfe and Fletcher

———, 1844. *The Medals of Creation; or, First Lessons in Geology, and in the Study of Organic Remains*, 2 vols continuously paginated, London: Bohn

———, 1846. *Thoughts on Animalcules; or, a Glimpse of the Invisible World Revealed by the Microscope*, London: John Murray

———, 1847. *Geological Excursions round the Isle of Wight*, London: Bohn

———, 1849. *Thoughts on a Pebble*, 8th ed., London: Reeve, Benham, and Reeve

———, 1850. *A Pictorial Atlas of Fossil Remains*, London: Bohn

———, 1851. *Petrifactions and Their Teachings; or, a Hand-Book to the Gallery of Organic Remains of the British Museum*, London: Bohn

Marcet, Jane, 2004. *Conversations on Chemistry* [facsimile reprint of 1st ed.], ed. Bernard Lightman, Bristol: Thoemmes

Marchand, Leslie A., ed., 1973–94. *Byron's Letters and Journals,* 13 vols, London: John Murray

Marsh, Joss, 1998. *Word Crimes: Blasphemy, Culture, and Literature in Nineteenth-Century England,* Chicago, Illinois: University of Chicago Press

[Martin, John], 1821. *A Description of the Pictures, Belshazzar's Feast, and Joshua,* 10th ed., London: n.p.

[———], 1822. *A Descriptive Catalogue of the Destruction of Pompeii and Herculaneum,* London: n.p.

[———], 1825. *The Deluge,* London: n.p.

[———], 1828a. *A Descriptive Catalogue of the Engraving of The Deluge,* London: n.p.

[———], 1828b. *Descriptive Catalogue of the Picture of The Fall of Nineveh,* London: n.p.

[———], 1832. *Descriptive Catalogue of the Engraving of The Fall of Babylon.* London: n.p.

Martin, Philip W., 1982. *Byron: A Poet before His Public,* Cambridge: Cambridge University Press

Matteson, Lynn R., 1981. "John Martin's 'The Deluge': A Study in Romantic Catastrophe", *Pantheon, 39,* 220–8

Matthew, H. C. G., and Brian Harrison, eds, 2004. *The Oxford Dictionary of National Biography: From the Earliest Times to the Year 2000,* 61 vols, Oxford: Oxford University Press (available online to subscribers at http://www.oxforddnb.com)

Mawe, J., 1821. *Familiar Lessons on Mineralogy and Geology,* London: J. Mawe

Mayer, David, III, 1969. *Harlequin in His Element: The English Pantomime, 1806–1836,* Cambridge, Massachusetts: Harvard University Press

Mayor, Adrienne, 2000. *The First Fossil Hunters: Paleontology in Greek and Roman Times,* Princeton, New Jersey: Princeton University Press

Meadows, A. J., ed., 1980. *Development of Science Publishing in Europe,* Amsterdam: Elsevier

Meisel, Martin, 1983. *Realizations: Narrative, Pictorial, and Theatrical Arts in Nineteenth-Century England,* Princeton, New Jersey: Princeton University Press

Merrill, Lynn L., 1989. *The Romance of Victorian Natural History,* Oxford: Oxford University Press

Mill, John, [c. 1855]. *The Fossil Spirit; A Boy's Dream of Geology,* 2nd ed., London: Darton and Co.

Miller, Hugh, 1835. *Scenes and Legends of the North of Scotland; or the Traditional History of Cromarty,* Edinburgh: A. and C. Black

[———], 1840. "Our First Year of Labour", *The Witness,* 15 April 1840

———, 1841. *The Old Red Sandstone; or New Walks in an Old Field,* Edinburgh: John Johnstone

———, 1847a. *The Old Red Sandstone,* 3rd ed., Edinburgh: John Johnstone

———, 1847b. *First Impressions of England and Its People,* Edinburgh: John Johnstone

———, 1848. "Geology of the Bass", in McCrie 1848, [51]–[139]

———, 1849. *Foot-Prints of the Creator: or, The Asterolepis of Stromness,* 2nd ed., London: Johnstone and Hunter

———, 1854. *My Schools and Schoolmasters; or, The Story of My Education,* Edinburgh: Constable

———, 1857. *The Testimony of the Rocks; or, Geology in Its Bearings on the Two Theologies, Natural and Revealed,* Edinburgh: Constable

———, 1858. *The Cruise of the Betsey; or, A Summer Holiday in the Hebrides. With Rambles of a Geologist*, ed. W. S. Symonds, Edinburgh: Constable

———, 1859. *Sketch-Book of Popular Geology; being a Series of Lectures Delivered before the Philosophical Institution of Edinburgh*, ed. Lydia Miller, Edinburgh: Constable

———, 1869. *Edinburgh and Its Neighbourhood, Geological and Historical; with the Geology of the Bass Rock*, ed. Lydia Miller, 3rd ed., Edinburgh: Nimmo

———, 1993. *My Schools and Schoolmasters*, ed. James Robertson, Edinburgh: B&W

———, 1994. *Scenes and Legends of the North of Scotland or The Traditional History of Cromarty* [based on 2nd ed.], ed. James Robertson, Edinburgh: B&W

———, 1995. *Hugh Miller's Memoir: From Stonemason to Geologist*, ed. Michael Shortland, Edinburgh: Edinburgh University Press

———, 2001. *The Testimony of the Rocks*, ed. Michael A. Taylor [facsimile reprint of Miller 1857], Cambridge: St Matthew

———, 2003. *The Cruise of the Betsey* [facsimile reprint of Miller 1858], ed. Michael A. Taylor, Edinburgh: NMS

Millhauser, Milton, 1954. "The Scriptural Geologists: An Episode in the History of Opinion", *Osiris*, 11, 65–86

———, 1956. "The Literary Impact of *Vestiges of Creation*", *Modern Language Quarterly*, 17, 213–26

Milner, Thomas, 1846. *The Gallery of Nature: A Pictorial and Descriptive Tour through Creation*, London: William S. Orr

———, 1880. *The Gallery of Nature*, new ed., London: W. and R. Chambers

Milton, John, 1853. *Paradise Lost*, new ed., London: Henry Washbourne

———, 1971. *Paradise Lost*, ed. Alastair Fowler, London: Longman

Mitchell, Robert, 1801. *Plans, and Views in Perspective, with Descriptions, of Buildings Erected in England and Scotland*, London: Robert Mitchell

Mitchell, Timothy, 1992. "Orientalism and the Exhibitionary Order", in Dirks 1992, 289–317

Mitchell, W. J. T., 1986. *Iconology: Image, Text, Ideology*, Chicago, Illinois: University of Chicago Press

———, 1998. *The Last Dinosaur Book: The Life and Times of a Cultural Icon*, Chicago, Illinois: University of Chicago Press

Moir, D. M., 1851. *Sketches of the Poetical Literature of the Past Half-Century*, Edinburgh: William Blackwood and Sons

Monckton, Norah, 1948. "Architectural Backgrounds in the Pictures of John Martin", *Architectural Review*, 104, 81–4

Montgomery, James, 1813. *The World before the Flood, a Poem, in Ten Cantos*, London: Longman *et al.*

Montgomery, Robert, 1839. *Satan: A Poem*, 5th ed., Glasgow: John Symington & Co.

Montgomery, Scott L., 1991. "Science as Kitsch: The Dinosaur and Other Icons", *Science as Culture*, 2, 7–58

Montulé, E., 1821. *A Voyage to North America, and the West Indies, in 1817*, London: Sir Richard Phillips and Co.

Moore, James R., 1986. "Geologists and Interpreters of Genesis in the Nineteenth Century", in Lindberg and Numbers 1986, 322–50

Moore, Thomas, 1915. *The Poetical Works*, ed. A. D. Godley, London: Oxford University Press

[Morley, Henry], 1851. "Our Phantom Ship on an Antediluvian Cruise", *Household Words*, 3, 492–6

Morrell, J. B., 1985. "Wissenschaft in Worstedopolis: Public Science in Bradford, 1800–1850", *BJHS*, 18, 1–23

———, 2005. *John Phillips and the Business of Victorian Science*, Aldershot: Ashgate

Morrell, Jack, and Arnold Thackray, 1981. *Gentlemen of Science: Early Years of the British Association for the Advancement of Science*, Oxford: Clarendon Press

Mortenson, Terry, 2004. *The Great Turning Point: The Church's Catastrophic Mistake on Geology—Before Darwin*, Green Forest, Arkansas: Master Books

Morus, Iwan Rhys, 1998. *Frankenstein's Children: Electricity, Exhibition, and Experiment in Early-Nineteenth-Century London*, Princeton, New Jersey: Princeton University Press

Myers, Greg, 1985. "Nineteenth-Century Popularizations of Thermodynamics and the Rhetoric of Social Prophecy", *Victorian Studies*, 29, 35–66

———, 1989. "Science for Women and Children: The Dialogue of Popular Science in the Nineteenth Century", in Christie and Shuttleworth 1989a, 171–200

———, 1990. *Writing Biology: Texts in the Social Construction of Scientific Knowledge*, Madison: University of Wisconsin Press

———, 1992. "Fictions for Facts: The Form and Authority of the Scientific Dialogue", *HS*, 30, 221–47

———, 1997. "Fictionality, Demonstration, and a Forum for Popular Science: Jane Marcet's *Conversations on Chemistry*", in Gates and Shteir 1997a, 43–60

Nash, Cristopher, ed., 1990. *Narrative in Culture: The Uses of Storytelling in the Sciences, Philosophy, and Literature*, London: Routledge

Neswald, Elizabeth R., 2006. *Thermodynamik als kultureller Kampfplatz: Zur Faszinationsgeschichte der Entropie 1850-1915*, Berlin: Rombach Verlag

Newlyn, Lucy, 1993. *Paradise Lost and the Romantic Reader*, Oxford: Clarendon Press

Newman, John Henry, 1859. *Lectures and Essays on University Subjects*, London: Longman *et al.*

Nichol, J. P., 1837. *Views of the Architecture of the Heavens: In a Series of Letters to a Lady*, Edinburgh: William Tait

Nicholson, H. Alleyne, 1877. *The Ancient Life-History of the Earth*, Edinburgh: Blackwood

Nicolson, Marjorie Hope, 1946. *Newton Demands the Muse: Newton's Opticks and the Eighteenth Century Poets*, Princeton, New Jersey: Princeton University Press

———, 1960. *Voyages to the Moon*, new ed., New York: Macmillan

———, 1963. *Mountain Gloom and Mountain Glory: The Development of the Aesthetics of the Infinite*, 2nd ed., New York: Norton

Niles, John D., 1991. "Pagan Survivals and Popular Belief", in Godden and Lapidge 1991, 126–41

Norman, David B., 1993. "Gideon Mantell's 'Mantel-piece': The Earliest Well-Preserved Ornithischian Dinosaur", *Modern Geology*, 18, 225–45

———, 2000. "Henry De la Beche and the Plesiosaur's Neck", *ANH*, 27, 137–48

Northrop, Henry Davenport, 1887. *Earth, Sea and Sky or Marvels of the Universe*, n.p.: n.p.

Numbers, Ronald L., 1992. *The Creationists*, Berkeley: University of California Press

O'Connor, Anne, 2007 in press. *Finding Time for the Old Stone Age: A History of Palaeolithic Archaeology and Quaternary Geology in Britain, 1860–1960*, Oxford: Oxford University Press

O'Connor, Ralph, 1999. "Mammoths and Maggots: Byron and the Geology of Cuvier", *Romanticism*, 5, 26–42

————, 2003a. "Hugh Miller and Geological Spectacle", in Borley 2003, 237–58

————, 2003b. "Thomas Hawkins and Geological Spectacle", *PGA*, 114, 227–41

————, 2003c. "The Poetics of Geology: A Science and Its Literature in Britain, 1802–1856", unpublished doctoral thesis, University of Cambridge

————, 2005a. "The Poetics of Earth Science: 'Romanticism' and the Two Cultures", *Studies in History and Philosophy of Science*, 36, 607–17

————, 2005b. "History or Fiction? Truth-Claims and Defensive Narrators in Icelandic Romance-Sagas", *Mediaeval Scandinavia*, 15, 101–69

————, 2006a. "Kirkdale Cave and the Poetry of William Buckland", *Studies in Speleology*, 14, 39–41

————, 2006b. *Icelandic Histories & Romances*, Stroud: Tempus

O'Curry, Eugene, 1873. *On the Manners and Customs of the Ancient Irish*, ed. W. K. Sullivan, 3 vols, London: Williams and Norgate

Oettermann, Stephan, 1997. *The Panorama: History of a Mass Medium*, trans. Deborah Lucas Schneider, New York: Zone

Olby, R. C., *et al.*, eds, 1990. *Companion to the History of Modern Science*, London: Routledge

Oldroyd, D. R., 1979. "Historicism and the Rise of Historical Geology", *HS*, 17, 191–213, 227–57

Opitz, Donald L., 2004. "Introduction", in Mary Roberts 2004, v–x

Orange, A. D., 1973. *Philosophers and Provincials: The Yorkshire Philosophical Society from 1822 to 1844*, York: Yorkshire Philosophical Society

————, 1975. "The Idols of the Theatre: The British Association and Its Early Critics", *AS*, 32, 277–94

Osborne, Roger, 1998. *The Floating Egg: Episodes in the Making of Geology*, London: Jonathan Cape

Osborne, Roger, and Alistair Bowden, 2001. *The Dinosaur Coast: Yorkshire Rocks, Reptiles and Landscape*, York: North York Moors National Park

Outram, Dorinda, 1984. *Georges Cuvier: Vocation, Science and Authority in Post-Revolutionary France*, Manchester: Manchester University Press

Owen, Richard, 1854. *Geology and Inhabitants of the Ancient World*, London: Crystal Palace Library

Page, Leroy Earl, 1963. "The Rise of the Diluvial Theory in British Geological Thought", unpublished doctoral thesis, University of Oklahoma

————, 1969. "Diluvialism and Its Critics in Great Britain in the Early Nineteenth Century", in Schneer 1969, 257–71

Paley, Morton D., 1986. *The Apocalyptic Sublime*, New Haven, Connecticut: Yale University Press

————, 1999. *Apocalypse and Millennium in English Romantic Poetry*, Oxford: Clarendon Press

Palmer, Elihu, 1819. *Principles of Nature*, new ed., London: Richard Carlile

Paradis, James G., 1996. "The Natural Historian as Antiquary of the World: Hugh Miller and the Rise of Literary Natural History", in Shortland 1996a, 122–50

————, 1997. "Satire and Science in Victorian Culture", in Lightman 1997a, 143–75

Paradis, James, and Thomas Postlewait, eds, 1981. *Victorian Science and Victorian Values: Literary Perspectives*, New York: New York Academy of Sciences

Parkinson, James, 1804–11. *Organic Remains of a Former World*, 3 vols, London: John Murray *et al.*

Parsons, Horatio A., 1835. *A Guide to Travellers Visiting the Falls of Niagara*, 2nd ed., Buffalo: Oliver G. Steele

Parsons, Nicholas T., 1988. *The Joy of Bad Verse*, London: Collins

Peale, Rembrandt, 1803. *An Historical Disquisition on the Mammoth, or, Great American Incognitum, an Extinct, Immense, Carnivorous Animal*, London: E. Lawrence

Penn, Granville, 1825. *A Comparative Estimate of the Mineral and Mosaical Geologies*, 2nd ed., 2 vols, London: James Duncan

Peters, Julie Stone, 2000. *Theatre of the Book 1480–1880: Print, Text, and Performance in Europe*, Oxford: Oxford University Press

Phillips, Samuel, 1854. *Guide to the Crystal Palace and Park*, London: Crystal Palace Library

"Philo", 1819. *A Short Narrative of the Creation and Formation of the Heavens and the Earth, &c. as Recorded by Moses in the Book of Genesis*, London: Longman *et al.*

Pidgeon, Edward, 1830. *The Fossil Remains of the Animal Kingdom*, in Cuvier *et al.* 1827–35, [XI] (*Supplementary Volume on the Fossils*)

Playfair, John, 1802. *Illustrations of the Huttonian Theory of the Earth*, Edinburgh: Cadell and Davies

———, 1805. "Biographical Account of the Late Dr James Hutton, F. R. S. Edin.", *Transactions of the Royal Society of Edinburgh*, 5, Part 3, 39–99

[———], 1811–12. "Transactions of the Geological Society", *Edinburgh Review*, 19, 207–29

[———], 1813–14. "Cuvier on the Theory of the Earth", *Edinburgh Review*, 22, 454–75

Pointon, Marcia R., 1970. *Milton & English Art: A Study in the Pictorial Artist's Use of a Literary Source*, Manchester: Manchester University Press

———, 1979. "Geology and Landscape Painting in Nineteenth-Century England", in Jordanova and Porter 1979, 84–116

Pollock, Martin, ed., 1983. *Common Denominators in Art and Science*, Aberdeen: Aberdeen University Press

[Porch, Thomas], 1833. *The Mysteries of Time; or, Banwell Cave. A Poem, in Six Cantos*, London: William Straker

Porter, Roy, 1976. "Charles Lyell and the Principles of the History of Geology", *BJHS*, 9, 91–103

———, 1977. *The Making of Geology: Earth Science in Britain 1660–1815*. Cambridge: Cambridge University Press

———, 1978a. "Gentlemen and Geology: The Emergence of a Scientific Career, 1660–1920", *Historical Journal*, 21, 809–36

———, 1978b. "Philosophy and Politics of a Geologist: G. H. Toulmin (1754–1817)", *Journal of the History of Ideas*, 39, 435–50

———, 1996. "Miller's Madness", in Shortland 1996a, 265–86

Potter, G. R., 1923. "Did Thomas Lovell Beddoes Believe in the Evolution of Species?", *Modern Philology*, 21, 89–100

[Powell, Baden], 1836. "Dr. Buckland's Bridgewater Treatise", *Magazine of Popular Science*, 2, 337–46

Price, Leah, 2000. *The Anthology and the Rise of the Novel: From Richardson to George Eliot*, Cambridge: Cambridge University Press

"Priscus", 1838. "Questions for Discussion:—Worldly Habits; Works of Fiction", *Christian Observer*, 38, 106–8

Proctor, Richard A., 1870. *Other Worlds than Ours: The Plurality of Worlds Studied under the Light of Recent Scientific Researches*, London: Longmans, Green, and Co.

Pugin, A. Welby, 1841. *Contrasts: or, a Parallel between the Noble Edifices of the Middle Ages, and Corresponding Buildings of the Present Day; Shewing the Present Decay of Taste*, 2nd ed. London: Charles Dolman

Purcell, Rosalind Wolff, and Stephen Jay Gould, 1992. *Finders, Keepers: Eight Collectors,* London: Hutchinson Radius

[Pyne, William Henry], "A Cockney Grey-Beard", 1821. "De Loutherbourg's Eidophusikon", *Literary Gazette,* 1821, 198–200, 216–18

Qureshi, Sadiah, 2004. "Displaying Sara Baartman, the 'Hottentot Venus'", *HS,* 42, 233–57

Rainger, Ronald, 1991. *An Agenda for Antiquity: Henry Fairfield Osborn & Vertebrate Paleontology at the American Museum of Natural History, 1890–1935,* Tuscaloosa: University of Alabama Press

Rappaport, Rhoda, 1982. "Borrowed Words: Problems of Vocabulary in Eighteenth-Century Geology", *BJHS,* 15, 27–44

———, 1997. *When Geologists Were Historians, 1665–1750,* Ithaca, New York: Cornell University Press

[Reade, John Edmund], 1829. *Cain the Wanderer: A Vision of Heaven: Darkness: and Other Poems,* London: Whittaker, Treacher, & Co.

Redmond, James, ed., 1987. *The Theatrical Space,* Themes in Drama, 9, Cambridge: Cambridge University Press

[Rennie, James], 1828. *Conversations on Geology,* London: Samuel Maunder

[———], 1840. *Conversations on Geology,* 3rd ed., London: J. W. Southgate

Richardson, Alan, 1994. *Literature, Education, and Romanticism: Reading as Social Practice, 1780–1832,* Cambridge: Cambridge University Press

———, 2001. *British Romanticism and the Science of the Mind,* Cambridge: Cambridge University Press

Richardson, G. F., 1838. *Sketches in Prose and Verse. (Second Series.),* London: Relfe and Fletcher

———, 1842. *Geology for Beginners,* London: Hippolyte Baillière

———, 1855. *An Introduction to Geology, and Its Associate Sciences,* ed. Thomas Wright, new ed., London: Bohn

Righter, Anne (= Anne Barton), 1962. *Shakespeare and the Idea of the Play,* London: Chatto and Windus

Ritvo, Harriet, 1987. *The Animal Estate: The English and Other Creatures in the Victorian Age,* Cambridge, Massachusetts: Harvard University Press

Roberts, Mary, 1837. *The Progress of Creation Considered, with Reference to the Present Condition of the Earth,* London: Smith, Elder & Co.

———, 2004. *The Conchologist's Companion* [facsimile reprint of 1st ed.], ed. Bernard Lightman, Bristol: Thoemmes

Roberts, Michael, 1998. "Geology and Genesis Unearthed", *Churchman,* 112, 225–55 (available online at http://scibel.gospelcom.net/content/scibelarticles. php?id=52)

Robertson, James, 2002. "Scenes, Legends and Storytelling in the Making of Hugh Miller", in Borley 2002, 17–25

Robson, John M., 1990. "The Fiat and Finger of God: The Bridgewater Treatises", in Helmstadter and Lightman 1990, 71–125

[Rodd, Thomas], "Philobiblos", 1820. *A Defence of the Veracity of Moses, in his Records of the Creation and General Deluge,* London: T. Rodd

Roger, Jacques, 1997. *Buffon: A Life in Natural History,* trans. Sarah Lucille Bonnefoi, Ithaca, New York: Cornell University Press

Romm, James S., 1992. *The Edges of the Earth in Ancient Thought: Geography, Exploration, and Fiction,* Princeton, New Jersey: Princeton University Press

Roselle, Daniel, 1968. *Samuel Griswold Goodrich, Creator of Peter Parley: A Study of His Life and Work,* Albany: University of New York Press

Rosman, Doreen M., 1984. *Evangelicals and Culture,* London: Croom Helm

Ross, Sydney, 1962. "*Scientist*: The Story of a Word", *AS*, 18, 65–85

Rowell, Geoffrey, 1974. *Hell and the Victorians: A Study of the Nineteenth-Century Theological Controversies Concerning Eternal Punishment and the Future Life,* Oxford: Clarendon Press

Rudwick, Martin J. S., 1963. "The Foundation of the Geological Society of London: Its Scheme for Co-operative Research and Its Struggle for Independence", *BJHS*, 1, 325–55

———, 1970. "The Strategy of Lyell's *Principles of Geology*", *Isis*, 61, 5–33

———, 1972. *The Meaning of Fossils: Episodes in the History of Palaeontology,* London: Macdonald

———, 1975. "Caricature as a Source for the History of Science: De la Beche's Anti-Lyellian Sketches of 1831", *Isis*, 66, 534–60

———, 1976. "The Emergence of a Visual Language for Geological Science, 1760–1840", *HS*, 14, 149–95

———, 1979. "Transposed Concepts from the Human Sciences in the Early Work of Charles Lyell", in Jordanova and Porter 1979, 67–83

———, 1985. *The Great Devonian Controversy: The Shaping of Scientific Knowledge among Gentlemanly Specialists,* Chicago, Illinois: University of Chicago Press

———, 1986. "The Shape and Meaning of Earth History", in Lindberg and Numbers 1986, 296–321

———, 1990–1. "Introduction", in C. Lyell 1990–1, I, vii–lviii

———, 1992. *Scenes from Deep Time: Early Pictorial Representations of the Prehistoric World,* Chicago, Illinois: University of Chicago Press

———, 1996. "Cuvier and Brongniart, William Smith, and the Reconstruction of Geohistory", *ESH*, 15, 25–36

———, 1997. *Georges Cuvier, Fossil Bones, and Geological Catastrophes: New Translations & Interpretations of the Primary Texts,* Chicago, Illinois: University of Chicago Press

———, 2001. "Jean-André de Luc and Nature's Chronology", in Lewis and Knell 2001, 51–60

———, 2004. *The New Science of Geology: Studies in the Earth Sciences in the Age of Revolution,* Aldershot: Ashgate

———, 2005a. *Lyell and Darwin, Geologists: Studies in the Earth Sciences in the Age of Reform,* Aldershot: Ashgate

———, 2005b. *Bursting the Limits of Time: The Reconstruction of Geohistory in the Age of Revolution,* Chicago, Illinois: University of Chicago Press

———, 2008 in press. *Worlds before Adam: The Reconstruction of Geohistory in the Age of Reform,* Chicago, Illinois: University of Chicago Press

Rupke, Nicholas, 1983a. "The Apocalyptic Denominator in English Culture of the Early Nineteenth Century", in Pollock 1983, 30–41

———, 1983b. *The Great Chain of History: William Buckland and the English School of Geology (1814–1849),* Oxford: Clarendon Press

———, 1990. "Caves, Fossils and the History of the Earth", in Cunningham and Jardine 1990, 241–59

———, 1997. "Oxford's Scientific Awakening and the Role of Geology", in Brock and Curthoys 1997, 543–62

———, 1998. " 'The End of History' in the Early Picturing of Geological Time", *HS*, 36, 61–90

Ruthven, K. K. 2001. *Faking Literature,* Cambridge: Cambridge University Press

St Clair, William, 2004. *The Reading Nation in the Romantic Period,* Cambridge: Cambridge University Press

Saussure, Horace-Bénedict de, 1779–96. *Voyages dans les Alpes,* 4 vols, Neuchâtel: Fauche

Saxon, A. H., 1968. *Enter Foot and Horse: A History of Hippodrama in England and France,* New Haven, Connecticut: Yale University Press

[Scafe, John], 1820a. *King Coal's Levee, or Geological Etiquette,* 4th ed, London: Longman *et al.*

[————], 1820b. *A Geological Primer in Verse,* London: Longman *et al.*

Schaffer, Simon, 1983. "Natural Philosophy and Public Spectacle in the Eighteenth Century", *HS,* 21, 1–43

————, 1990. "Genius in Romantic Natural Philosophy", in Cunningham and Jardine 1990, 82–98

————, 1996. "Babbage's Dancer and the Impresarios of Mechanism", in Spufford and Uglow 1996, 53–80

Schneer, Cecil J., ed., 1969. *Toward a History of Geology,* Cambridge, Massachusetts: MIT Press

Schoch, Richard W., 1998. *Shakespeare's Victorian Stage: Performing History in the Theatre of Charles Kean,* Cambridge: Cambridge University Press

Schock, Peter A., 1995. "The 'Satanism' of *Cain* in Context: Byron's Lucifer and the War against Blasphemy", *Keats-Shelley Journal,* 44, 182–215

Scotland, Nigel, 1995. *John Bird Sumner: Evangelical Archbishop,* Leominster: Gracewing

Scott, Walter, 1890. *The Journal of Sir Walter Scott from the Original Manuscript at Abbotsford,* 2 vols, Edinburgh: David Douglas

"A Scriptural Geologist", 1839. "The Bishop of Calcutta, the Rev. H. Melvill, and Dr. Chalmers, on Scriptural Geology", *Christian Observer,* 39, 25–31

[Scrope, George Poulett, and William Broderip], 1836. "Dr. Buckland's *Bridgewater Treatise*", *Quarterly Review,* 56, 31–64

Secord, Anne, 1994a. "Corresponding Interests: Artisans and Gentlemen in Nineteenth-Century Natural History", *BJHS,* 27, 383–408

————, 1994b. "Science in the Pub: Artisan Botanists in Early Nineteenth-Century Lancashire", *HS,* 32, 269–315

————, 1996. "Michael Shortland (ed.), Hugh Miller's Memoir", *BJHS,* 29, 105–6

————, 2002. "Botany on a Plate: Pleasure and the Power of Pictures in Promoting Early Nineteenth-Century Scientific Knowledge", *Isis,* 93, 28–57

————, 2008 forthcoming. *Artisan Naturalists,* Chicago, Illinois: University of Chicago Press

Secord, James A., 1982. "King of Siluria: Roderick Murchison and the Imperial Theme in Nineteenth-Century British Geology", *Victorian Studies,* 25, 413–42

————, 1986. *Controversy in Victorian Geology: The Cambrian-Silurian Dispute,* Princeton, New Jersey: Princeton University Press

————, 1991. "Edinburgh Lamarckians: Robert Jameson and Robert E. Grant", *Journal of the History of Biology,* 24, 1–18

————, 1997. "Introduction", in C. Lyell 1997, ix–xliii

————, 2000a. *Victorian Sensation: The Extraordinary Publication, Reception, and Secret Authorship of* Vestiges of the Natural History of Creation, Chicago, Illinois: University of Chicago Press

————, 2000b. "Progress in Print", in Frasca-Spada and Jardine 2000, 369–89

————, 2003a. "From Miller to the Millennium", in Borley 2003, 328–37

————, 2003b. "Introduction", in [Clark] 2003, v–x

————, 2004. "Monsters at the Crystal Palace", in de Chadarevian and Hopwood 2004, 138–69

Sedgwick, Adam, 1826–33a. "[Presidential Address]" [1830], *Proceedings of the Geological Society of London,* 3rd series, 1, 187–212

————, 1826–33b. "[Presidential Address]" [1831], *Proceedings of the Geological Society of London,* 3rd series, 1, 281–316

————, 1833. *A Discourse on the Studies of the University,* Cambridge: John W. Parker

Sellers, Charles Coleman, 1947. *Charles Willson Peale,* 2 vols, Philadelphia, Pennsylvania: American Philosophical Society

————, 1980. *Mr. Peale's Museum: Charles Willson Peale and the First Popular Museum of Natural Science and Art,* New York: Norton

Semonin, Paul, 2000. *American Monster: How the Nation's First Prehistoric Creature Became a Symbol of National Identity,* New York: New York University Press

Seznec, Jean, 1964. *John Martin en France,* London: Faber

Shaffer, Elinor S., ed., 1998. *The Third Culture: Literature and Science,* Berlin: Walter de Gruyter

Shakespeare, William, 1993. *The Complete Oxford Shakespeare,* ed. Stanley Wells and Gary Taylor, new ed., 3 vols continuously paginated, London: BCA

Shapin, Steven, 1983. "'Nibbling at the teats of science': Edinburgh and the Diffusion of Science in the 1830s", in Inkster and Morrell 1983, 151–78

————, 1990. "Science and the Public", in Olby *et al.* 1990, 990–1007

Sharrock, Roger, 1962. "The Chemist and the Poet: Sir Humphry Davy and the Preface to *Lyrical Ballads*", *NRRSL,* 17, 57–76

Shatto, Susan, 1976. "Byron, Dickens, Tennyson, and the Monstrous Efts", *Yearbook of English Studies,* 6, 144–55

Sheets-Pyenson, Susan, 1981. "A Measure of Success: The Publication of Natural History Journals in Early Victorian Britain", *Publishing History,* 9, 21–36

————, 1985. "Popular Science Periodicals in Paris and London: The Emergence of a Low Scientific Culture, 1820–1875", *AS,* 42, 549–72

Shelley, Percy Bysshe, 1989–2000. *The Poems of Shelley,* ed. Geoffrey Matthews and Kelvin Everest, 2 vols, London: Longman

Sherman, Daniel J., and Irit Rogoff, eds, 1994. *Museum Culture: Histories, Discourses, Spectacles,* London: Routledge

Shiach, Morag, 1989. *Discourse on Popular Culture: Class, Gender and History in Cultural Analysis, 1730 to the Present,* Cambridge: Polity Press

Shinn, Terry, and Richard Whitley, eds, 1985. *Expository Science: Forms and Functions of Popularisation,* Dordrecht: Reidel

Shortland, Michael, 1994. "Darkness Visible: Underground Culture in the Golden Age of Geology", *HS,* 32, 1–61

————, ed., 1996a. *Hugh Miller and the Controversies of Victorian Science,* Oxford: Clarendon Press

————, 1996b. "Hugh Miller's Contribution to the *Witness:* 1840–1856", in Shortland 1996a, 287–300

Siegfried, Robert, and Robert H. Dott, Jr., eds, 1980. *Humphry Davy on Geology: The 1805 Lectures for the General Audience,* Madison: University of Wisconsin Press

Sillars, Stuart, 1995. *Visualisation in Popular Fiction, 1860–1960: Graphic Narratives, Fictional Images,* London: Routledge

Simpson, J. A., and E. S. C. Weiner, eds, 1991. *The Oxford English Dictionary,* 2nd ed., compact edition, Oxford: Oxford University Press

Sims-Williams, Patrick, 1986. "The Visionary Celt: The Construction of an Ethnic Preconception", *Cambridge Medieval Celtic Studies*, 11, 71–96

Siskin, Clifford, 1998. *The Work of Writing: Literature and Social Change in Britain, 1700–1830*, Baltimore, Maryland: Johns Hopkins University Press

Smiles, Samuel, 1878. *Robert Dick, Baker, of Thurso, Geologist and Botanist*, 10th ed., London: John Murray

Smith, Bernard, 1985. *European Vision and the South Pacific*, 2nd ed., New Haven, Connecticut: Yale University Press

Smith, Jonathan, 1994. *Fact and Feeling: Baconian Science and the Nineteenth-Century Literary Imagination*, Madison: University of Wisconsin Press

Snow, C. P., 1993. *The Two Cultures* [reprint of 2nd ed.], Cambridge: Cambridge University Press

Sommer, Marianne, 2003. "The Romantic Cave? The Scientific and Poetic Quests for Subterranean Spaces in Britain", *ESH*, 22, 172–208

———, 2004. "'An amusing account of a cave in Wales': William Buckland (1784–1856) and the Red Lady of Paviland", *BJHS*, 37, 53–74

———, 2008 forthcoming. *Bones and Ochre: The Curious Afterlife of the Red Lady of Paviland*, Cambridge, Massachusetts: Harvard University Press

Sotheby, William, 1834. *Lines Suggested by the Third Meeting of the British Association for the Advancement of Science*, London: G. and W. Nicol

[Southwell, Charles], 1841. "Theory of Regular Gradation I", *Oracle of Reason*, 1, 5–6

Spearing, A. C., 1976. *Medieval Dream-Poetry*, Cambridge: Cambridge University Press

Spenser, Edmund, 1977. *The Faerie Queene*, ed. A. C. Hamilton, London: Longman

Špinar, Zdeněk V., 1972. *Life Before Man*, London: Thames and Hudson

Spufford, Francis, and Jenny Uglow, eds, 1996. *Cultural Babbage: Technology, Time and Invention*, London: Faber

Stableford, Brian, 2003. "Science Fiction before the Genre", in James and Mendlesohn 2003, 15–31

Stafford, Barbara Maria, 1984. *Voyage into Substance: Art, Science, Nature, and the Illustrated Travel Account, 1760–1840*, Cambridge, Massachusetts: MIT Press

Stafford, Barbara Maria, and Frances Terpak, 2001. *Devices of Wonder: From the World in a Box to Images on a Screen*, New York: Getty

Stafford, Fiona J., 1988. *The Sublime Savage: A Study of James Macpherson and the Poems of Ossian*, Edinburgh: Edinburgh University Press

———, 1994. *The Last of the Race: The Growth of a Myth from Milton to Darwin*, Oxford: Clarendon Press

Steffan, Truman Guy, ed., 1968. *Lord Byron's* Cain, Austin: University of Texas Press

Stevenson, Sara, 2002. *The Personal Art of David Octavius Hill*, New Haven, Connecticut: Yale University Press

Stiling, Rodney L., 1999. "Scriptural Geology in America", in Livingstone *et al.* 1999, 177–92

Story, William W., 1863. *Roba di Roma*, 2 vols, London: Chapman and Hall

Stott, Rebecca, 1999. "Thomas Carlyle and the Crowd: Revolution, Geology and the Convulsive 'Nature' of Time", *JVC*, 4, 1–24

Sullivan, W. K., 1873. *Introduction*, vol. I of O'Curry 1873

Sumner, John Bird, 1816. *A Treatise on the Records of the Creation, and on the Moral Attributes of the Creator*, 2 vols, London: J. Hatchard

Sutherland, Elizabeth, and Marian McKenzie Johnston, 2002. *Lydia: Wife of Hugh Miller of Cromarty*, East Linton: Tuckwell

Taquet, Philippe, and Kevin Padian, 2004. "The Earliest Known Restoration of a Pterosaur and the Philosophical Origins of Cuvier's *Ossemens Fossiles*", *Comptes Rendus Palevol*, 3, 157–75

Tayler, W. Elfe, 1855. *Geology: Its Facts and Its Fictions*, London: Houlston and Stoneman

Taylor, Michael A., 1994. "The Plesiosaur's Birthplace: The Bristol Institution and Its Contribution to Vertebrate Palaeontology", *Zoological Journal of the Linnean Society*, 112, 179–96

———, 1997. "Before the Dinosaur: The Historical Significance of the Fossil Marine Reptiles", in Callaway and Nicholls 1997, xix–xlvi

———, 2002. "Joseph Clark III's Reminiscences about the Somerset Fossil Reptile Collector Thomas Hawkins (1810–1889): 'Very Near the Borderline between Eccentricity and Criminal Insanity' ", *Somerset Archaeology and Natural History*, 146, 1–10

———, 2003. "Introduction", in Miller 2003, A11–A62

———, 2007. *Hugh Miller: Stonemason, Geologist, Writer*, Edinburgh: NMS

Taylor, Michael A., and Hugh S. Torrens, 1986. "Saleswoman to a New Science: Mary Anning and the Fossil Fish *Squaloraja* from the Lias of Lyme Regis", *Proceedings of the Dorset Natural History and Archaeological Society*, 108, 135–48

"T. E.", 1830. "Dr. Ure's Geology", *Magazine of Natural History*, 3, 90–2

Tennyson, Alfred, 1969. *The Poems of Tennyson*, ed. Christopher Ricks, London: Longmans

Tennyson, Hallam, 1897. *Alfred Lord Tennyson: A Memoir by His Son*, 2 vols, London: Macmillan

Terhune, Alfred McKinley, and Annabelle Burdick Terhune, eds, 1980. *The Letters of Edward FitzGerald*, 4 vols, Princeton, New Jersey: Princeton University Press

Terrall, Mary, 2000. "Natural Philosophy for Fashionable Readers", in Frasca-Spada and Jardine 2000, 239–54

[Thackeray, William Makepeace], "M. A. Titmarsh", 1846. *Notes of a Journey from Cornhill to Grand Cairo*, London: Chapman and Hall

Thackray, Arnold, 1974. "Natural Knowledge in Cultural Context: The Manchester Model", *American Historical Review*, 79, 672–709

Thackray, J. C., 1976. "James Parkinson's *Organic Remains of a Former World* (1804–1811)", *Journal of the Society for the Bibliography of Natural History*, 7, 451–66

———, 2003. *To See the Fellows Fight: Eye Witness Accounts of Meetings of the Geological Society of London and Its Club, 1822–1868*, London: British Society for the History of Science

Thompson, Thomas, 1835. "An Attempt to Ascertain the Animals Designated in the Scriptures by the Names Leviathan and Behemoth", *Magazine of Natural History*, 8, 193–7, 307–21

Thomson, James, 1855–61. *Poetical Works*, ed. Robert Bell, 2 vols, London: Parker

———, 1987. *The Seasons and The Castle of Indolence*, ed. James Sambrook, 3rd ed., Oxford: Clarendon Press

Todd, Ruthven, 1946. *Tracks in the Snow: Studies in English Science and Art*, London: Grey Walls

Tolkien, J. R. R., 1974. *The Return of the King: Being the Third Part of The Lord of the Rings*, new ed., London: George Allen and Unwin

———, 1983. *The Monsters and the Critics and Other Essays*, ed. Christopher Tolkien, London: George Allen and Unwin

Topham, Jonathan R., 1992. "Science and Popular Education in the 1830s: The Role of the *Bridgewater Treatises*", *BJHS*, 25, 397–430

———, 1993. "'An Infinite Variety of Arguments': The *Bridgewater Treatises* and British Natural Theology in the 1830s", doctoral thesis, University of Lancaster

———, 1998. "Beyond the 'Common Context': The Production and Reading of the Bridgewater Treatises", *Isis*, 89, 233–62

———, 2000. "Scientific Publishing and the Reading of Science in Nineteenth-Century Britain: A Historiographical Survey and Guide to Sources", *Studies in the History and Philosophy of Science*, 31, 559–612

———, 2004a. "The *Mirror of Literature, Amusement and Instruction* and Cheap Miscellanies in Early Nineteenth-Century Britain", in Cantor *et al.* 2004a, 37–66

———, 2004b. "Periodicals and the Making of Reading Audiences for Science in Early Nineteenth-Century Britain: The *Youth's Magazine*, 1828–37", in Henson *et al.* 2004, 57–69

Torrens, Hugh, 1995. "Mary Anning (1799–1847) of Lyme; 'the greatest fossilist the world ever knew'", *BJHS*, 28, 257–84

———, 1998. "Geology and the Natural Sciences: Some Contributions to Archaeology in Britain 1780–1850", in Brand 1998, 35–59

———, 1999. "Politics and Paleontology: Richard Owen and the Invention of Dinosaurs", in Farlow and Brett-Surman 1999, 175–90

———, 2002. *The Practice of British Geology, 1750–1850*, Aldershot: Ashgate

———, 2006a. "The Life and Times of Hastings Elwin or Elwyn (1777–1852) and His Critical Role in Founding the Bath Literary and Scientific Institution in 1823", *Geological Curator*, 8, 141–68

———, 2006b. "Notes on 'The Amateur' in the Development of British Geology", *PGA*, 117, 1–8

Torrens, Hugh S., and John A. Cooper, 1985. "George Fleming Richardson (1796–1848)—Man of Letters, Lecturer and Geological Curator", *Geological Curator*, 4, 249–72

Torrens, Hugh S., and Michael A. Taylor, 1987–94. "Geological Collectors and Museums in Cheltenham 1810–1988: A Case History and Its Lessons", *Geological Curator*, 5, 175–213

Treadwell, James, 1993. "Blake, John Martin, and the Illustration of *Paradise Lost*", *WI*, 9, 363–82

Trimmer, Joshua, 1841. *Practical Geology and Mineralogy*, London: John W. Parker

Turner, Frank M., 1978. "The Victorian Conflict between Science and Religion: A Professional Dimension", *Isis*, 69, 356–76

Turner, G., 1799. "Memoir on the Extraneous Fossils denominated Mammoth Bones", *Transactions of the American Philosophical Society*, 4, 510–18

Turner, Sharon, 1832. *The Sacred History of the World, as Displayed in the Creation and Subsequent Events to the Deluge*, London: Longman *et al.*

Ulrich, John M., 2006. "Thomas Carlyle, Richard Owen, and the Paleontological Articulation of the Past", *JVC*, 11, 30–58

Ure, Andrew, 1829. *A New System of Geology*, London: Longman *et al.*

Vincent, David, 1989. *Literacy and Popular Culture: England 1750–1914*, Cambridge: Cambridge University Press

Volney, Constantin-François Chasseboeuf, comte de, 1979. *A New Translation of Volney's Ruins*, ed. Robert D. Richardson, Jr., 2 vols, New York: Garland

Walker, D. P., 1964. *The Decline of Hell: Seventeenth-Century Discussions of Eternal Torment*, London: Routledge

Wallach, Alan P., 1968. "Cole, Byron, and the *Course of Empire*", *Art Bulletin,* 50, 375–9

Warner, Eric, ed., 1992. *Virginia Woolf: A Centenary Perspective,* New York: St Martin's

Wawn, Andrew, 1982. "*Gunnlaugs saga ormstungu* and the Theatre Royal Edinburgh 1812: Melodrama, Mineralogy and Sir George Mackenzie", *Scandinavica,* 21, 139–51

———, 2000. *The Vikings and the Victorians: Inventing the Old North in Nineteenth-Century Britain,* Cambridge: Brewer

Weindling, Paul, 1980. "Science and Sedition: How Effective Were the Acts Licensing Lectures and Meetings, 1795–1819?", *BJHS,* 13, 139–53

Wennerbom, A. J., 1999. "Charles Lyell and Gideon Mantell, 1821–1852: Their Quest for Elite Status in English Geology", unpublished doctoral thesis, University of Sydney

Wheeler, Michael, 1990. *Death and the Future Life in Victorian Literature and Theology,* Cambridge: Cambridge University Press

Whewell, William, 1852. "On the General Bearing of the Great Exhibition", in Anon. 1852, 3–34

White, Edward, 1876. *Life in Christ,* 2nd ed., London: Elliot Stock

White, Paul, 2002. "Cross-Cultural Encounters: The Co-Production of Science and Literature in Mid-Victorian Periodicals", in Luckhurst and McDonagh 2002, 75–95

———, 2005. 'Ministers of Culture: Arnold, Huxley and Liberal Anglican Reform of Learning', *HS,* 43, 115–38

Whitley, Richard, 1985. "Knowledge Producers and Knowledge Acquirers: Popularisation as a Relation between Scientific Fields and Their Publics", in Shinn and Whitley 1985, 3–28

Wiener, Joel H., 1983. *Radicalism and Freethought in Nineteenth-Century Britain: The Life of Richard Carlile,* Westport, Connecticut: Greenwood Press

Wilcox, Scott B., 1988. "Unlimiting the Bounds of Painting", in Hyde 1988, 13–44

Wilkinson, Henry, 1824. *Cain, a Poem, Intended to be Published in Parts, Containing an Antidote to the Impiety and Blasphemy of Lord Byron's* Cain. Part I, London: Baldwin, Cradock, and Joy

[Wilks, Samuel], "S. C. W." 1834a. "The Fossil Shell", *Christian Observer,* 34, 219–29

[———], 1834b. "Review of Cole's Letter to Sedgwick on Geology", *Christian Observer,* 34, 369–87

Williams, David, 1982. *Cain and* Beowulf: *A Study in Secular Allegory,* Toronto: University of Toronto Press

Williams, Raymond, 1983. *Keywords: A Vocabulary of Culture and Society,* 2nd ed., London: Fontana

Wilson, Edward O., 1998. *Consilience: The Unity of Knowledge,* London: Little, Brown and Company

Wilson, Leonard G., 1972. *Charles Lyell: The Years to 1841: The Revolution in Geology,* New Haven, Connecticut: Yale University Press

———, 1998. *Lyell in America: Transatlantic Geology, 1841–1853,* Baltimore, Maryland: Johns Hopkins University Press

Wilton, Andrew, 1980. *Turner and the Sublime,* London: British Museum Publications

———, 1990. *Painting and Poetry: Turner's* Verse Book *and His Work of 1804–1812,* London: Tate Gallery

Wiseman, Nicholas, 1836. *Twelve Lectures on the Connexion between Science and Revealed Religion*, 2 vols, London: Joseph Booker

Wonders, Karen, 1993. *Habitat Dioramas: Illusions of Wilderness in Museums of Natural History*, Uppsala: Uppsala University

Wood, Gillen D'Arcy, 2001. *The Shock of the Real: Romanticism and Visual Culture, 1760–1860*, London: Palgrave

Wood, R. Derek, 1993. "The Diorama in Great Britain in the 1820s", *History of Photography*, 17, 284–95 (available with additional images online at http://www.midleykent.fsnet.co.uk)

Wooding, Jonathan M. ed., 2000. *The Otherworld Voyage in Early Irish Literature: An Anthology of Criticism*, Dublin: Four Courts

Wordsworth, Jonathan, 1982. *William Wordsworth: The Borders of Vision*, Oxford: Clarendon Press

Wordsworth, William, 1944–72. *The Poetical Works*, ed. E. de Selincourt and H. Darbishire, 5 vols, Oxford: Clarendon Press

———, 1972. *The Prelude: A Parallel Text*, ed. J. C. Maxwell, new ed., London: Penguin

Wyatt, John, 1995. *Wordsworth and the Geologists*, Cambridge: Cambridge University Press

Yanni, Carla, 1999. *Nature's Museums: Victorian Science and the Architecture of Display*, Baltimore, Maryland: Johns Hopkins University Press

Yearley, Steven, 1985. "Representing Geology: Textual Structures in the Pedagogical Presentation of Science", in Shinn and Whitley 1985, 79–101

Yeo, Richard, 1984. "Science and Intellectual Authority in Mid-Nineteenth-Century Britain: Robert Chambers and *Vestiges of the Natural History of Creation*", *Victorian Studies*, 28, 5–31

———, 1993. *Defining Science: William Whewell, Natural Knowledge, and Public Debate in Early Victorian Britain*, Cambridge: Cambridge University Press

———, 2001. *Science in the Public Sphere: Natural Knowledge in British Culture, 1800–1860*, Aldershot: Ashgate

Young, Edward, 1989. *Night Thoughts*, ed. Stephen Cornford, Cambridge: Cambridge University Press

Young, George, 1838. *Scriptural Geology; or An Essay on the High Antiquity Ascribed to the Organic Remains Imbedded in Stratified Rocks*, London: Simpkin, Marshall, and Co.

Young, George, and John Bird, 1822. *A Geological Survey of the Yorkshire Coast*, Whitby: George Clark

———, 1828. *A Geological Survey of the Yorkshire Coast*, 2nd ed., Whitby: R. Kirby

Yule, John David, 1976. 'The Impact of Science on British Religious Thought in the Second Quarter of the Nineteenth Century', unpublished doctoral thesis, University of Cambridge

Ziolkowski, Theodore, 1990. *German Romanticism and Its Institutions*, Princeton, New Jersey: Princeton University Press

# Credits

Grateful acknowledgement is made to those who have kindly granted permission to quote from their manuscripts: Roderick Gordon; the President and Fellows of Magdalen College, Oxford.

Grateful acknowledgement is made to those who have kindly granted permission to reproduce copyright images in their possession. Figs. 0.1, 1.1, 1.4, 1.5, 2.6, 3.1, 3.2, 3.3, 3.5, 3.6, 3.7, 3.8, 5.1, 5.2, 7.7, 7.8, 7.12, 7.14, 7.15, 7.16, 7.18, 7.19, 7.25, 8.1, 8.5, 8.6, 9.3, 9.6, 9.9, 10.3, 10.4, E.1, Plate 2: Syndics of Cambridge University Library. 1.2, 1.3, 7.1, 7.3: Guildhall Library, City of London. 2.1, 2.3, 2.4, 2.9: Oxford University Museum of Natural History. 2.2, 4.3: Department of Geology, National Museum of Wales. 2.5, 2.11, 2.12, 4.1, 4.2, 6.3, 6.4, 6.5, 6.6, 6.7, 7.6, 8.2, 8.3, 8.4, 8.7, 8.8, 8.9, 8.10, 8.11, 8.12, 9.1, 9.2, 9.4, 9.5, 9.7, 9.8, Plates 1, 3, 4, 5: author's collection. 2.7: Martin Rudwick. 2.8: National Portrait Gallery, London. 2.10: Lyme Regis Museum. 3.4, 7.10, 7.11, 7.17, 7.20: Trustees of the British Museum. 7.2: © University of Aberdeen. 7.4: City of Westminster Archives Centre. 7.5: University of Exeter Library (Bill Douglas Centre). 7.9, 7.13, 7.22, 7.23, 7.24: Bodleian Library, University of Oxford. 7.21: Trustees of Sir John Soane's Museum. 10.1: Scottish National Photography Collection, SNPG. 10.2: City of Edinburgh Museums and Galleries (City Art Centre). 10.5, Plate 6: Alexander Turnbull Library, Wellington, New Zealand. Plate 7: John O'Connor. Plate 8: Belshazzar's Feast (oil on canvas) Martin, John (1789–1854), © Yale Center for British Art, Paul Mellon Collection, USA, The Bridgeman Art Library.

# Index

Page numbers in italics refer to illustrations, and their captions are referred to using the form *21n*. In author-entries and artist-entries, works are listed at the end of the entry (in chronological rather than alphabetical order).

techniques borrowed from, 197, 366
Ararat, Mount, 303–4, 397
*Arcana of Science* (annual), 234
archaeology: geology as, 148; Palaeolithic, 438
Arctic, panoramas of the, 269, 274
Ariosto, Ludovico, 132
Aristophanes, 345
Aristotle, 17, 135, 348. *See also* eternalism
Artis, Edmund, 121n. 12
artisans. *See* working classes
Ashe, Thomas, 36; and American Indian folklore, 39–46, 158; as author of scurrilous autobiography of Princess Caroline, 36; as fossil collector, 36–9; fossil theft by, 39; as mammoth impersonator, 45; *Memoirs of Mammoth* (1806), 11, 36–46, 125, 230, 256, 338, 359; *Memoirs and Confessions* (1815), 36, 38–9, 45
Ashmolean Museum (Oxford), 75, 76
association theory, 297–300, 316–18, 440
Assyria. *See* Babylon, the fall of; Nineveh
astronomy: gendering of, 153; geology borrowing prestige of, 62, 88, 130, 382, 394, 439; geology's prestige borrowed by, 437–8; and heaven and hell, 306–7, 419, 420; as imaginative resource for geological writers, 67–8, 87–8, 181, 373, 382–3, *383*; and natural theology, 22, 68, 102–4; popularization of, 21–2, 205, 206n. 49, 430n. 152, 437–8; practised by angels, 430; and the sublime, 21–2, 151–3, 236. *See also* outer space; plurality of worlds
*As You Like It* (Shakespeare), 5, 184, 301, 377
atheism, 60
*Athenaeum* (periodical): on biblical literalism, 209, 210–11; on geology vogue, 195; on Miller, 287–8, 404
Atherstone, Edwin, 290
Attila, the Spiritual, 417
audiences for science: academic, 71–4, 75, 192; active or passive, 7, 10, 49–50, 256–8; female, 105–6, 153, 192, 234–41; juvenile, 145–6, 152–8, 234–41; multiple, 7–8, 24–6, 119, 167, 171, 333–4; polite, 217–61, 345–7; private, 26, 81–3, 91–5, 102, 105–6, 110; and public participation in scientific debate, 99–102, 204–5,

211; widening, 192–3, 198, 217–18, 242. *See also* middle classes; popularization; "rational amusement"; readers; upper classes; working classes
Augusta, Josef, 435
Augustine, Saint, 63
Austen, Jane, 284; *Emma* (1816), 65
"Author of 'A Portrait of Geology'", 206
authorship: and control over meaning, 7–8, 234, 357; economics of, 119, 346; feminine, 202, 204–5, 250–1, 252–3, 435
autobiographical mode: chatty, 362, 393–6, 401; fantastic, 251–2, 253–4; as frame for whole book, 232, 248, 253; functions of, 38–9, 409–10; heroic, 38–9, 45, 230, 246–7, 256, *257*; and talking fossils, 253, 344–5
*Autobiography of the Earth* (Hutchinson), 248
avalanches. *See* disasters, representations of
*Awful Changes. See under* De la Beche
*Az ember tragédiája* (Mádach), 252, 253

Babel, Tower of: representations, 286, 307, 308–9, *310*, 311, *plate 8*
baboon, smoking, 397
Babylon, the fall of: depicted by Martin, 286–7, 310–1, *310*, *plate 8*; depicted theatrically, 305, 312–15, *313*, *314*, 362; moral of, 300, 308, 310–1, *310*n., 315; as type of other cataclysms, 307–9. *See also* Babel, Tower of
Bacon, Francis, 74, 141, 202n. 29. *See also* nature: as book or archive
Bailey, P. J., 252, 253, 353–4
Bakewell, Robert, 27, 46; relationship to Geological Society, 46, 68. See also *Introduction to Geology* (Bakewell)
Baldred, Saint, 403
*Baldrs draumar*, 108
ballads: and cheap print, 224; about dragons, *238*, 355; sung at Geological Survey, 81–2, 112n. 137
Balzac, Honoré de, 152, 322–3
Barker, Robert, 267, *267*, 285
Barthes, Roland, 10; on text-to-image relation, 291
Bartholin, 108
Barton, Anne, 377
Barton, Bernard, 290